Lecture Notes in Computer Science 14242

Founding Editors

Gerhard Goos
Juris Hartmanis

Editorial Board Members

The series Lecture Notes in Computer Science (LNCS), including its subseries Lecture Notes in Artificial Intelligence (LNAI) and Lecture Notes in Bioinformatics (LNBI), has established itself as a medium for the publication of new developments in computer science and information technology research, teaching, and education.

LNCS enjoys close cooperation with the computer science R & D community, the series counts many renowned academics among its volume editors and paper authors, and collaborates with prestigious societies. Its mission is to serve this international community by providing an invaluable service, mainly focused on the publication of conference and workshop proceedings and postproceedings. LNCS commenced publication in 1973.

Stefan Wesarg · Esther Puyol Antón ·
John S. H. Baxter · Marius Erdt ·
Klaus Drechsler · Cristina Oyarzun Laura ·
Moti Freiman · Yufei Chen · Islem Rekik ·
Roy Eagleson · Aasa Feragen · Andrew P. King ·
Veronika Cheplygina ·
Melani Ganz-Benjaminsen · Enzo Ferrante ·
Ben Glocker · Daniel Moyer · Eikel Petersen
Editors

Clinical Image-Based Procedures, Fairness of AI in Medical Imaging, and Ethical and Philosophical Issues in Medical Imaging

12th International Workshop, CLIP 2023
1st International Workshop, FAIMI 2023
and 2nd International Workshop, EPIMI 2023
Vancouver, BC, Canada, October 8 and October 12, 2023
Proceedings

 Springer

Editors
Stefan Wesarg (ID)
Fraunhofer-Institute for Computer Graphics
Research (IGD)
Darmstadt, Germany

John S. H. Baxter (ID)
Université de Rennes
Rennes, France

Esther Puyol Antón (ID)
King's College London
London, UK

Marius Erdt (ID)
Singapore, Singapore

Additional Editors *see next page*

ISSN 0302-9743   ISSN 1611-3349 (electronic)
Lecture Notes in Computer Science
ISBN 978-3-031-45248-2   ISBN 978-3-031-45249-9 (eBook)
https://doi.org/10.1007/978-3-031-45249-9

This Springer imprint is published by the registered company Springer Nature Switzerland AG
The registered company address is: Gewerbestrasse 11, 6330 Cham, Switzerland

Paper in this product is recyclable.

Additional Editors

Klaus Drechsler
Aachen University of Applied Sciences
Aachen, Germany

Moti Freiman (iD)
Technion – Israel Institute
of Technology
Haifa, Israel

Islem Rekik (iD)
Imperial College London
London, UK

Aasa Feragen (iD)
Technical University of Denmark
Kgs Lyngby, Denmark

Veronika Cheplygina (iD)
University of Copenhagen
Copenhagen, Denmark

Enzo Ferrante (iD)
Universidad Nacional del Litoral
Santa Fe, Argentina

Daniel Moyer (iD)
Vanderbilt University
Nashville, TN, USA

Cristina Oyarzun Laura
Fraunhofer-Institute for Computer
Graphics Research (IGD)
Darmstadt, Germany

Yufei Chen
Tongji University
Shanghai, China

Roy Eagleson (iD)
Western University
London, ON, Canada

Andrew P. King (iD)
King's College London
London, UK

Melani Ganz-Benjaminsen (iD)
University of Copenhagen
Copenhagen, Denmark

Ben Glocker (iD)
Imperial College London
London, UK

Eikel Petersen (iD)
Technical University of Denmark
Kgs. Lyngby, Denmark

CLIP Preface

The *12th International Workshop on Clinical Image-based Procedures: Towards Holistic Patient Models for Personalised Healthcare (CLIP 2023)* was held in Vancouver, Canada on October 12, 2023. Again, it was organized in conjunction with the International Conference on Medical Image Computing and Computer Assisted Intervention – MICCAI 2023.

Following the CLIP's long tradition of translational research, the goal of the developments presented in our workshop was to bring basic research methods closer to clinical practice. One of the key aspects regarding the applicability of the presented methods is the creation of *holistic patient models* as an important step towards personalised healthcare. As a matter of fact, the clinical picture of a patient does not uniquely consist of medical images. It is rather the combination of medical image data from multiple modalities with other patient data, e.g., omics, demographics or electronic health records, that is desirable. Since 2019 CLIP has put a special emphasis on this area of research.

This year, CLIP switched back to normal. After being organized as a hybrid event last year, the workshop was held as an in-person event again. Considering the huge variety of different topics among the MICCAI satellite events, one could observe that the number of workshops has increased once more. As a result, CLIP 2023 had to be shortened to a half-day event – like many other MICCAI workshops. Having that in mind, the number of 8 submissions is a good result – especially since the quality of the submissions has increased. This was reflected by the paper scores assigned during the reviewing process. Therefore, we decided to introduce a novel short paper category (3 papers) in addition to the accepted 5 full papers. All submitted papers were peer-reviewed double-blind by at least 3 experts.

All accepted papers were presented by their authors during the workshop. The attendees voted for the best one, which received the Best Paper Award of CLIP 2023. In addition to the oral presentations provided by the authors of the accepted papers, CLIP 2023 featured a keynote given by Mathias Unberath from the Department of Computer Science at Johns Hopkins University. His talk about recent efforts around digital twins in medicine was well received by the audience. We would like to thank Mathias for enriching CLIP 2023 with his talk!

Furthermore, we take this opportunity to thank also our program committee members, authors and attendees who helped to make CLIP 2023 a big success.

October 2023

Stefan Wesarg
Moti Freiman
Marius Erdt
Klaus Drechsler
Cristina Oyarzun Laura
Yufei Chen

CLIP Organization

Organizing Committee

Yufei Chen	Tongji University, China
Klaus Drechsler	Aachen University of Applied Sciences, Germany
Marius Erdt	Singapore
Moti Freiman	Technion, Israel
Cristina Oyarzun Laura	Fraunhofer IGD, Germany
Stefan Wesarg	Fraunhofer IGD, Germany

Program Committee

Shafa Balaram	Institute for Infocomm Research, A*STAR, Singapore
Jan Egger	Graz University of Technology, Austria
Moritz Fuchs	Technical University of Darmstadt, Germany
Katarzyna Heryan	University of Science and Technology, Poland
Martin Hoßbach	Clear Guide Medical, USA
Yogesh Karpate	ChiStats Labs Private Limited, India
Purnima Rajan	Clear Guide Medical, USA
Tillmann Rheude	Berlin Institute of Health at Charité, Germany
Andreas Wirtz	Fraunhofer IGD, Germany
Lukas Zerweck	Fraunhofer ITMP, Germany
Stephan Zidowitz	Fraunhofer MEVIS, Germany

FAIMI Preface

During the last few years, research in the area of fairness in machine learning has highlighted the potential risks associated with biased systems in various application scenarios. A large body of research studies has shown that machine learning systems can be biased in terms of demographic attributes like gender, ethnicity, age or geographical distribution, presenting unequal behaviour on disadvantaged or underrepresented subpopulations. Even though fairness in machine learning has been extensively studied in decision-making scenarios like job hiring, credit scoring and criminal justice, it wasn't until recently that researchers started to study and characterize bias and design mitigation strategies for systems in medical image computing (MIC) and computer assisted interventions (CAI).

Aiming to continue and expand the discussion, the *Fairness of AI in Medical Imaging* (FAIMI) MICCAI 2023 workshop was devoted to creating awareness about potential fairness issues that can emerge in the context of machine learning. Moreover, our goal was also to bring together researchers from the MIC, CAI, machine learning and fairness communities who use and develop models for the analysis of biomedical images and encourage discussions about bias assessment and mitigation strategies. To this end, our workshop was divided into three sessions (1) a presentation from an expert keynote speaker; (2) oral presentations provided by the authors of accepted papers; and (3) poster presentations. All accepted papers were presented as posters by their authors during the workshop and the attendees also voted for the recipient of the Best Paper Award.

To select the peer-reviewed papers we leveraged the CMT tool. We applied a double-blind review process and had each submitted paper reviewed by three independent reviewers. The papers were assigned to reviewers taking into account (and avoiding) potential conflicts of interest and recent work collaborations between peers. Of the 20 papers submitted, 19 papers were accepted for publication. The best four regular papers were invited to give oral presentations.

This is the first time that MICCAI has had a workshop on Fairness in AI for medical imaging, and we would like to thank all authors, reviewers and organizers for their time, efforts, contributions and support in making FAIMI 2023 a successful event.

October 2023

Esther Puyol-Antón
Aasa Feragen
Andrew P. King
Enzo Ferrante
Veronika Cheplygina
Melanie Ganz
Ben Glocker
Daniel Moyer
Eike Petersen

FAIMI Organization

Program and Organizing Committee

Veronika Cheplygina	IT University Copenhagen, Denmark
Aasa Feragen	Technical University of Denmark, Denmark
Enzo Ferrante	CONICET, Universidad Nacional del Litoral, Argentina
Melanie Ganz-Benjaminsen	University of Copenhagen, Denmark
Ben Glocker	Imperial College London, UK
Andrew King	King's College London, UK
Daniel Moyer	Vanderbilt University, USA
Eike Petersen	Technical University of Denmark, Denmark
Esther Puyol-Antón	HeartFlow and King's College London, UK

Program Committee

Amelia Jiménez-Sánchez	University of Copenhagen, Denmark
Annika Reinke	German Cancer Research Center, Germany
Bishesh Khanal	Nepal Applied Mathematics and Informatics Institute for Research, Nepal
Cian Scannell	Eindhoven University of Technology, The Netherlands
Cosmin Bercea	Technical University of Munich, Germany
Emma Stanley	University of Calgary, Canada
Hilde Weerts	Eindhoven University of Technology, The Netherlands
Mahesan Niranjan	University of Southampton, UK
Maria A. Zuluaga	EURECOM, France
Pola Schwöbel	Amazon Web Services, Germany
Rodrigo Echeveste	CONICET/Universidad Nacional del Litoral, Argentina
Stephen Pfohl	Google Research, USA
Tareen Dawood	King's College London, UK
Tiarna Lee	King's College London, UK

EPIMI Preface

The second *Ethical & Philosophical Issues in Medical Imaging* (EPIMI) workshop continued to investigate questions that underlie medical imaging research at the most fundamental level. These investigations bridge the purely technical considerations traditionally seen in medical imaging research with humanistic ones and expand their scope beyond *what* is done towards *why* it is done. By doing so, we hoped to more thoroughly understand our basic assumptions in performing medical imaging research and move towards dramatic answers to questions we never before thought to ask.

This instance of EPIMI concentrated on topics surrounding open science, turning a critical lens on the subject. Open science has been thought of as a mechanism for ensuring reproducible and honest research that is accessible to all. However, deeper investigation shows that mere openness is not a panacea for the scientific community. Souza *et al.* critique open research in artificial intelligence on the basis of global equity, providing a perspective that nuances how we can view fairness in medical machine learning research. Rosenblatt *et al.* critique the relationship between open datasets and trust in said datasets, showing how adversarial examples could be used to subtly change open datasets to disguise data fraud. These investigations show that even the concept of open science needs to be refined to best meet the needs of our community.

Papers were selected using a double-blind review process with author rebuttal through CMT. Each paper was evaluated by at least two independent reviewers in terms of the topicality of its content, the clarity of its presentation, the theoretical quality of its arguments, and the empirical quality of the evidence used to support them. Authors were then given an opportunity to critique their review and suggest changes to their final paper. Of the five papers submitted, two were selected for the EPIMI 2023 workshop, giving an acceptance rate of 40%. Both accepted papers were invited to give an oral presentation followed by an extended discussion period.

Lastly, we would to thank all those who submitted papers, the reviewers who critiqued them, and all those who contributed to the discussion at EPIMI 2023.

October 2023

John S. H. Baxter
Islem Rekik
Roy Eagleson
Luping Zhou

EPIMI Organization

Program and Organizing Committee

John S. H. Baxter   Université de Rennes, France
Islem Rekik   Imperial College London, UK
Roy Eagleson   Western University, Canada
Luping Zhou   University of Sydney, Australia

Steering Committee

Elisabetta Lalumera   Università di Bologna, Italy
Pierre Jannin   Université de Rennes, France
Terry Peters   Western University, Canada
Dinggang Shen   Shanghai Technical University, China

Contents

FAIMI

EPIMI

CLIP

Automated Hand Joint Classification of Psoriatic Arthritis Patients Using Routinely Acquired Near Infrared Fluorescence Optical Imaging

Lukas Zerweck[1,3]($\boxtimes$) (iD), Stefan Wesarg[2,3] (iD), Jörn Kohlhammer[2,3] (iD), and Michaela Köhm[1,3,4] (iD)

[1] Fraunhofer Institute for Translational Medicine and Pharmacology ITMP, Frankfurt am Main, Germany
lukas.zerweck@itmp.fraunhofer.de
[2] Fraunhofer Institute for Computer Graphics Research IGD, Darmstadt, Germany
[3] Fraunhofer Cluster of Excellence Immune-Mediated Diseases CIMD, Frankfurt am Main, Germany
[4] Division of Rheumatology, Goethe-University Frankfurt, Frankfurt am Main, Germany

Abstract. Near infrared fluorescence optical imaging (NIR-FOI) is a relatively new imaging modality to diagnose arthritis in the hands. The acquired data has two spatial dimensions and one temporal dimension, which visualizes the time dependent distribution of an administered color agent. In accordance with previous work, we hypotesize that the distribution process allows a joint-wise classification into *inflammatory affected* and *unaffected*.

In this work, we present the first approach to objectively classify hand joint NIR-FOI image stacks by designing, training, and testing a neural network. Previously presented model architectures for spatio-temporal classification do not yield satisfying results when trained on NIR-FOI data. A recall value of 0.812 of the over- and a recall value of 0.652 of the underrepresented class is achieved, the model's robustness tested against small variations and its attention visualized in activation maps. Even though these results leave room for further improvement, they also indicate, that the model architecture can capture the latent features of the data. We are confident, that more available data will lead to a robust classification model and can support medical doctors in using NIR-FOI as a diagnostic tool for PsA.

Keywords: Near Infrared Fluorescence Optical Imaging ·
Spatio-Temporal Data Classification · Neural Network · Psoriatic
Arthritis

1 Introduction

Psoriasis arthritis (PsA) affects 0.3-1% of the general population [3] and may lead to permanent structural joint damage [2] if left untreated. An early diagno-

S. Wesarg et al. (Eds.): CLIP/FAIMI/EPIMI 2023, LNCS 14242, pp. 3–11, 2023.
https://doi.org/10.1007/978-3-031-45249-9_1

sis and start of treatment can slow down disease progression and can reduce costs [2]. However, clear indicators of early stages for diagnosing PsA, e.g. biomarkers, are yet to be discovered.

In recent years, near infrared fluorescence optical imaging (NIR-FOI) emerged as a potential diagnostic tool in the field of rheumatology to detect the existence and early stages of arthritis in the hand joints and is, especially for PsA, part of ongoing research. However, up until today, the acquired data is evaluated semi-quantitatively by capturing each investigated joint's inflammatory status in a NIR-FOI specific score, from which FOIAS is most frequently used [10].

In this work, we present the first approach using a neural network for an objective evaluation of NIR-FOI spatio-temporal imaging data by extracting an image stack per joint and classifying the entire three-dimensional spatio-temporal data stack into *inflammatory affected* and *unaffected*.

We use the FOIAS score, which directly evaluates the imaging data, as ground truth label to show, that the suggested model architecture can capture the latent features of the given data. In future work, we will tackle further research questions, of which the early detection of PsA is the main focus.

2   Background

Related Work. In this work, we present our idea for NIR-FOI joint stack classification, which is described in detail in Subsect. 3. There are different approaches for classifying spatio-temporal data in the literature. We mention the approaches we compared to the suggested idea, but do not present comparison results.

Due to the temporal connection between different slices, the approaches are often based on recurrent neural networks (RNN). Two RNN based approaches, which are used for video classification, are gated recurrent units (GRU) [1] and Long Short-Term Memory units (LSTM) [6]. The usage of GRUs is suggested by the official Keras website [9], while in Halder et al. [4] a CNN-BiLSTM is presented to capture the temporal relations. In Mao et al. [8] a graph network for video classification is proposed. However, the data in Mao et al. [8] differ from ours in the sense that an overall video semantic, based on different camera angles and views, needs to be found. We implemented all these approaches and compared them to the presented model in this work. However, none of these yielded satisfying classification results.

Thus, we divided our model into two simpler calculations. Extraction of a temporal embedding, which is then classified by a convolutional neural network. Using a temporal convolutional network [7] for the time embedding extraction, lead to better results than the RNN based approaches, but does not outperform our model.

Since all of the approaches mentioned above showed good results in the past, we currently do not have a satisfactory explanation for the low performance classifying the data used in this work. One difference to previously suggested models is the low resolution of the used data for this work.

Data. NIR-FOI is an imaging modality tailored to the color agent indocyanine green. It acquires 360 images at one image per second, resulting in two spatial dimensions (x, y) and one temporal dimension (t). Even though a general distribution process is described in previous work [10], the color agent distribution varies greatly between data sets. Thus, a neural network is trained to learn the latent features of *affected* and *unaffected* joints.

The data sets used for this work were acquired during a multi-centre study, which fulfilled Good Clinical Practice Guidelines in accordance with the Declaration of Helsinki. It was approved by the ethics committee of the University Hospital Frankfurt am Main and all participants gave signed consent to be included in the study and to the usage of their data for research purposes. In total, 659 data sets from 27 patients, with 104 joints labeled as *affected* and 555 joints labeled as *unaffected*, are included.

3 Method

All described methods are implemented using Python 3.3.8 and Tensorflow 2.3.0. To train the neural network for classifying joints as *affected* and *unaffected*, each data set needs to go through a pre-processing pipeline.

Firstly, 26 joint stacks are extracted from each patient's image stack, based on a calculated two-dimensional segmentation map. Each stack is adjusted to a size of width = 50, height = 50, and channel = 360 by spatial interpolation. The wrists and interphalangeal joint of both thumbs are not considered, due to their different positioning with regard to the measurement device (CCD chip). Then, each stack goes through a sequence of three steps for each training epoch: Addition of white noise, z-score normalization, and further augmentation (e.g. random flipping, random rotation, and more).

To test the neural network's robustness, during model testing, described in Subsect. 4, white - and salt-and-pepper noise are added to the data sets.

Model. The entire model structure is visualized in Fig. 1. The model has two main paths. While one side, including the "time blocks" and "space blocks", is learning the latent features (latent path), the other path is serving as a big skip connection (inspired by a residual block [5]) (skip path).

For the latent path, firstly five "time blocks" with a decreasing number of channels is performed. In each "time block" a one-dimensional convolution is calculated, which serves as a fully connected layer in temporal dimension. The output of each "time block" is concatenated to a 50×50x5 latent space: The time embedding. Continuously decreasing the number of channels, while keeping each step's embedding, combines the high- and low-level time features. Then, the time embedding runs through three "space blocks", to capture the time embedding's spatial features. The final tensor of shape 12×12x512 is reduced to a 512-element vector by global average pooling to be concatenated with the result of the skip path.

The idea of the skip path is to capture the input data set in a few values and infuse it into the fully connected layers at the end of the model, to make learning

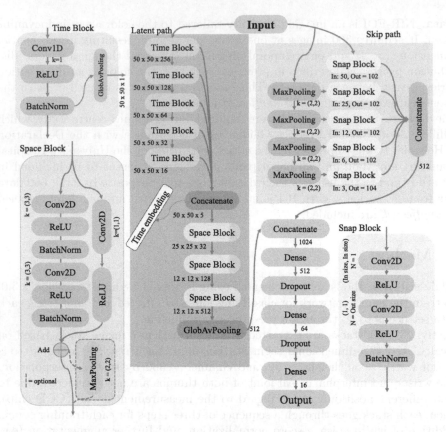

Fig. 1. Model structure. The figure visualizes the entire model architecture with its two main paths: The latent path capturing latent features in temporal and spatial dimension, and the skip path to make learning easier. Additionally, the different block types are shown (Time-, Space- and Snap Block).

easier. It consists of five "snap blocks", which all have the original input in its initial or spatially pooled dimensions. A "snap block" captures the whole stack in a single value by performing a two-dimensional convolution, with the filter size as big as the spatial dimensions and *valid* padding. This single value is convoluted with a 1×1 filter either 102 or 104 times, to adjust the vector size. The five results of the "snap blocks" are concatenated to a vector of size 512 and then concatenated with the latent path output.

Finally, the concatenated vector of size 1024 runs through four dense layers with decreasing size and two dropout layers.

Model Training. The available data is randomly split with a ratio of 70% (abs.: 416) training -, 20% (abs.:132) validation - and 10% (abs.: 66) test data, while considering the class imbalance. However, the validation data is only used

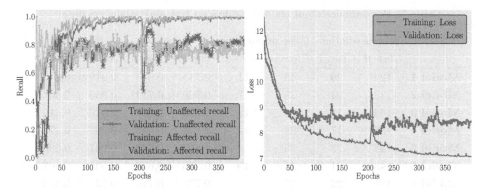

Fig. 2. Recall for trainings- and validation data for both classes and all epochs.

Fig. 3. Trainings- and validation loss for all epochs.

for visualizing the model's performance on non-training data during training and is not used for fine tuning. Thus, validation and test data can be considered equal, during model evaluation. The model is trained with the following setup: Epochs: 400, batch size: 64, optimizer: Adam, loss: binary cross-entropy. The learning rate is not stable and decreases over time starting at 10^{-4} and ends at $2 \cdot 10^{-5}$ with two intermediate increases.

Model Robustness Testing. As mentioned before the amount of available data sets for the class *affected* is relatively small. Thus, different combinations of training- and test data sets can lead to very different results. To get a reliable result, the neural network is trained multiple times from scratch with the same training conditions but with a random assignment of samples to training - and test data set.

4 Results

The results of all training runs are summarized in Table 1. The model with the gray background is used for further investigation. Due to the high class imbalance, the accuracy, as well as precision value, have limited explanatory power. Instead, the recall value for both classes over all 400 epochs is visualized in Fig. 2. Additionally, the loss is visualized in Fig. 3. In both figures, the increase in learning rate causes a fluctuation in recall and loss value. However, the recovery of both values after a small amount of epochs indicates, that the model is optimized towards the global and is not stuck in a local minimum. The recall and loss values are relatively stable for the validation data, which indicates a maximal performance with the given data. Additionally, the validation loss does not increase after many epochs of training. Thus, no numerical signs of overfitting can be observed.

Table 1. Performance of all trained models on test and validation data.

	True prediction		False prediction		Recall		Precision	
	Class 0	Class 1	Class 0	Class 1	Class 0	Class 1	Class 0	Class 1
Model 1	124	20	43	11	0.743	0.645	0.919	0.317
Model 2	132	24	35	7	0.790	0.774	0.950	0.407
Model 3	142	18	25	13	0.850	0.581	0.916	0.419
Model 4	138	17	29	14	0.826	0.548	0.908	0.370
Model 5	144	23	23	8	0.862	0.742	0.947	0.500
Model 6	143	21	24	10	0.856	0.677	0.935	0.467
Model 7	130	22	37	9	0.778	0.710	0.935	0.373
Model 8	128	19	39	12	0.766	0.613	0.914	0.328
Model 9	140	18	27	13	0.838	0.581	0.915	0.400
	Σ	Σ	Σ	Σ				
	1221	182	282	97	0.812	0.652	0.926	0.392

Investigating Model Results. To get a deeper understanding of the trained neural network, two evaluation steps are performed. Firstly, the model's focus on the input joint stacks is visualized. Secondly, its robustness against small data changes is tested.

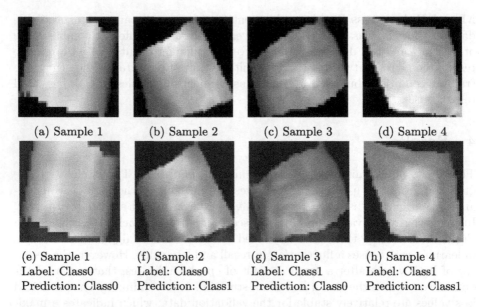

(a) Sample 1   (b) Sample 2   (c) Sample 3   (d) Sample 4

(e) Sample 1   (f) Sample 2   (g) Sample 3   (h) Sample 4
Label: Class0   Label: Class0   Label: Class1   Label: Class1
Prediction: Class0   Prediction: Class1   Prediction: Class0   Prediction: Class1

Fig. 4. For the images in (a) - (d) the standard deviation along the temporal dimension is calculated. (e) - (h) show the overlay of these stand deviation images with the corresponding activation map. A sample for each possible classification result is visualized.

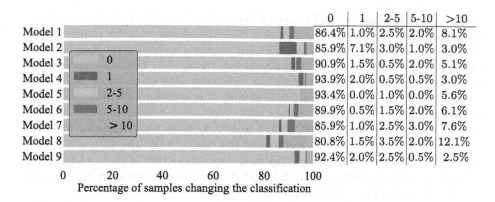

	0	1	2-5	5-10	>10
Model 1	86.4%	1.0%	2.5%	2.0%	8.1%
Model 2	85.9%	7.1%	3.0%	1.0%	3.0%
Model 3	90.9%	1.5%	0.5%	2.0%	5.1%
Model 4	93.9%	2.0%	0.5%	0.5%	3.0%
Model 5	93.4%	0.0%	1.0%	0.0%	5.6%
Model 6	89.9%	0.5%	1.5%	2.0%	6.1%
Model 7	85.9%	1.0%	2.5%	3.0%	7.6%
Model 8	80.8%	1.5%	3.5%	2.0%	12.1%
Model 9	92.4%	2.0%	2.5%	0.5%	2.5%

Percentage of samples changing the classification

Fig. 5. Summary of the robustness test. Each column contains the percentage of data sets, for which the classification is **switched** as many times as the bar color or column title states (e.g. the label/column 0 displace the percentage of data sets, which switch 0 times the classification during all 100 classifications).

For four samples activation maps are shown in Fig. 4, to visualize the influence of different parts of the initial joint stack on the classification. To capture the dynamic distribution progress in one image and thus, enable an evaluation of the activation maps, the standard deviation along the temporal dimension is calculated, and visualized in Fig. 4 (a)-(d). While the neural network focuses on meaningful parts for correctly classified samples (Fig. 4 (e) and (h)), it seems to focus on less meaningful areas for wrongly classified samples (Fig. 4 (f) and (g)). Especially, in Fig. 4 (g) the model seems to miss the bright spot and rather focuses on the joint's edge.

To test the validity of the final result, all test and validation data sets are predicted 100 times, with randomly added white - and salt-and-pepper noise individually for each sample and test. A sample switching between different predictions due to added noise represents an uncertain neural network and can indicate overfitting (the approach is inspired by adversarial attacks). The smallest number of 0 switches can be observed for model 8 with a switch rate of 80.8%. Model 5, the one highlighted in Table 1, has a switch rate of 93.4% and thus, appears robust against smaller changes in the data (the results are summarized in Fig. 5). Both evaluation steps support the conclusion of a not overfitted model.

5   Discussion and Future Work

In this work, we present a neuronal network architecture, to classify joints, based on NIR-FOI spatio-temporal imaging data. Approaches from the literature to classify the data used in this work do not yield satisfying results, for which we currently do not have a substantiated explanation.

With the presented idea, an average recall of 0.812 of the over- and an average recall of 0.652 of the underrepresented class is achieved. These values clearly state, that the trained model performs better than random guessing, especially considering the class imbalance, but still has a lot of potential for improvement. The development over all epochs of both recall values, as well as the loss, indicate, that a learning is achieved, without overfitting the model. This conclusion is also supported by the repetitive testing, for which the vast majority of samples do not change their classification when applying minor changes to the input image (compare Fig. 5).

The visualized activation maps, shown in Fig. 4, allow the conclusion that the model can capture the latent features of the data. While the activation maps of correct classified samples support this conclusion (Fig. 4 (e) and (h)), the wrongly classified samples also indicate that more data is necessary to eliminate wrong attention (Fig. 4 (f) and (g)).

In summary, the presented pipeline shows the potential to be a robust classifier, for the extracted NIR-FOI joint stacks. Thus, the presented neural network architecture will be trained on more data to develop an automated evaluation system to support clinicians in diagnosing PsA with the support of NIR-FOI.

Acknowledgment. This project has received funding from the Innovative Medicines Initiative 2 Joint Undertaking (JU) under grant agreement No 101 007 757 (Hippocrates). The JU receives support from the European Union's Horizon 2020 research and innovation program and EFPIA.

References

1. Ballas, N., Yao, L., Pal, C., Courville, A.: Delving deeper into convolutional networks for learning video representations. arxiv.org/abs/1511.06432 (2015)
2. D'Angiolella, L.S., et al.: Cost and cost effectiveness of treatments for psoriatic arthritis: a systematic literature review. PharmacoEconomics **36**(5), 567–589 (2018). https://doi.org/10.1007/s40273-018-0618-5
3. Gladman, D.D.: Psoriatic arthritis: epidemiology, clinical features, course, and outcome. Ann. Rheum. Dis. **64**(suppl 2), ii14–ii17 (2005). https://doi.org/10.1136/ard.2004.032482
4. Halder, R., Chatterjee, R.: CNN-BiLSTM model for violence detection in smart surveillance. SN Comput. Sci. **1**(4) (2020). https://doi.org/10.1007/s42979-020-00207-x
5. He, K., Zhang, X., Ren, S., Sun, J.: Deep residual learning for image recognition. In: Proceedings of the IEEE Conference on Computer Vision and Pattern Recognition (CVPR) (2016)
6. Hochreiter, S., Schmidhuber, J.: Long short-term memory. Neural Comput. **9**(8), 1735–1780 (1997). https://doi.org/10.1162/neco.1997.9.8.1735
7. Lea, C., Flynn, M.D., Vidal, R., Reiter, A., Hager, G.D.: Temporal convolutional networks for action segmentation and detection. arxiv.org/abs/1611.05267 (2016)
8. Mao, F., Wu, X., Xue, H., Zhang, R.: Hierarchical video frame sequence representation with deep convolutional graph network. In: ECCV 2018 Workshops, pp. 262–270 (2019). https://doi.org/10.1007/978-3-030-11018-5_24
9. Paul, S.: Video classification with a CNN-RNN architecture (2021). https://keras.io/examples/vision/video_classification/. Accessed 25th Apr 2023

10. Werner, S.G., et al.: Indocyanine green-enhanced fluorescence optical imaging in patients with early and very early arthritis: a comparative study with magnetic resonance imaging. Arthritis Rheum. **65**(12), 3036–3044 (2013). https://doi.org/10.1002/art.38175

Automatic Neurocranial Landmarks Detection from Visible Facial Landmarks Leveraging 3D Head Priors

Oded Schlesinger[✉], Raj Kundu, Stefan Goetz, Guillermo Sapiro,
Angel V. Peterchev, and J. Matias Di Martino

Duke University, Durham, NC, USA
oded.schlesinger@duke.edu

Abstract. The localization and tracking of neurocranial landmarks is essential in modern medical procedures, e.g., transcranial magnetic stimulation (TMS). However, state-of-the-art treatments still rely on the manual identification of head targets and require setting retroreflective markers for tracking. This limits the applicability and scalability of TMS approaches, making them time-consuming, dependent on expensive hardware, and prone to errors when retroreflective markers drift from their initial position. To overcome these limitations, we propose a scalable method capable of inferring the position of points of interest on the scalp, e.g., the International 10–20 System's neurocranial landmarks. In contrast with existing approaches, our method does not require human intervention or markers; head landmarks are estimated leveraging visible facial landmarks, optional head size measurements, and statistical head model priors. We validate the proposed approach on ground truth data from 1,150 subjects, for which facial 3D and head information is available; our technique achieves a localization RMSE of 2.56 mm on average, which is of the same order as reported by high-end techniques in TMS. Our implementation is available at https://github.com/odedsc/ANLD.

Keywords: Automatic landmark detection · Supervised learning · TMS

1 Introduction

Recent work in TMS achieved remarkable efficacy in treating patients with severe depression [2]. This approach has been recently cleared by the U.S. Food and Drug Administration (FDA) to target specific brain regions. Other work [9] tracked and estimated head pose while exploiting visible facial landmarks gaining attention for its potential toward accurate and scalable novel marker-less neuronavigation paradigms. Limiting factors for the adoption of such personalized TMS treatments include the need for expensive hardware (e.g., neuronavigation tools), the precise tracking of scalp keypoints, and the requirements of obtaining the subject fcMRI. Our work takes a step toward overcoming some

© The Author(s), under exclusive license to Springer Nature Switzerland AG 2023
S. Wesarg et al. (Eds.): CLIP/FAIMI/EPIMI 2023, LNCS 14242, pp. 12–20, 2023.
https://doi.org/10.1007/978-3-031-45249-9_2

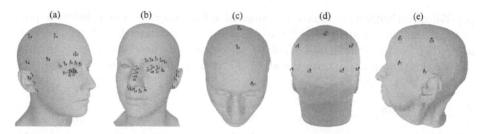

Fig. 1. We leverage head model priors (3DMM) and MRI scans to predict the location of points on the scalp (10–20 System landmarks - blue points) from visible facial landmarks (subset of 68 MultiPIE landmarks scheme - red points). Examples of annotated landmarks on: UHM mean shape (a, b), IXI subject (c, d), ADNI subject (e). (Color figure online)

of these limitations, introducing a computer vision based methodology, leveraging recent advances in facial 3D Morphable Models (3DMM) and full-head MRI datasets to study the relationships between facial landmarks and cranial topology, as illustrated in Fig. 1. We train neural networks with these relationships to infer target keypoints coordinates.

Related Work. Manual annotation and tracking of cranial landmarks are prone to displacement errors, which may lead to incorrect medical diagnosis or inadequate treatment [1]. Torres et al. [14] presented a deep-learning landmark detection method focused on 3D infant head surface shapes. Xiao et al.'s [15] introduced a method that relies on head surface sampling to infer the 10–20 System landmark locations. Gilani et al. [5] established a dense correspondence between faces for detecting facial landmarks using full 3D face data. Ploumpis et al. [11,12] introduced the Universal Head Model (UHM), a 3DMM that combines head, face, ears, eyes, and mouth models, capturing fine-detailed human head structure from subjects across ages, gender, and ethnicity.

One of the limitations of Torres et al.'s [14] proposed method is that it requires a complete 3D head shape. Xiao et al.'s [15] method requires a digital pointer and the acquisition of the neurocranium surface to annotate the neurocranial 10–20 System landmark locations. Gilani et al.'s [5] method requires using a complete surface to detect landmarks and do so only within the same surface.

In contrast to previous methods, our proposed detection method utilizes easily acquired features (visible facial landmarks) to predict other features (neurocranial landmarks) originating in another, unobserved and hard-to-acquire surface, and does not require special tools or having the subject fcMRI. Hence, it could help with detecting these landmarks which are essential for various critical interventions. A possible use-case of our method includes the extraction of such features using cameras during interventions, as exemplified by Matsuda et al. [9].

Main Contributions. (i) We validate for the first time an ensemble of local 3DMM [11,12] and test its global accuracy using ground truth human 3D data;

(ii) We quantitatively evaluate the mutual information between different portions of the human anatomy (iii) We develop and validate data-driven models to predict from visible facial keypoints, and optionally head size measurements, the 3D position of the International 10–20 System neurocranial landmarks.

2   Methods

3DMM is a statistical tool which normally consists of a 3D mesh representing the mean shape of the sampled data, and a set of K eigenvalues and eigenvectors. The eigenvectors share a dense correspondence with the mean shape, which enables the representation of the variety in the model's class and the generation of high-fidelity synthetic instances and examples within it.

We considered facial features consistent with the 68 MultiPIE landmarks scheme defined by Gross et al. [6]. For the neurocranium region, we used the International 10–20 System [8], which is commonly used for electroencephalogram (EEG) electrode placement and for coil localization in TMS procedures. For our predictions to be robust and independent of facial expression, we chose the 1–3 and 15–48 landmarks to represent facial structure. These landmarks are visually distinguishable and located in rigid regions of the face, making them easy to acquire and stable across facial expressions. We selected geodesic distances to represent cranial size measurements such as the head circumference and the distances between landmarks which are commonly used as reference points during the 10–20 System landmarks manual annotation, namely the inion, nasion, and pre-auricular points (see 'A1' and 'A2' in Fig. 1(a) and (b) ear area).

2.1   Datasets and Preprocessing

Learning 3DMM-Based Shape Priors. The UHM mean shape has been already annotated with the 68 MultiPIE facial landmarks. The 10–20 System landmarks were defined following Jasper et al. [8] by picking the closest point to each landmark. We used the UHM to create a synthetic dataset. Each synthetic head instance was generated by adding deformations to the UHM mean shape. In order to generate well-grounded synthetic head instances, we considered each of the model's $K = 500$ eigenvectors and eigenvalues. The products of each of these was multiplied by a weight sampled from the normal distribution with mean 0 and standard deviation of 0.35. This standard deviation was selected based on empirical observations, generating synthetic instances that resemble those of real-life human heads in both MRI datasets (as described below). Following the described process, we gathered 25,000 instances to create the synthetic dataset.

MRI Scans. Evaluations are conducted using two publicly available datasets, IXI [7] and ADNI1 Complete 1Yr 1.5T collection (ADNI) [10], consisting of the head and the face of 581 and 639 subjects, respectively.

In order to obtain subject's facial and neurocranial features, we first preprocess the head MRI image by segmenting the cranium outer shape and meshing

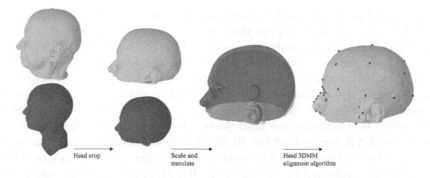

Fig. 2. Different stages of the subject's head (light gray) landmark annotation process using the head 3DMM (slate gray). When both meshes are aligned, we use the 3DMM landmark annotations to annotate the subject's head mesh. Using facial landmarks (red points) and cranial size features we infer cranial landmarks (blue points) coordinates. (Color figure online)

it using the open source software package SimNIBS 3.2.5. The meshed scans are cropped so that we only use parts of the head where the selected face landmarks are located, as illustrated in Fig. 2. The data of 38 IXI and 32 ADNI subjects was not processed due to technical issues or having unusual image artifacts. Consequently, 543 IXI and 607 ADNI valid subjects were included in the study.

Following that, we align the subject and 3DMM point clouds. This is a joint optimization problem consists of two main elements: rigid point clouds alignment, and template deformation. First, we translate the subject's head point cloud and scale it to coarsely fit the UHM mean shape general proportions. Then, we morph the 3DMM to match the subject's aligned head point cloud, utilizing Generalized ICP [13] for rigid points alignment and matching correspondence set establishment, and BFGS algorithm [4] for finding the morphable model weights which minimize the distances between corresponding points in the point clouds. After the alignment step, we utilize the established correspondence set between both point clouds and the pre-existing UHM landmark annotations to annotate each of the facial and 10–20 System landmarks on the subject's head using its correspondent on the fitted 3DMM head. Thus, all landmarks are annotated within the subject's point cloud and not on the fitted 3DMM head point cloud.

Further implementation details can be found in the associated code repository.

2.2   Models Training and Evaluation

To learn a data-driven representation capable of mapping facial keypoints into head landmarks, we implement a neural network model with PyTorch, namely a multilayer perceptron (MLP) with 3 hidden fully connected layers. Between the layers, there are Leaky ReLU activation functions, batch normalization, and dropout layers. This model input consists of the spatial coordinates of the selected facial landmarks and the three geodesic distances described above. Its output is the spatial coordinates of the desired 10–20 System landmark position.

We rely on the assumption that parts of the cranium are related and there is some cranial structural regularity. Therefore, landmark features originating in some cranial regions may help infer landmark positions in other regions. We also evaluate the UHM by assessing how well it captures this regularity and the relationships between different head parts. We train the models mentioned in Sects. 3.1 and 3.3, for each of the 10–20 System landmarks, with a batch size of 8 and Adam optimizer for 500 iterations with squared L2 norm loss (MSE). Predictions are evaluated by measuring the RMSE of the predicted position displacement. For proper and more robust evaluations of the models' performance, we employ 5-fold cross-validation and report the mean RMSE over folds. Using these models, it is possible to obtain the spatial coordinates of neurocranial landmarks, without acquiring the shapes of the entire face or head.

3   Experimental Results

3.1   Neurocranial Landmark Coordinates Prediction

We predict neurocranial landmark positions using the above-mentioned facial landmark positions and geodesic distances. The predictions appearing in the following experiments were conducted with the same folds for each one of the datasets. We trained three types of neural network models: the Synthetic, IXI, and ADNI models, which were trained on each one of the datasets, respectively.

Displacement Errors. For both MRI landmark datasets, landmarks' displacement RMSE of a 5-fold cross-validation model, are significantly lower than those of the Synthetic model, as illustrated in Fig. 3. The UHM was not created and designed specifically for this task, and displacement errors are somewhat affected by annotating landmarks using it. Yet it serves as a baseline for comparing our method while using subjects' data with statistical head model priors, utilizing these priors only, thus demonstrating and emphasizing the essence of real data in training.   Both MRI models achieved similar results across different neurocranial landmarks when tested with the same training scheme, and share similar patterns.

Landmark Coordinates Mutual Information. We are interested in assessing the relationships between landmark positions from an information theory perspective. We quantitatively evaluated these by computing their mutual information values. Mutual information share similar patterns and values across both MRI datasets, which reinforces the validity of these relationships as captured by our annotation method. Full mutual information computation code and a detailed table are provided in the associated GitHub repository.

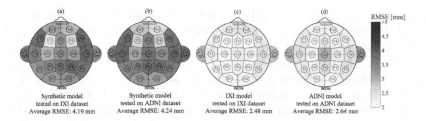

Fig. 3. Displacement RMSE across 10–20 System landmarks predicted from facial landmarks and three geodesic distances measured across subjects' head surface.

3.2   3DMM Validation

The UHM [11,12] was constructed based on the fusion of 3DMM models which represent different parts of the head, and were constructed separately using different subjects. Therefore, UHM may not maintain and preserve the relationships between different head parts. We investigate this by comparing the error between aligning it to the datasets subjects' entire head, as mentioned before, and aligning it using the face region only. We also compare the error over the face when fitting using the entire head versus using only the face.

The metric we use for these comparisons is the Chamfer Distance, which is invariant to the number of points in the point clouds. Essentially, we evaluate the fitting performance between two point clouds, P_1 and P_2, using:

$$CD(P_1, P_2) = \frac{1}{|P_1|} \sum_{p_1 \in P_1} \min_{p_2 \in P_2} \|p_1 - p_2\|_2 + \frac{1}{|P_2|} \sum_{p_2 \in P_2} \min_{p_1 \in P_1} \|p_1 - p_2\|_2. \quad (1)$$

The alignment results across all subjects in both MRI datasets (see Table 1) demonstrate that Chamfer distances, when evaluated over one part of the head, are not significantly higher than when evaluated using the same part that was used for alignment. Hence, it can be deduced that the UHM does preserve the relationships in the human head structure.

Table 1. Chamfer distances RMSE comparison over MRI datasets. Alignment being done utilizing the entire head or only the face region, and measured over both regions.

			RMSE [mm]	
			Head	Face
IXI	Aligned with	Head	8.32	10.03
		Face	10.21	9.47

			RMSE [mm]	
			Head	Face
ADNI	Aligned with	Head	7.5	7.92
		Face	7.98	5.93

3.3   Ablation Study

Based on our experimental results, there exists a regularity according to which the neurocranial features can be predicted based on other features. We evaluate how these are maintained in subjects originating from different datasets, composed of different populations. For this study we test each of the MRI models, IXI and ADNI, on the other dataset. When facial landmarks data acquisition is done with no advanced medical imaging devices, some errors are added to the landmark positions, those are normally on the scale of 1 mm [3]. We model those errors by adding Gaussian noise in random directions, with mean of 1 mm and standard deviation of 0.5 mm to the facial landmark coordinates. The RMSE of these is on average 0.31 mm higher across landmarks, meaning, the added noise degraded the results by a smaller degree than its own magnitude. Cranial size measurements, represented by geodesic distances, might not be accessible to all users. Therefore, we also evaluate the results of models which were only given facial landmarks as input. The results of these ablation study experiments are illustrated in Fig. 4.

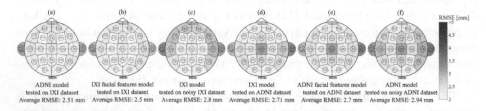

Fig. 4. Ablation study results. Illustrated scalps represent the displacement RMSE across 10–20 System landmarks. MRI models (a, d) show good results also when tested on the other MRI dataset. Missing cranial size measurements (b, e) slightly degrade prediction results. Noisy features acquisition (c, f) demonstrates the models' robustness.

4   Discussion and Conclusions

This paper proposes a scalable method that allows the efficient and practical localization of neurocranial landmarks, which is part of state-of-the-art medical procedures, and achieves an average localization RMSE of 2.56 mm on average over two MRI datasets. The proposed method's displacement errors are not equally distributed over the predicted landmarks when tested with human subjects. Landmarks located near the face or at the back of the head produced better results than those located near the ears. This trend repeats in all the experiments conducted, showing the consistency of our method under different conditions.

A potential limitation of our work includes the prediction models not being trained or tested directly with data collected in intervention settings, although these were evaluated and validated by numerous real subjects' data originating from two different populations. Additionally, preprocessing is reliant on MRI subjects not having distinct facial or cranial deformities. Displacement errors for models tested on the IXI dataset are consistently lower than those of the

ADNI dataset, likely because IXI dataset images include fewer small deformities that allowed for a more accurate reconstruction of subjects' head surface.

During the development of this method, we leverage an existing local 3DMM ensemble. To this end, we align it to the subject's head using two different regions and compare the results over them. We demonstrate that despite the expected decrease in performance while utilizing or evaluating partial data, the model is able to capture the shape of the subject's head accurately.

In contrast to other methods, our detection method only requires facial features that are easy to obtain; some head measurements can be added to further improve its precision. Ablation study shows our method is robust to perturbations which are common in the facial landmarks annotation task.

We have explored the relationships between different cranial features, and demonstrated a method that utilizes those relationships to infer neurocranial landmark positions, outperforming other common used methods. This method helps to lower the cost and increase access to critical interventions such as TMS, as well as other applications such as EEG and surgery for deep brain stimulation.

Acknowledgments. Research reported in this publication was supported by the Duke Institute for Brain Sciences and the National Institute of Mental Health of the National Institutes of Health under Award Number R01MH129733. Additional support from NSF, ONR, and NGA, is acknowledged. The content is solely the responsibility of the authors and does not necessarily represent the official views of the funding agencies. The authors thank Dr. Dennis Turner for helpful discussions of neurosurgical applications.

References

1. Atcherson, S.R., Gould, H.J., Pousson, M.A., Prout, T.M.: Variability of electrode positions using electrode caps. Brain Topogr. **20**(2), 105–111 (2007)
2. Cole, E.J., et al.: Stanford neuromodulation therapy (SNT): a double-blind randomized controlled trial. Am. J. Psychiatry **179**(2), 132–141 (2022)
3. Fagertun, J., et al.: 3d facial landmarks: inter-operator variability of manual annotation. BMC Med. Imaging **14**(1), 1–9 (2014)
4. Fletcher, R.: Practical Methods of Optimization. Wiley, Hoboken (2013)
5. Gilani, S.Z., Mian, A., Shafait, F., Reid, I.: Dense 3d face correspondence. IEEE Trans. Pattern Anal. Mach. Intell. **40**(7), 1584–1598 (2017)
6. Gross, R., Matthews, I., Cohn, J., Kanade, T., Baker, S.: Multi-pie. Image Vis. Comput. **28**(5), 807–813 (2010)
7. IXI dataset. https://brain-development.org/ixi-dataset/
8. Jasper, H.H.: The ten-twenty electrode system of the international federation. Electroencephalogr. Clin. Neurophysiol. **10**, 370–375 (1958)
9. Matsuda, R.H., Souza, V.H., Kirsten, P.N., Ilmoniemi, R.J., Baffa, O.: MarLe: markerless estimation of head pose for navigated transcranial magnetic stimulation. Phys. Eng. Sci. Med. **46**, 1–10 (2023)
10. Petersen, R.C., et al.: Alzheimer's disease neuroimaging initiative (ADNI): clinical characterization. Neurology **74**(3), 201–209 (2010)
11. Ploumpis, S., et al.: Towards a complete 3d morphable model of the human head. IEEE Trans. Pattern Anal. Mach. Intell. **43**(11), 4142–4160 (2020)

12. Ploumpis, S., Wang, H., Pears, N., Smith, W.A., Zafeiriou, S.: Combining 3d morphable models: a large scale face-and-head model. In: Proceedings of the IEEE/CVF Conference on Computer Vision and Pattern Recognition, pp. 10934–10943 (2019)
13. Segal, A., Haehnel, D., Thrun, S.: Generalized-ICP. In: Robotics: Science and Systems, Seattle, WA, vol. 2, p. 435 (2009)
14. Torres, H.R., et al.: Anthropometric landmark detection in 3d head surfaces using a deep learning approach. IEEE J. Biomed. Health Inform. **25**(7), 2643–2654 (2020)
15. Xiao, X., et al.: Semi-automatic 10/20 identification method for MRI-free probe placement in transcranial brain mapping techniques. Front. Neurosci. **11**, 4 (2017)

Subject-Specific Modelling of Knee Joint Motion for Routine Pre-operative Planning

Jeffry Hartanto[1(✉)], Wee Kheng Leow[1], Andy Khye Soon Yew[2],
Joyce Suang Bee Koh[2], and Tet Sen Howe[2]

[1] Department of Computer Science, National University of Singapore, Singapore,
Singapore
{jhartanto,leowwk}@comp.nus.edu.sg
[2] Department of Orthopaedics, Singapore General Hospital, Singapore, Singapore
andy.yew.k.s@sgh.com.sg, {joyce.koh.s.b, howe.tet.sen}@singhealth.com.sg

Abstract. To avoid undesired surgical outcomes, a subject-specific computational model of the knee joint is needed for the pre-operative planning stage to predict possible surgical outcomes for a specific subject. Existing models are not suitable for routine clinical practice as they either expose a subject to excessive radiation through multiple computed tomography (CT) scans or require complex auxiliary data obtained using motion capture, electromyography or multiple magnetic resonance imaging scans. Unlike the existing works, this paper presents a subject-specific knee joint model that requires only one CT scan of a patient's knee, which is the minimum amount of information acquired for the diagnosis of a patient's condition in routine clinical practice. Inspired by the functional anatomy of patellofemoral joint in which patella motion is constrained by femoral groove, this paper proposes a novel contact surface model that generates joint motion by encoding the interactions between the surfaces of two contacting bones through several motion paths. Test results show that the proposed model is sufficiently accurate compared to the existing models, which shows promise in its application for routine pre-operative planning of knee surgery.

Keywords: Subject-specific modelling · Knee joint motion model · Routine pre-operative planning

1 Introduction

Knee joint is prone to injuries, such as torn ligaments, that often restrict its range of motion [12,21]. To restore normal knee joint motion, complex surgical interventions are needed, whose outcome depends on the surgeons' experience as well as pre-operative planning. A slight mistake in surgery can result in abnormal knee joint motion. To avoid that, modelling of normal knee joint motion of a specific subject is useful for pre-operative planning of knee surgery such as reconstruction of torn ligament. In particular, the modeled joint motion can be used to help surgeons to identify the optimal attachment sites of a graft that replaces the torn ligament.

Existing knee joint models can be categorized into three categories, namely static models [9,20,22], kinematic models [4,6,10,11] and dynamic musculoskeletal (DM) models [5,7,13,18]. Static models [9,20,22] model the poses of knee bones at prescribed flexion angles by directly constructing 3D bone poses at prescribed flexion angles from computed tomography (CT) or magnetic resonance imaging (MRI) images of a patient's knee. They are simple and accurate but cannot model the full range of knee motion because they either expose a subject to excessive radiation (multiple CTs) or are too costly (multiple MRIs). Kinematic models [4,6,10,11] model the poses of knee bones over a range of flexion angles without considering forces. Methods for kinematic models capture 3D geometries of knee bones at neutral pose from CT or MRI images. In addition, they use knee motion data obtained from low-resolution MRI sequence, fluoroscopic video or motion capture (mocap) system to determine bone poses over a range of flexion angles. These additional computations incur additional errors on modelling accuracy such as mocap errors [3]. DM models [5,7,13,18] model the poses of knee bones and forces over a range of flexion angles. Methods for dynamic models usually apply the kinematic method with mocap data to obtain bone poses, and adopt various muscle models to model muscle forces and joint moments. Methods that use mocap data have the same drawback as kinematic models, whereas other methods that do not use mocap data usually require high computational cost.

The existing knee joint models are subject specific. But, they are not applicable for routine clinical practice mainly due to their excessive auxiliary data requirement. Thus, they have been used primarily for medical research and post-surgery assessments. To address the challenges, this paper proposes a subject-specific knee joint model that requires bone surface models constructed from one CT scan of a patient's knee at neutral pose. This is the minimum amount of information acquired for the diagnosis of a patient's condition in routine clinical practice. Therefore, it is suitable for routine clinical practice. Our work may improve the outcomes of knee surgeries and reduce post-surgery recovery time such as reconstruction of torn ligament.

2   Method

Inspired by the functional anatomy of femoral groove on constraining the patella's movement in patellofemoral (PF) joint motion [8], a novel *contact surface model* is proposed to use the contact surfaces between the bones to characterize the motion between them. This contact surface model is applied separately to the PF joint and the tibiofemoral (TF) joint to generate patella poses and tibia poses with respect to the femur. The generated tibia poses are used to measure the knee flexion angle with respect to the femur. Then, the generated patella poses are matched to the tibia poses temporally according to the constraint on the changes of patellar tendon between them, completing the knee motion model. Details of part of the model are covered in the following subsections.

2.1 Contact Surface Model of PF and TF Joint

Contact surface models of PF and TF joints generate patella and tibia poses with respect to the femur based on their respective contact surfaces with the femur. Formally, the patella/tibia pose generation problem is defined as follows: given 3D surface models of patella/tibia B and femur F at neutral pose, then generate the poses of B with respect to F for the full range of joint motion. The generated bone poses should satisfy the following constraints:

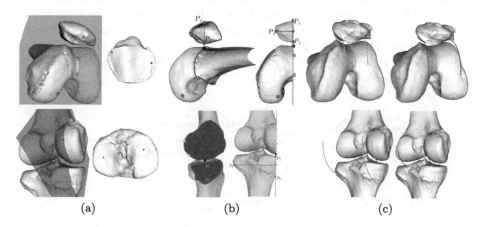

(a)   (b)   (c)

Fig. 1. Schematic of contact surface model of PF and TF joints. (a) Landmarks (dots/cube) and motion plane (green). (b) Intersection curves (blue) and distinctive points (P_1-P_3). (c) Motion paths (green) and refinement paths (blue). (Color figure online)

1. The bone is constrained to move in a plane that intersects the femur, which ensures that the bone moves along its contact surface on the femur.
2. The bone is constrained to maintain appropriate gap between itself and the femur. This gap accounts for the thickness of soft tissues between the contacting bones.

One way to generate bone poses that satisfy these constraints is to generate motion paths that define the bone poses with respect to the femur. Then, the bone poses can be obtained by rigidly transforming the bone model along the paths using Similarity Transformation Algorithm [1]. In this way, generations of motion paths for PF and TF joints are slightly different due to their different contact surfaces with respect to the femur, which will be described in the next subsection. On the other hand, the generations of tibia and patella poses are identical and straightforward.

Generation of Motion Paths. Given the 3D surface models of patella/tibia B and femur F, three **motion landmarks** are placed at distinctive features on F. For PF joint, they are placed at the start, mid, and end of femoral groove (Fig. 1a green) where the patella engages the femoral groove during knee flexion-extension motion. For TF joint, they are placed at the most protruded features along the medial femoral condyle (Fig. 1a green) where the tibia rolls and glides on the femur during knee flexion-extension motion. These landmarks are used to define a plane that is related to the motion of B with respect to F, i.e., in-plane translation and flexion of bone B that accounts for four degrees of freedom (DoF). To model tilting and internal/external rotation of the bone B over the femoral condyle, which accounts of the other two DoF, two **refinement landmarks** are placed on each side of B (Fig. 1a blue), which accounts of the other two DoF. For both joints, these landmarks are placed at the sides where the gaps between the two surfaces are small, i.e., indicating contact. Finally, a **termination landmark** (Fig. 1a cube) is placed on F to denote the lowest position of B with respect to F, i.e., final position of the patella/tibia with respect to the femur.

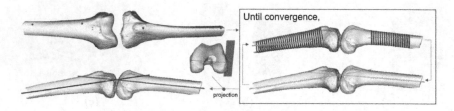

Fig. 2. Schematic for computation of knee flexion angles.

A motion plane (Fig. 1a) that fits the motion landmarks is intersected with B and F to generate intersection curves (Fig. 1b). The distance between the closest points on the two intersection curves describes the thickness of soft tissues between B and F. Moreover, the closest point on F relates to the starting point of B w.r.t. F. At least three distinctive points are needed to define a pose in 3D. For that, the surface normal at the closest point on F intersects B at two points, a point P_1 at distance d_1 and a point P_2 at distance d_2 (Fig. 1b). The third distinctive point P_3 is obtained by intersecting B with a line that intersects the mid-point between P_1 and P_2 along the direction of motion plane's unit surface normal at a distance d_3. Three motion paths that correspond to the three distinctive points P_1–P_3 are computed based on the points along the intersection curve and points' surface normal at distances d_1–d_3 (Fig. 1c). For PF joint, the contact shifts to femoral condyle at late knee flexion, which can be captured by extending the third motion path in a similar way to the generation of the first motion path. Then, the first two motion paths are extended based on the extended third motion path through triangulation. Next, the estimation of refinement paths for the refinement landmarks are easily obtained by triangulation based on the three motion paths. To maintain a constant gap, each point in the estimated refinement paths is placed along the direction of P_1 and P_2 from its closest point on F at distance r_1 and r_2, respectively, where r_1 and r_2 are the distances from the refinement landmarks to their respective closest points on F (Fig. 1c).

2.2   Computation of Knee Flexion Angle

Computationally, the knee flexion angle of a TF joint pose is measured as the angle between the centrelines of the femoral and tibial shafts called *shaft axes* (Fig. 2). To obtain the shaft axes, two extent landmarks are placed on each bone model to indicate the shaft part of the model. Then, an iterative optimization method is applied separately on each bone model to iteratively find an axis that best estimates the shaft axis (Fig. 2). The method first applies PCA on the bone model, and uses the first eigenvector for good initialization as the largest variance of the bone model is along the shaft axis. Then, it computes intersection curves by intersecting the bone model with each cross-sectional plane normal to the shaft axis between the two extent landmarks. Next, it updates the shaft axis as a line that best fits the centroids of the intersection curves. This method iterates until convergence occurs, which is when the shaft axes computed in two consecutive iterations are collinear. Since the shaft axes are unlikely to be coplanar in 3D, two landmarks are placed on the femoral bone model to indicate the anatomical flexion axis that defines a flexion plane [2] (Fig. 2). Finally, the knee flexion angle is measured as the angle between the shaft axes projected onto the flexion plane. This method is applied to each generated TF pose to obtain flexion angles for the generated TF joint poses.

2.3   Matching Tibia and Patella Poses

To obtain knee joint motion, the separately generated tibia poses T_i at various flexion angles θ_i and generated patella poses P_j need to be matched. Physically, tibia and patella are connected through patellar tendon. Thus, an intuitive way to match the generated tibia and patella poses is by temporally matching them according to a constraint on the patellar tendon length changes, which is a *constrained temporal matching problem*. However, the patellar tendon length change pattern of each subject is unknown and differs across different subjects. So, a generic length change pattern $l^*(\theta)$ that quadratically fits the computed tendon lengths of some subjects is adapted to each individual subject. In clinical practice, only the neutral knee pose at angle θ_1 is available for a subject. Thus, the generic pattern that best fits a subject is defined as $l'(\theta) = l^*(\theta) + l(\theta_1) - l^*(\theta_1)$.

The tibia pose T_1 and the patella pose P_1 of the neutral knee pose is a known match $(1,1)$. Let $d(i,j)$ denote *measured* patellar tendon length between T_i and P_j, which is computed as geodesic distance between a pair of clinically-identified attachment sites of patellar tendon. Then, the objective of the matching problem is to find matching pairs (i,j), for each i, that minimize the following error:

$$E = \sum_{i=1}^{m} \|d(i,j) - l'(\theta_i)\|, \tag{1}$$

subject to the following constraints: (1) *Boundary condition*: $(1,1)$ is a match. (2) *One-to-one matching*: each T_i matches only one P_j. (3) *Monotonically increasing matching*: for matches (i,j) and (i',j'), then $j > j' \Leftrightarrow i > i'$.

Dynamic programming (DP) is the natural choice for this problem as the recursive equation of DP is formulated by looking backward. Suppose (i,j) is a match, and $D(i,j)$ denote the total error of the optimal sequence from $(1,1)$ to (i,j). Since patellar tendon is physically stiff, the best matching patella pose for tibia pose T_i must be within a subsequent local window of patella poses of previous tibia pose. Therefore, with a window size w, $D(i,j)$ can be formulated recursively as:

$$D(1,1) = \|d(1,1) - l'(\theta_1)\|,$$
$$\text{For } 2 \leq i \leq m, i \leq j \leq \min\left((i-1)w+1, n\right),$$
$$D(i,j) = \|d(i,j) - l'(\theta_i)\| + \min_{j'} D(i-1,j'), \text{ for } \max(1, j-w) \leq j' < j. \tag{2}$$

Note that the last generated tibia pose T_m and the last generated patella pose P_n are not necessarily an optimal match. The optimal match j^* of m is given by

$$D(m,j^*) = \min_{j'} D(m,j'), \text{ for } m \leq j' \leq \min\left((m-1)w+1, n\right). \tag{3}$$

3   Experiments and Discussions

CT scans of five healthy cadaver knees (S1–S5) were positioned using a jig and captured from neutral knee pose of around $0°$ to $120°$ at intervals of $30°$. These include the left knees of subjects S1, S2 and S4, and the right knees of subjects S3 and S5. The acquired CT scans were used to construct 3D surface models of the knee joint using Visualization Toolkit v5.8 [17]. First, each CT scan was manually segmented using a thresholding method to extract distinct bone regions. Next, the segmented CT scans were used to construct 3D knee surface models using marching cubes method [15]. Lastly, post-processing such as surface smoothing was applied to the constructed 3D surface models. As a result, 3D knee surface models of each subject at five different poses were obtained.

The knee joint motion model was applied to each subject's 3D knee bone models at neutral pose. For each subject, 200 tibia poses and 600 patella poses were generated by the model. Larger number of patella poses than the number of tibia poses is needed to ensure that each tibia pose has an optimal match to a particular patella pose. The matching patella pose of each tibia pose was determined by adapting $l^*(\theta)$ of subjects S1–S5. For each subject, the acquired bone models at neutral pose was used as input to the motion model, whereas bone models at the other four poses were used for evaluation. The experiment was run on a PC with Intel i7-5930K at $3.5\,\text{GHz}$, $32\,\text{GB}$ RAM and NVIDIA GeForce GTX 980. The total computational time needed to generate specific knee joint poses at an interval of about $0.5°$ flexion angle was less than an hour.

3.1 Evaluation of Generated Patella and Tibia Poses

This evaluation measures the accuracy of the separately generated tibia and patella poses with respect to the four acquired poses. Since the clinically measured flexion angles of the acquired poses are inaccurate, it is unreasonable to directly compare the generated poses at specific measured flexion angle. Instead, for each acquired pose, the best matching generated pose is identified in terms of the Hausdorff distance to the acquired pose. Then, the differences between the matching generated poses and acquired poses are measured in terms of the 3D translation and rotation from the matching generated poses. The translation and rotation errors are each measured as a vector with respect to the acquired pose using Similarity Transformation Algorithm [1].

Table 1. Modelling errors of generated patella poses (a) and generated tibia poses (b). Refer to the main text for explanation of the symbols, i.e., +, 0, and −.

	Pose	Translation				Rotation			
		2	3	4	5	2	3	4	5
(a)	S1	1.24 + 0 −	0.71 + + −	0.17 − 0 0	0.91 + + +	0.06 0 0 0	0.01 0 0 0	0.00 0 0 0	0.00 0 0 0
	S2	0.84 + + +	0.35 + 0 +	0.42 + 0 +	1.78 + − +	0.00 0 0 0	0.00 0 0 0	0.06 0 0 0	0.15 0 0 0
	S3	0.42 + 0 +	0.43 0 + +	0.53 − − +	0.41 − + 0	0.01 0 0 0	0.01 0 0 0	0.01 0 0 0	0.03 0 0 0
	S4	0.65 0 − +	0.80 + − −	0.48 + + 0	0.64 + 0 +	0.03 0 0 0	0.06 0 0 0	0.14 0 0 0	0.15 0 0 −
	S5	1.47 − + +	1.92 − + +	2.05 − + +	2.11 0 − +	0.00 0 0 0	0.00 0 0 0	0.00 0 0 0	0.01 0 0 0
(b)	S1	1.38 − 0 −	0.91 0 0 −	1.65 + + −	3.04 + + −	0.00 0 0 0	0.02 0 0 0	0.12 0 − 0	0.29 + − 0
	S2	1.29 0 0 −	1.51 0 + −	2.10 − + −	2.07 − + −	0.00 0 0 0	0.00 0 0 0	0.00 0 0 0	0.00 0 0 0
	S3	0.14 0 + 0	0.87 − + −	1.57 − + −	1.85 − + −	0.00 0 0 0	0.00 0 0 0	0.02 0 0 0	0.05 0 0 0
	S4	1.05 − + +	1.77 − + −	2.55 − + −	2.45 − + −	0.03 0 0 0	0.00 0 0 0	0.00 0 0 0	0.02 0 0 0
	S5	0.49 + 0 0	0.66 + + −	1.32 − + −	2.45 − + −	0.00 0 0 0	0.00 0 0 0	0.01 0 0 0	0.04 0 0 0

For tabulating the error vectors, they are resolved into signed-magnitudes along three principle directions, namely medial-lateral (M-L), anterior-posterior (A-P), and inferior-superior (I-S). Three symbols are used to denote each signed magnitude. The symbol + indicates significant magnitude (> 0.1 mm for translation, > 0.1° for rotation) in the medial, anterior, and inferior directions. The symbol − indicates significant magnitude in the lateral, posterior, and superior directions. 0 indicates insignificant magnitude in any of principle directions.

Table 1 shows the modelling errors of the generated patella poses and generated tibia poses. The ranges of translation errors are 0.17–2.11 mm for patella poses and 0.14–3.04 mm for tibia poses, and both of them have varying translation error pattern across different subjects and flexion angles. Most of the rotation errors of both patella and tibia poses are less than 0.1°. When comparing the modelling errors of the two joint motions, the translation and rotation errors of the generated patella poses are generally smaller than those of the generated tibia poses. One possible reason for such differences is that TF joint motion might be more prominently guided by muscles and ligaments, whereas the PF joint motion is mainly constrained by the femoral groove.

In comparison with the existing models, [6,16,19] reported patella translation errors that range from 0.37 mm to 0.88 mm, and rotation errors that range from 0.3° to 1.75°. For tibia, [14,19] reported translation errors that range from 0.5 mm to 2.4 mm, and rotation errors that range from 0.5° to 3.3°. So, the rotation errors of our model are significantly smaller, whereas the translation errors are comparable for some subjects. Therefore, this evaluation shows that the amounts of modelling errors of both the generated tibia and patella poses are sufficiently small for pre-operative planning of knee surgery.

Table 2. Modelling errors of matching generated patella poses. Refer to the main text for explanation of the symbols, i.e., +, 0, and −.

Pose	Translation				Rotation			
	2	3	4	5	2	3	4	5
S1	1.24 + − +	1.66 + + −	4.31 + + −	7.21 + + −	0.05 0 0 0	0.02 0 0 0	0.04 0 0 0	0.17 + 0 0
S2	1.66 + − +	0.63 + + 0	1.34 + + +	2.41 + + +	0.01 0 0 0	0.01 0 0 0	0.09 0 0 0	0.29 + + −
S3	1.92 + + −	1.10 + + −	1.21 − + 0	1.85 0 + −	0.00 0 0 0	0.01 0 0 0	0.00 0 0 0	0.01 0 0 0
S4	6.24 0 + −	4.92 + + −	4.63 + + −	4.95 + + 0	0.66 + + −	0.54 + + −	0.47 + 0 −	0.50 + + −
S5	4.01 − − +	4.63 − − +	3.11 − − +	2.63 0 − +	0.02 0 0 0	0.08 0 0 0	0.05 0 0 0	0.02 0 0 0

3.2 Evaluation of Tibia and Patella Pose Matching

This section first evaluates the computed flexion angles as it is used to obtain desired patellar tendon length $l'(\theta_i)$. The computed flexion angle linearly increases at a rate of about 0.5° per generated tibia pose, which indicates that there is no abrupt change on the generated TF joint motion. For all subjects, the generated tibia poses start at around 20° computed flexion angle and end at slightly above 120° computed flexion angle.

This section also evaluates the total errors for matching tibia and patella poses with DP at various window sizes w, namely 3, 9 and 15. Among the different window sizes, DP obtains the smallest D with window size $w \geq 9$ for all subjects, i.e., from 7.30 mm to 10 mm across the five subjects. These results indicate that $w = 3$ might be too small to obtain optimal result, whereas $w = 5$ is unnecessarily large. Therefore, DP with $w = 9$ is used for the subsequent evaluations.

Ideally, the matched generated poses and their respective best matching acquired poses should be the same. In reality, there are small differences between them for most subjects mainly due to the differences between the adapted and acquired tendon length change pattern. To quantify the difference, translation and rotation errors from the matching generated patella poses to their respective best matched acquired poses are measured. The translation and rotation are each measured as a vector with respect to the acquired pose using Similarity Transformation Algorithm [1]. The error vectors are tabulated in the same way as in Table 1. Table 2 shows the modelling errors of the matching generated patella poses. The translation errors ranging from 0.63 mm to 7.21 mm, and some rotation errors are greater than 0.1°. This result indicates that the errors are more prominent on the translation than the rotation. The smallest and largest average modelling errors are obtained with subject S3 and S4, respectively. The modelling errors are larger than those in Table 1, which is potentially due to the difference between the adapted and acquired tendon length change pattern.

4   Conclusion

This paper proposes a subject-specific knee joint motion model for routine pre-operative planning. Unlike the existing works, the subject-specific knee joint motion model requires only one CT scan of a patient's knee at full extension, which is the minimum amount of information in knee surgery planning. Thus, it is suitable for routine clinical practice. Essentially, it employs a novel contact surface model, which describes the interactions between the surfaces of two contacting bones, to generate patella and tibia joint poses of the subject separately. Then, it applies an iterative optimization method and dynamic programming to compute flexion angles and temporally synchronize the generated tibia and patella poses, respectively. Comprehensive experimental evaluation shows that the model is sufficiently accurate compared to those reported in the previous studies. Consequently, this work could potentially be extended to model motions of other joints, and be used for routine pre-operative planning of orthopaedic surgeries. This work has several limitations. Primarily, the accuracy of the model may be inconsistent due to subject variability in identifying the landmarks, which could be reduced by developing an algorithm that automatically identifies these distinctive bone surfaces. Beside that, the model omits the change of contact between the contacting bones during joint motion, which could be addressed by modelling the variation of cartilage thickness and contact surface. Lastly, evaluation of the proposed model on a larger number of subjects is needed for better validation before deploying such a solution into routine clinical practice.

References

1. Arun, K., Huang, T., Blostein, S.: Least-squares fitting of two 3-d point sets. IEEE Trans. Pattern Anal. Mach. Intell. **5**, 698–700 (1987)
2. Asano, T., Akagi, M., Nakamura, T.: The functional flexion-extension axis of the knee corresponds to the surgical epicondylar axis: in-vivo analysis using a biplanar image-matching technique. Arthroplasty **20**(8), 1060–1067 (2005)
3. Cereatti, A., et al.: Standardization proposal of soft tissue artefact description for data sharing in human motion measurements. J. Biomech. **62**, 5–13 (2017)
4. Chen, H.C., Wu, C.H., Wang, C.K., Lin, C.J., Sun, Y.N.: A joint-constraint model-based system for reconstructing total knee motion. IEEE Trans. Biomed. Eng. **61**(1), 171–181 (2014)
5. Crouch, D., Huang, H.: Simple EMG-driven musculoskeletal model enables consistent control performance during path tracing tasks. In: Conference Proceedings of the IEEE Engineering in Medicine and Biology Society, pp. 1–4 (2016)
6. Fellows, R., et al.: Magnetic resonance imaging for in-vivo assessment of three-dimensional patellar tracking. J. Biomech. **38**(8), 1643–1652 (2005)
7. Gerus, P., et al.: Subject-specific knee joint geometry improves predictions of medial tibiofemoral contact forces. J. Biomech. **46**(16), 2778–2786 (2013)
8. Hamill, J., Knutzen, K., Derrick, T.: Biomechanical Basis of Human Movement, 4th edn. Wolters Kluwer Health, Philadelphia (2015)
9. Jeong, W., et al.: An analysis of the posterior cruciate ligament isometric position using an in-vivo 3-dimensional computed tomography-based knee joint model. Knee Surg. Sports Traumatol. Arthrosc. **26**(10), 1333–1339 (2010)
10. Kainz, H., Modenese, L., Lloyd, D., Maine, S., Walsh, H., Carty, C.: Joint kinematic calculation based on clinical direct kinematic versus inverse kinematic gait models. J. Biomech. **49**(9), 1658–1669 (2016)
11. Kedgley, A., McWalter, E., Wilson, D.: The effect of coordinate system variation on in-vivo patellofemoral kinematic measures. Knee **22**(2), 88–94 (2015)
12. Kyung, H.S., Kim, H.J.: Medial patellofemoral ligament reconstruction: a comprehensive review. Knee Surg. Relat. Res. **27**(3), 133–140 (2015)
13. Lenhart, R., Kaiser, J., Smith, C., Thelen, D.: Prediction and validation of load-dependent behavior of the tibiofemoral and patellofemoral joints during movement. Ann. Biomed. Eng. **43**(11), 2675–2685 (2015)
14. Li, J.S., et al.: Prediction of in-vivo knee joint kinematics using a combined dual fluoroscopy imaging and statistical shape modeling technique. J. Biomech. Eng. **136**(12), 124503 (2014)
15. Lorensen, W., Cline, H.: Marching cubes: a high resolution 3D surface construction algorithm. In: Proceedings of the 14th SIGGRAPH, pp. 163–169 (1987)
16. Otake, Y., Esnault, M., Grupp, R., Kosugi, S., Sato, Y.: Robust patella motion tracking using intensity-based 2D–3D registration on dynamic bi-plane fluoroscopy: Towards quantitative assessment in MPFL reconstruction surgery. In: Proceedings of Medical Imaging, pp. 105–110 (2016)

17. Schroeder, W., Lorensen, B., Martin, K.: The Visualization Toolkit: An Object-Oriented Approach of 3D Graphics. Kitware (2004)
18. Seth, A., Sherman, M., Reinbolt, J., Delp, S.: OpenSim: a musculoskeletal modeling and simulation framework for in silico investigations and exchange. Procedia IUTAM **2**, 212–232 (2011)
19. Sharma, G., et al.: Radiological method for measuring patellofemoral tracking and tibiofemoral kinematics before and after total knee replacement. Bone Jt. Res. **1**(10), 263–271 (2012)
20. Song, S., Pang, C.H., Kim, C., Kim, J., Choi, M., Seo, Y.J.: Length change behavior of virtual medial patellofemoral ligament fibers during in-vivo knee flexion. Am. J. Sports Med. **43**(5), 1165–1171 (2015)
21. Yamauchi, D., Sato, N., Morita, Y.: An experimental study on the relationship between the components and movement of the human knee using an android model - measurement of internal-external rotation and anterior-posterior tibial translation -. In: Proceedings of ICCAS, pp. 811–815 (2015)
22. Yoo, Y.S., et al.: Changes in the length of the medial patellofemoral ligament: an in-vivo analysis using 3-dimensional computed tomography. Am. J. Sports Med. **40**(9), 2142–2148 (2012)

Towards Fine-Grained Polyp Segmentation and Classification

Yael Tudela[1]([✉]) [iD], Ana García-Rodríguez[2] [iD], Glória Fernández-Esparrach[2] [iD], and Jorge Bernal[1] [iD]

[1] Computer Vision Center, Barcelona, Spain
{ytudela,jbernal}@cvc.uab.cat
[2] Hospital Clinic de Barcelona, Barcelona, Spain
anagrod4@gmail.com, mgfernan@clinic.cat

Abstract. Colorectal cancer is one of the main causes of cancer death worldwide. Colonoscopy is the gold standard screening tool as it allows lesion detection and removal during the same procedure. During the last decades, several efforts have been made to develop CAD systems to assist clinicians in lesion detection and classification. Regarding the latter, and in order to be used in the exploration room as part of resect and discard or leave-in-situ strategies, these systems must identify correctly all different lesion types. This is a challenging task, as the data used to train these systems presents great inter-class similarity, high class imbalance, and low representation of clinically relevant histology classes such as serrated sessile adenomas.

In this paper, a new polyp segmentation and classification method, Swin-Expand, is introduced. Based on Swin-Transformer, it uses a simple and lightweight decoder. The performance of this method has been assessed on a novel dataset, comprising 1126 high-definition images representing the three main histological classes. Results show a clear improvement in both segmentation and classification performance, also achieving competitive results when tested in public datasets. These results confirm that both the method and the data are important to obtain more accurate polyp representations.

Keywords: Medical image segmentation · Colorectal Cancer · Vision Transformer · Classification

1 Introduction

Colorectal cancer is one of the main causes of cancer death worldwide. Its mortality rate depends on the stage it is detected on, hence the importance of screening programs to detect its precursor lesion, the polyp, before it develops into cancer. Colonoscopy is the gold standard screening tool as it allows lesion detection and removal within the same exploration. During the last decades, artificial intelligence has played a key role to assist clinicians in several domains. Regarding colonoscopy, the majority of works are focused on lesion detection to reduce

S. Wesarg et al. (Eds.): CLIP/FAIMI/EPIMI 2023, LNCS 14242, pp. 32–42, 2023.
https://doi.org/10.1007/978-3-031-45249-9_4

polyp miss rate, but in the last few years polyp classification has received a lot of attention from the research community. This is aligned with clinicians needs, as having an accurate in-vivo diagnosis would allow them to efficiently implement strategies such as resect and discard [10] or leave-in-situ which would result in an increase in procedure efficiency. To achieve this, an accurate lesion histology prediction is required, which is not an easy task due to both the scarce availability of publicly annotated databases with histology and the great class imbalance they present. Additionally, these datasets seldom include other types of lesions, such as serrated sessile adenomas, which also convey clinical relevance as they might evolve into cancer.

Regarding CAD systems for polyp classification, neural networks have already been used to tackle semantic polyp segmentation, aiming at providing a histology prediction of the detected lesion. U-net architecture [19] has been a method with high relevance for biomedical image segmentation. One of the reasons for this is related to the high flexibility of encoder-decoder architecture and the easy integration with different strategies for aggregating low-level and high-level features to properly define polyp areas within an image. Regarding polyp segmentation, the work of [13] introduces ResUnet++, a model based on U-net that uses spatial and channel attention to improve segmentation results. In [9], the authors use a partial decoder for a coarse segmentation and then use Reverse Attention to improve the miss-aligned segmentation.

In recent years, research focus has moved towards attention-based models. This is due to the Vision Transformer revolution started by [8], which showed a consistent improvement on several vision tasks. This also applies to polyp segmentation: in [7] authors showed how they improved results on multiple metrics across datasets by using Pyramid-ViT (PVT) as an encoder.

Nevertheless, it has to be noted that all previous works, while achieving good performance, are not completely valid for an actual clinical use as they tackle the problem in an incomplete way. Unlike fine-grained semantic segmentation, they face the task as a salient object segmentation problem which does not provide information about the histology of the lesion, thus not allowing them to assist in resect and discard or leave-in-situ strategies.

Fine-grained visual classification differs from image classification task in that it aims to identify hard-to-distinguish object classes such as bird species (Birdsnap dataset [1] distinguishes between 500 species) or vehicle models (Stanford Dataset [14] contains 196 car models) whilst the second uses coarse-grained classes like birds or vehicles. In the context of semantic polyp segmentation, a differentiation is made between adenomatous, serrated sessile adenomatous, and hyperplastic polyps, with the former being the most likely to develop into cancer.

In this work Swin-Expand is introduced, which is a new simple yet effective fine-grained polyp segmentation method. It proposes a simple encoder-decoder method based on Swin-Transformer and a margin loss that forces the network to produce more discriminative features enforcing the model to obtain more accurate class segmentation maps.

The presented methodology is validated in the PolypSegm-ASH dataset, a novel dataset that contains 1225 high-definition images with both pixel-wise polyp mask and histology class annotations. This work also presents a comparative analysis of the performance of several segmentation strategies in public datasets.

The contributions of this paper can be summarized as follows: (1) PolypSegm-ASH: a dataset for fine-grained polyp segmentation task; (2) Swin-Expand: A novel yet simple method to obtain segmentation maps that obtain good results in both class-aware and in salient-object segmentation methods; (3) Additional benchmarks on multiple state-of-the-art methods on class agnostic polyp segmentation setting.

2   Method

The proposed architecture is shown in Fig. 1. Four modules can be clearly differentiated: (1) An encoder for extracting images features at different scales; in this case, Swin-Transformer; (2) a modified Feature Pyramid Network [15] to enrich the features; (3) a decoder for aggregating multi-scale features and recover and restore resolution and (4) segmentation head that takes the decoder features and generates the final segmentation maps.

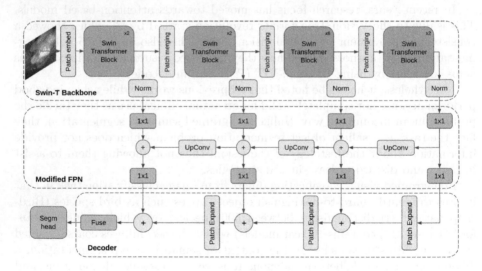

Fig. 1. Overview of the proposed Swin-Expand architecture.

2.1   Swin Transformer Encoder

Swin Transformer processes images $\mathbf{x} \in \mathbb{R}^{H \times W \times 3}$ by splitting them into non-overlapping tokens with 4×4 resolution which are projected to C-dimensional embeddings. Those tokens go under four stages, each one consisting of several Swin-transformer blocks and a Patch-Merging block that aggregates information from 2×2 neighbor tokens, reducing the tokens on each stage. In this work, the features of each stage are taken just before the Patch-Merging block. Afterwards, a LayerNorm (LN) is applied to them, and the resulting output is used as the input for the next block.

The Swin-Transformer has been chosen as the encoder for two main reasons. Firstly, the hierarchical representation of the features is achieved by default by the model. Secondly, the use of (Shifted-)Window-Self-Attention ((S)W-SA) proves to be less computationally intensive than other encoders such as ViT.

2.2 Multi-Scale Feature Enhancement

A modified version of Feature Pyramid Network (FPN) was implemented to obtain more meaningful features on each hierarchical level of the encoder. FPN adds a top-down stage over the encoder, taking multi-scale features and aggregating information across different scales to produce more meaningful features. It consists of a series of lateral connections on each of the encoder feature levels and several up-sampling operations which are used to combine the feature maps with the ones obtained in previous levels. Then, each newly combined feature is again re-projected to a new dimension for their further aggregation.

In this study, the input and output embedding dimensions were preserved so the input size is the same as the output from the FPN. Lateral connections are performed by 1×1 convolution; then the up-sampling is done by a bilinear 2x interpolation at each level followed by another 1×1 convolution. The up-sampled features are finally passed through a 1×1 convolution followed by a GeLU activation [11] that restores the original embedding size.

2.3 Patch-Expanding Decoder

The proposed decoder is composed of a top-down pyramid similar to FPN and it is composed of Patch-Expand blocks, as used in Swin-Unet [4]. Patch-Expand performs the analogous operation of Patch-Merging from Swin-Transformer. Patch-Expand block consists of a linear layer and a LayerNorm. Linear layers expand the embedding from C to 2C on each token. Then, new embeddings are rearranged to create four new tokens of dimension C/2. Finally, LayerNorm is applied to the new tokens. This decoder gradually recovers resolution while keeping meaningful information of each hierarchical scale using lightweight operations.

Each step can be defined as follows:

$$P_{t+1} = Expand_t(P_t + P_{t-1}), \tag{1}$$

where P_t and $Expand$ refers to patches and Patch-Expand at level t respectively. The proposed decoder is composed of three Patch-Expand blocks that integrate gradually each scale followed by a 1×1 convolution, increasing the channels before the head.

2.4   Upsample Head

After Patch-Expand decoder, the method obtains a set of tokens with the same size as the features at the start of stage 1 of the encoder (patches of resolution $\frac{H}{4}*\frac{W}{4}$). Segmentation maps are further refined by using standard UpConvBn blocks (which consist of a bilinear up-sample by 2x, followed by a 3×3 convolution and BatchNorm) before the prediction layer. An ablation study to determine the number of UpConvBn blocks needed to perform correctly was carried out: results showed that only 1 block is needed to obtain good segmentation maps.

2.5   Loss Function

A combination of Cross-Entropy and Tversky Loss is used to optimize the proposed method, following previous works [21]. Given the considerable overlap of the dataset classes and the non-trivial nature of their discrimination, a loss that less heavily penalizes incorrect classifications is sought. A Multi-Label margin loss is employed to tackle this problem. Multi-label losses tend to produce better scores on hard examples, then making optimization easier and achieving a better separation of the different classes.

Fig. 2. Sample images and annotations from PolypSegm-ASH. Red: Adenomatous; Green: Serrated Sessile Adenomatous; Blue: Hyperplastic. (Color figure online)

3   PolypSegm-ASH Dataset

This dataset was built with data acquired at the Hospital Clinic of Barcelona, Spain. It is composed of a total of 1126 HD polyp images. There are a total of 473 unique polyps, with a variable number of different shots per polyp (minimum: 2, maximum: 24, median: 10). Special attention was paid to ensure that images from the same polyp show different conditions. An external frame-grabber and a white light endoscope were used to capture raw images. The dataset contains images with two different resolutions: 1920×1080 and 1350×1080.

Images were pixel-wise annotated by clinicians and separated into three classes: adenomatous, serrated sessile adenomatous and hyperplastic polyps. Figure 2 displays one example of each class from the train split. Three stratified splits without polyp ID overlapping were presented with 788, 113 and 224 images for train, validation and test respectively. Additional statistics are provided in Table 1, where a notable imbalance can be observed between the three classes.

All the experiments code along with the dataset and its official splits can be found in https://github.com/yaeltudela/polyp_ash

Table 1. Detailed content of the PolypSegm-ASH dataset.

	Train	Valid	Test	All
Classes				
Adenoma	535	77	148	760
Serrated	122	17	37	176
Hyperplastic	131	19	40	190
Unique polyps	329	53	91	473
Localization				
Rectum/Sigma	339	57	105	501
Other	449	56	120	625
Size				
<= 5 mm	327	41	94	462
6 mm–10 mm	174	40	87	301
>= 10 mm	287	32	44	363

These are the main features of the novel PolypSegm-ASH Dataset with respect to existing ones:

- **The first fine-grained segmentation dataset.** Contains pixel-wise annotations of adenomatous, serrated sessile adenomatous and hyperplastic.
- **Annotated by expert clinicians.** Class labels of each lesion were obtained from expert histopathologists after pathological analysis. Pixel-wise annotations were made by expert clinicians on the field and externally validated.

4   Results

Datasets and Evaluation Metrics. The methods are validated on the novel PolypSegm-ASH test-set and in the following well-established binary polyp segmentation: CVC-ClinicDB [3], CVC-ColonDB [2], KVasir-SEG [12], and Etis-LaribDB [20]. For the binary datasets the splits proposed in [9] are used.

The methods have been evaluated using the standard semantic segmentation metric, mean Intersection over Union ($mIoU$), in both scenarios: fine-grained and binary. These metrics are referred to as $mIoU_{cls}$ and $mIoU_{bin}$ respectively.

Additionally, image-level predictions were derived by calculating class probabilities with a very naive method. Predictions $p \in \mathbb{R}^C \times H \times W$ have been taken and arg-max has been applied across dimension C. Subsequently, the histogram of the arg-maxed prediction has been computed, and the histogram has been normalized according to the H and W dimensions. Finally, the class probability of C = 0, corresponding to the background, was removed.

Image-level metrics were used in order to derive the confusion matrix and to compute the Matthews Correlation Coefficient (MCC), which provides complete information by using all the values of the derived confusion matrix.

Implementation Details. All the methods compared in this work were implemented in PyTorch [18] framework and trained on one NVIDIA RTX 2070 SUPER 8 Gb GPU. All the methods share the same Swin-Transformer [16] encoder, following the official implementation. Unless stated otherwise, the following configuration was used across all methods: encoder with *tiny* configuration and pre-trained weights in ImageNet22k [6].

AdamW [17] optimization was used with weight decay set to 0.05, gradient clipping set to 0.5. A base learning rate of 1e−4 was applied, along with a linear scheduler featuring a brief warm-up of 250 steps. The models were trained for a maximum of 5000 steps (approximately 50 epochs on the training split) using a batch size of 8, and early stopping (*patience* = 5) was employed to prevent overfitting

Images were resized to 224 × 224 and standard practices for data augmentation were followed. Further regularization was achieved by using RandAugment [5] on the *light* configuration along with motion blur and defocus augmentations, which have been shown to enhance the robustness of models handling colonoscopy images. The total training time of all the presented experiments for the proposed method took approximately 19 h. An output rate of around 98 FPS was achieved, making the method ready to be used in a real-time CAD system.

Table 2. Results on PolypSegm-ASH test. $mIoU_{cls}$ and $mIoU_{bin}$ are pixel-level metrics while MCC is an image-level metric derived from pixel-level predictions. Relative improvement is shown in parenthesis.

Method	Loss	Params (M)	MCC	$mIoU_{cls}$	$mIoU_{bin}$
Swin-FPN	CE	31.33	0.4414	0.5466	0.8490
Segformer	CE	32.42	0.3989	0.4961	0.8112
Swin-UperNet	CE	37.56	0.3913	0.5066	0.8315
Swin-Unet	CE	41.38	0.3821	0.3821	0.8218
Swin-Expand	CE	**30.31**	0.5308	0.5634	0.8525
Swin-FPN	Margin	31.33	0.5171 (17.15)	0.5754 (5.27)	0.8607
Segformer	Margin	32.42	0.4717 (18.25)	0.5486 (10.58)	0.8453
Swin-UperNet	Margin	37.56	0.4967 (26.94)	0.5620 (10.94)	0.8536
Swin-Unet	Margin	41.38	0.3717 (2.51)	0.4849 (26.90)	0.8309
Swin-Expand	Margin	**30.31**	**0.6114 (15.18)**	**0.6024 (6.92)**	**0.8635**

4.1 Experiments on PolypSegm-ASH

Swin-Expand was evaluated against various state-of-the-art methods used as baselines on different medical image segmentation datasets. To ensure a fair comparison, all the methods were using the same optimization framework. The results, presented in Table 2, demonstrate that all methods exhibit improved performance with Multi-Label Margin loss, showing a consistent increase in all metrics.

4.2 Experiments on Binary Polyp Segmentation

Table 3 presents results on a binary setting, where a comparison is made with other state-of-the-art methods across 4 publicly available datasets. The evaluation employs the DICE score and mIoU as metrics. As the methods were not re-implemented, the results presented are those reported in their original works. In the case of the method presented in this paper, it was trained on PolypSegm-ASH and evaluated on the remaining datasets. Binary mask predictions were created by the aggregation of predictions from all classes excluding the background one, mirroring the approach taken by other methods where they only differ from background and polyp.

Competitive results across all datasets were achieved by Swin-Expand. A fine-tuned version of the method (designated as Swin-Expand-FT), where the initial method was fine-tuned over 25k steps with the same training data used by the other methods, is also provided. The experimental model exhibited strong generalization across all datasets and surpassed previous work in 3 of them, despite a relatively short fine-tuning period. These results suggest that by training on more complex tasks, a performance boost can be achieved through the learning of complex representations.

Table 3. Comparison with other state-of-the-art methods on four benchmark datasets for binary polyp segmentation. The number of images are denoted in parenthesis.

	ClinicDB(62)		ColonDB(380)		KVasir-SEG(100)		ETIS-Larib(196)	
Method	DICE	mIoU	DICE	mIoU	DICE	mIoU	DICE	mIoU
PraNet	0.8990	0.8490	0.7120	0.6400	0.8980	0.8400	0.6280	0.5670
PolypPVT	0.9308	0.8828	0.8075	0.7185	0.9123	0.8630	0.7867	0.7097
LDNet	**0.9431**	**0.9421**	0.7842	**0.8339**	0.9070	**0.9101**	0.7434	0.8226
Swin-Expand	0.8074	0.7775	0.7800	0.6918	0.8972	0.8511	0.7761	0.8533
Swin-Expand-FT	0.8899	0.8592	**0.8406**	0.7916	**0.9222**	0.8810	**0.8097**	**0.8696**

4.3 Ablation Study. Effect of Up-Samples Before Predictions

An ablation study was conducted to evaluate the impact of the number of UpConvBn blocks on the performance of the presented model. As shown in

Table 4, using just one unique block (obtaining mask resolution of $\frac{H}{2} \times \frac{W}{2}$) yields good results while keeping low model parameters. Despite an improvement observed with an additional block, the substantial increase in parameter count contradicts the objective of maintaining a lightweight model. Consequently, only one UpConvBn block was retained.

Table 4. Ablation study on the number of ConvUpsample blocks. Column "parameter" refers only to the Upsample-head module of the proposed architecture.

# blocks	Resolution	Parameters (M)	$mIoU_{cls}$	MCC	$mIoU_{bin}$
–	56 × 56	0.103	0.5393	0.5157	0.8358
1	112 × 112	0.820 (6.30x)	0.5634	0.5308	0.8525
2	224 × 224	1.296 (1.54x)	0.5792	0.5412	0.8658

5   Conclusion

In this work Swin-Expand is proposed. It is a fine-grained polyp segmentation method that uses Swin-Transformer as an encoder, a modified FPN to enrich features, and a lightweight decoder based on Patch-Expand blocks to produce accurate segmentation maps for polyp segmentation. Experimental results demonstrate competitive results in both fine-grained image segmentation and salient object segmentation.

Although the presented methodology is already capable of being part of a CAD system and assists clinicians in diagnosis, there is still room for improvement, especially on metrics related to under-represented lesions. In future work, the exploration of weakly supervised learning approaches is planned aiming to develop a more robust and generalizable method by incorporating other polyp datasets without lesion histology information. Additionally, the acquisition of NBI data is desired to study if its utilization impacts the performance of the method in a manner analogous to in-vivo optical diagnosis.

Acknowledgements. This work was supported by Grant Numbers: PID2020-120311 RB-I00 and RED2022-134964-T, funded by MCIN/AEI/10.13039/501100011033.

References

1. Berg, T., Liu, J., Lee, S.W., Alexander, M.L., Jacobs, D.W., Belhumeur, P.N.: Birdsnap: large-scale fine-grained visual categorization of birds. In: 2014 IEEE Conference on Computer Vision and Pattern Recognition, pp. 2019–2026 (2014). https://doi.org/10.1109/CVPR.2014.259
2. Bernal, J., Sánchez, J., Vilariño, F.: Towards automatic polyp detection with a polyp appearance model. Pattern Recognit. **45**(9), 3166–3182 (2012). https://doi.org/10.1016/j.patcog.2012.03.002

3. Bernal, J., Sánchez, F.J., Fernández-Esparrach, G., Gil, D., Rodríguez, C., Vilariño, F.: WM-DOVA maps for accurate polyp highlighting in colonoscopy: validation vs. saliency maps from physicians. Comput. Med. Imaging Graph. **43**, 99–111 (2015). https://doi.org/10.1016/j.compmedimag.2015.02.007
4. Cao, H., et al.: Swin-Unet: Unet-like pure transformer for medical image segmentation. In: Karlinsky, L., Michaeli, T., Nishino, K. (eds.) ECCV 2022. LNCS, vol. 13803, pp. 205–218. Springer, Cham (2022). https://doi.org/10.1007/978-3-031-25066-8_9
5. Cubuk, E.D., Zoph, B., Shlens, J., Le, Q.: RandAugment: practical automated data augmentation with a reduced search space. In: Larochelle, H., Ranzato, M., Hadsell, R., Balcan, M., Lin, H. (eds.) Advances in Neural Information Processing Systems, vol. 33, pp. 18613–18624. Curran Associates, Inc. (2020)
6. Deng, J., Dong, W., Socher, R., Li, L.J., Li, K., Fei-Fei, L.: ImageNet: a large-scale hierarchical image database. In: 2009 IEEE Conference on Computer Vision and Pattern Recognition, pp. 248–255. IEEE (2009)
7. Dong, B., Wang, W., Fan, D.P., Li, J., Fu, H., Shao, L.: Polyp-PVT: polyp segmentation with pyramid vision transformers. arXiv preprint arXiv:2108.06932 (2021)
8. Dosovitskiy, A., et al.: An image is worth 16x16 words: transformers for image recognition at scale. arXiv preprint arXiv:2010.11929 (2020)
9. Fan, D.-P., et al.: PraNet: parallel reverse attention network for polyp segmentation. In: Martel, A.L., et al. (eds.) MICCAI 2020. LNCS, vol. 12266, pp. 263–273. Springer, Cham (2020). https://doi.org/10.1007/978-3-030-59725-2_26
10. Hassan, C., Pickhardt, P.J., Rex, D.K.: A resect and discard strategy would improve cost-effectiveness of colorectal cancer screening. Clin. Gastroenterol. Hepatol. **8**(10), 865–869.e3 (2010). https://doi.org/10.1016/j.cgh.2010.05.018. https://www.sciencedirect.com/science/article/pii/S1542356510005434
11. Hendrycks, D., Gimpel, K.: Gaussian error linear units (gelus). arXiv preprint arXiv:1606.08415 (2016)
12. Jha, D., et al.: Kvasir-SEG: a segmented polyp dataset. In: Ro, Y.M., De Neve, W. (eds.) MMM 2020. LNCS, vol. 11962, pp. 451–462. Springer, Cham (2020). https://doi.org/10.1007/978-3-030-37734-2_37
13. Jha, D., et al.: ResUNet++: an advanced architecture for medical image segmentation. In: 2019 IEEE International Symposium on Multimedia (ISM), pp. 225–2255 (2019). https://doi.org/10.1109/ISM46123.2019.00049
14. Krause, J., Stark, M., Deng, J., Fei-Fei, L.: 3d object representations for fine-grained categorization. In: 4th International IEEE Workshop on 3D Representation and Recognition (3dRR-13), Sydney, Australia (2013)
15. Lin, T.Y., Dollár, P., Girshick, R., He, K., Hariharan, B., Belongie, S.: Feature pyramid networks for object detection. In: Proceedings of the IEEE Conference on Computer Vision and Pattern Recognition, pp. 2117–2125 (2017)
16. Liu, Z., et al.: Swin transformer: hierarchical vision transformer using shifted windows. In: Proceedings of the IEEE/CVF International Conference on Computer Vision, pp. 10012–10022 (2021)
17. Loshchilov, I., Hutter, F.: Decoupled weight decay regularization. arXiv preprint arXiv:1711.05101 (2017)
18. Paszke, A., et al.: PyTorch: an imperative style, high-performance deep learning library. In: Advances in Neural Information Processing Systems (2019)
19. Ronneberger, O., Fischer, P., Brox, T.: U-Net: convolutional networks for biomedical image segmentation. In: Navab, N., Hornegger, J., Wells, W.M., Frangi, A.F. (eds.) MICCAI 2015. LNCS, vol. 9351, pp. 234–241. Springer, Cham (2015). https://doi.org/10.1007/978-3-319-24574-4_28

20. Silva, J., Histace, A., Romain, O., Dray, X., Granado, B.: Toward embedded detection of polyps in WCE images for early diagnosis of colorectal cancer. Int. J. Comput. Assist. Radiol. Surg. **9**(2), 283–293 (2014). https://doi.org/10.1007/s11548-013-0926-3
21. Sushma, B., Raghavendra, C.K., Prashanth, J.: CNN based U-Net with modified skip connections for colon polyp segmentation. In: 2021 5th International Conference on Computing Methodologies and Communication (ICCMC), pp. 1762–1766 (2021). https://doi.org/10.1109/ICCMC51019.2021.9418037

Automated Orientation and Registration of Cone-Beam Computed Tomography Scans

Luc Anchling[1,2(✉)], Nathan Hutin[1,2], Yanjie Huang[1], Selene Barone[1,5],
Sophie Roberts[8], Felicia Miranda[1,7], Marcela Gurgel[1], Najla Al Turkestani[1,6],
Sara Tinawi[1], Jonas Bianchi[1,4], Marilia Yatabe[1], Antonio Ruellas[9],
Juan Carlos Prieto[3], and Lucia Cevidanes[1]

[1] University of Michigan, Ann Arbor, MI, USA
lanchlin@umich.edu
[2] CPE Lyon, Lyon, France
[3] University of North Carolina, Chapel Hill, USA
[4] University of the Pacific, San Francisco, USA
[5] Magna Graecia University of Catanzaro, Catanzaro, Italy
[6] King Abdulaziz University, Jeddah, Saudi Arabia
[7] Bauru Dental School, University of Sao Paulo, Bauru, SP, Brazil
[8] Department of Orthodontics, University of Melbourne, Melbourne, Australia
[9] Federal University of Rio de Janeiro, Rio de Janeiro, Brazil

Abstract. Automated clinical decision support systems rely on accurate analysis of three-dimensional (3D) medical and dental images to assist clinicians in diagnosis, treatment planning, intervention, and assessment of growth and treatment effects. However, analyzing longitudinal 3D images requires standardized orientation and registration, which can be laborious and error-prone tasks dependent on structures of reference for registration. This paper proposes two novel tools to automatically perform the orientation and registration of 3D Cone-Beam Computed Tomography (CBCT) scans with high accuracy ($<3°$ and $<2\,\mathrm{mm}$ of angular and linear errors when compared to expert clinicians). These tools have undergone rigorous testing, and are currently being evaluated by clinicians who utilize the 3D Slicer open-source platform. Our work aims to reduce the sources of error in the 3D medical image analysis workflow by automating these operations. These methods combine conventional image processing approaches and Artificial Intelligence (AI) based models trained and tested on de-identified CBCT volumetric images. Our results showed robust performance for standardized and reproducible image orientation and registration that provide a more complete understanding of individual patient facial growth and response to orthopedic treatment in less than 5 min.

Keywords: Deep Learning · Standardized Orientation · Medical Image Registration · 3D CBCT scans · Image Processing

S. Wesarg et al. (Eds.): CLIP/FAIMI/EPIMI 2023, LNCS 14242, pp. 43–58, 2023.
https://doi.org/10.1007/978-3-031-45249-9_5

1 Introduction

In the current three-dimensional (3D) medical longitudinal image analysis workflow (Fig. 1), two steps are essential: orientation and registration. While the former consists of aligning the images to a common and standardized frame of reference, the latter enx == tails the visual task of reducing disparities between corresponding points in two images. This is achieved by determining a geometric transformation that enables their effective superimposition.

Cone-beam computed tomography (CBCT) is a relatively low radiation dose imaging technique that is widely used in Dentistry due to its low cost and its detailed three-dimensional (3D) information about various craniofacial tissues and structures [14]. However, because of the inconsistency in imaging acquisition protocols, patient position during image acquisition, and settings used on scanners in different clinical centers, the volumetric CBCT scans are not acquired in a reproducible standardized position [24]. The patient head posture is also irregular across various acquisitions with variability in the head pitch, roll, and yaw position. Having a standardized orientation and head position is essential for any 3D imaging analysis workflow, as inconsistent orientation can affect both the downstream image registration task and the quality of measurements throughout the image analysis workflow [5].

While standardized orientation is of primary importance to harmonize craniofacial imaging records, registration of images at multiple time points is also a crucial step in the medical image analysis workflow [2]. It is a prerequisite for several medical image applications [18], such as growth observation [28], follow-up [26] or simulations [19] of treatments. Depending on the area affected by the treatment, growth, and/or clinical condition, the alignment often should utilize specific stable craniofacial structures of reference: image changes over time should be analyzed relative to these references rather than considering the entire image best fit.

These two fundamental steps in the image analysis workflow can be laborious, error-prone, and time-consuming for expert clinicians, averaging up to 45 min. The challenges for robust automated orientation and registration are numerous due to the variability of the morphology in the various craniofacial structures in CBCT.

Our proposed tool implements all of the functionalities of both Automated Standardized Orientation (ASO), to automatically orient and position the scan, and Automated Registration (AReg), to register two time point CBCT scans according to the region of reference chosen, in a user-friendly plugin available in the 3D Slicer [10] software. The goal of this work is to propose a novel, free, open-source automatic workflow for medical image analysis that also includes automatic multi-anatomical skeletal structures segmentation (Fig. 1).

2 Materials

A total of 465 CBCT scans of the head in Digital Imaging and Communications in Medicine (DICOM) format were used in this work. The images were acquired

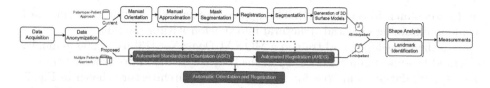

Fig. 1. Comparison between the current and the proposed medical image analysis workflow with the different steps.

at different clinical centers with different scanners, image acquisition protocols, and fields of view. All DICOM files were anonymized removing all identifiable personal information using the 3D Slicer Batch Anonymizer module. The University of Michigan Institutional Review Board (IRB) HUM00238037 waived the requirement for informed consent and granted IRB exemption. The machine learning models for the orientation (coarse orientation) and stable regions of reference (ROFs also called mask segmentation) were trained using the same CBCT scans. Clinical experts used two open-source software packages, ITK-SNAP 3.8 [27] and 3D Slicer 5.2 [10], to interactively segment masks for each of the volumetric images and align the head in a common spatial orientation. This served as the ground truth for training our machine-learning models. A dataset consisting of 290 CBCT scans was used for training the orientation and different mask (cranial base, maxilla, and mandible) models, with manual segmentation performed on 135 CBCT scans. The remaining scans were reserved for testing the ASO (121 scans) and AReg (54 scans) tools.

3 Proposed Method

The different methods for automatic orientation and registration of 3D CBCT scans rely on a combination of algorithmic and deep-learning techniques to perform both the image orientation and the registration automatically.

3.1 Automated Standardized Orientation (ASO)

Our method starts with a rough alignment step performed by a trained Machine Learning (ML) model and a landmark-based image alignment method. The latter uses automatic landmark identification to predict the points of reference.

Coarse Orientation. To improve the precision of the orientation process with the ASO tool, we require a rough head orientation of the scan as an initialization, before fine-tuning it with the landmark-based registration.

To standardize each 3D volumetric scan and establish a resilient ML model capable of identifying scan head orientation, an initial image preprocessing phase is executed. This step involves rescaling each image to a reduced resolution. The rescaling procedure encompasses isometric dimensions for both image size (128^3)

and voxel dimensions $(1.5\,\text{mm}^3)$. This approach ensures computational efficiency while maintaining essential structural information within the images.

We then developed and trained a MONAI DenseNet [13] using 290 manually oriented by expert clinicians CBCT scans that we divided into 70% for training, 10% for validation, and 20% for testing using the architecture shown in Fig. 7. A random rotation transform was utilized as data augmentation to enlarge the dataset. The training hyperparameters were defined as follows: a batch size of 25, a learning rate of $1e-4$, and a maximum angle of $\pi/2$ for random rotations. The loss function was the Cosine Similarity, which determines the angular difference between two vectors.

Automatic Landmark Identification (ALI-CBCT) [12]. Automatic landmark identification is based on a multi-scale combination of a MONAI DenseNet and a fully-connected layers approach agent that is trained to make an initial estimate of the landmark position in the lower resolution scan and refined it using the higher resolution.

The landmarks used for the ASO step are part of the different craniofacial structures. Depending on the orientation that is desired, either relative to the cranial base or the maxilla, various landmarks are then used for the process, as presented in Table 1.

Table 1. Different planes with selected landmarks for each orientation type (landmark definition are shown in Table 5).

Structure	Horizontal Plane (Landmarks)	Vertical Plane (Landmarks)
Cranial Base	Frankfort Horizontal (ROr, LOr, RPo, LPo)	Midsagittal Plane (Ba, S, N)
Maxilla	Occlusal Plane (UR6O, UL6O, UR1O)	Midsagittal Plane (ANS, PNS, IF)

Landmark Based Image Alignment. The alignment of the source image to a target image with a standard orientation (cranial base in Fig. 2 and maxilla in Fig. 6) is fine-tuned using the landmarks described above, which improves the precision of the orientation and position of the CBCT scan with the ASO tool.

To automatically match the identified landmarks with a group of reference landmarks marked by a clinician, we applied the Iterative Closest Point (ICP) algorithm [17]. To obtain better accuracy, this required an initial alignment step [3], which we achieved by aligning three randomly chosen landmarks to minimize the distance between the two sets of landmarks.

All the transformations computed during this process are concatenated and the final ASO transformation is then applied to the CBCT volume.

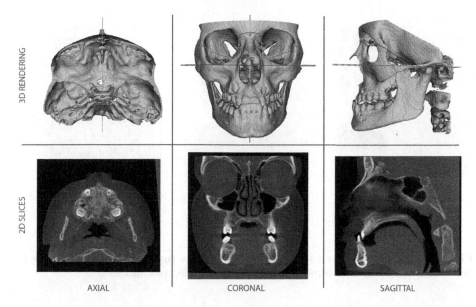

Fig. 2. Cranial Base standardized orientation with Frankfurt horizontal and Midsagittal plane. With both the 3D rendering and the different slice views.

3.2 Automated Registration (AReg)

Our automatic alignment is based on a regional voxel-based registration (VBR) approach which automatically aligns mask segmentations that contains only stable ROFs within the cranial base, the mandible, or the maxilla.

Automatic Multi-anatomical Skull Structure Segmentation for Masks (AMASSS-CBCT) [11]. AMASSS relies on a MONAI UNEt TRansformers initially trained to segment different craniofacial structures (hard and soft tissues). However, for the proposed registration approaches we then trained AMASSS models to segment stable ROFs within each craniofacial structure which are defined following works of reference for each region, cranial base [6], mandible [22], and maxilla [20] are described in Fig. 3. For this purpose, different mask segmentations were trained with 135 anonymized CBCTs for each of the three ROFs.

Flexible Voxel-Based Registration. Our algorithm utilized the Python library optimized for medical image registration for the desired voxel-based method: SimpleElastix [16].

To perform the regional VBR, we applied the automatically-obtained segmentation as a mask to the fixed image, to only keep the important information from the scan within the delimited area. To have an accurate transformation between the moving and the fixed image, we initially run a VBR between the full images

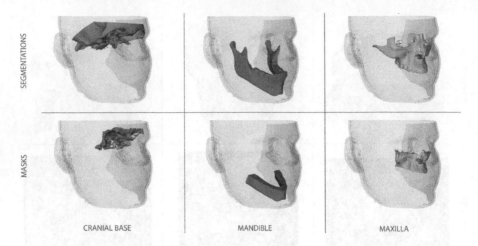

SEGMENTATIONS

MASKS

CRANIAL BASE   MANDIBLE   MAXILLA

Fig. 3. Comparison between segmentation and masks (used for the registration) for different craniofacial structures (cranial base, mandible, and maxilla).

and then with the masked image as a fine-tuning step. The optimized parameters used for the two registration steps in this application for craniomaxillofacial CBCT imaging are described in Table 4. The tool then generates registered label maps and surface meshes using the AMASSS function.

3.3 Evaluation Metrics

The precision of the ASO and the AReg approaches was tested using 3D point-to-point measurements between our automated method and the manually oriented/registered scans by expert clinicians. The different quantifications were performed in the Automated Quantitative 3D Components (AQ3DC) tool in 3D Slicer, which is the automated version of the Q3DC tool previously validated for craniofacial quantification [7,8]. Two types of measurements were computed: linear (in millimeters) and angular (in degrees). For the former, the linear distance was calculated between each selected point for each component: right-left, anterior-posterior, superior-inferior, and the 3D Distance. For the latter, two lines were used two calculate the angular errors of the yaw, pitch, and roll of the facial structures (Table 6 for the orientation and Table 7 for the registration).

Orientation. The datasets used to test the precision of the automated orientation approach consisted of 121 CBCT scans. The ASO outputs were compared to previously manually oriented scans, 62 for the cranial base and 59 for the maxillary orientation. For the cranial base, only full-face large field-of-view (FOV) scans were used, while for the maxilla, the sample also included small FOV scans of the maxilla, which may be indicated depending on the clinical application.

Registration. Two approaches were compared to test AReg. The first approach tested only the flexible VBR algorithm by using manually-approximated scans and manually-segmented regions of reference as input for the registration of different time points. The second approach tested the full AReg pipeline for automated mask segmentation and registration. Both angular and linear changes were assessed for 27 patients comparing the two approaches with blinded user experts' measurements.

3.4 Implementation

Our Python-based algorithm requires multiple libraries for the different image-processing tasks accomplished throughout the proposed method: SimpleITK [15], VTK [23], SimpleElastix [16]. To implement these tools, we used the Medical Open Network for Artificial Intelligence (MONAI) library, which is a PyTorch-based framework for medical image analysis. MONAI offers several advantages for our work, such as high performance, modularity, and interoperability with other libraries.

4   Results

The experiments have been carried out on a system with an NVIDIA Quadro RTX 6000/8000 GPU and an Intel(R) Xeon(R) Gold 6226R CPU.

4.1   Orientation

Table 2 shows the angular changes for the different types of orientation tested. The average computational time per patient to orient them with our novel proposed tool was 65 s.

Table 2. The angular differences for the two craniomaxillofacial orientation regions.

Type	Yaw	Pitch	Roll
Cranial Base	1.53° ± 1.03°	1.68° ± 1.23°	0.64° ± 0.53°
Maxillary	1.35° ± 1.19°	2.60° ± 2.42°	1.30° ± 1.47°

4.2   Registration

Table 3 shows the yaw, pitch, and roll difference relative to the user expert for each regional registration approach. The various linear displacements are described for the cranial base in Table 8, maxilla in Table 9, and mandible in Table 10. Also, the graphic display of the registered surface meshes with the different approaches is shown in Fig. 4. The average registration computational time per patient was 4 min.

Table 3. The angular differences between each of the two new approaches and the previously validated experts' clinician approach for the three craniomaxillofacial regions were assessed.

Type	Yaw	Pitch	Roll
Cranial Base			
Approach 1	0.06° ± 0.16°	0.17° ± 0.19°	0.07° ± 0.12°
Approach 2	0.23° ± 0.23°	0.56° ± 0.71°	0.24° ± 0.31°
Maxillary			
Approach 1	0.46° ± 0.55°	1.38° ± 1.34°	0.30° ± 0.27°
Approach 2	0.51° ± 0.52°	1.63° ± 1.66°	0.44° ± 0.45°
Mandible			
Approach 1	0.32° ± 0.29°	1.19° ± 1.34°	0.28° ± 0.26°
Approach 2	0.29° ± 0.34°	1.65° ± 1.76°	0.31° ± 0.30°

Approach 1: only the flexible VBR; Approach 2: fully automated registration

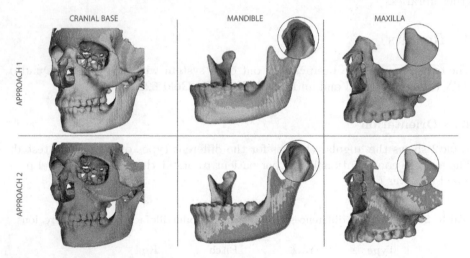

Fig. 4. Comparison between the two automated methods for registration with the clinician experts' registration (yellow). The first approach (green) only tests the registration step using masks made by clinicians as references. The second approach (orange) automatically generates masks and performs the registration. (Color figure online)

5   Discussion

To the best of our knowledge, this is the first study that introduces a free open-source automated tool integrating multiple steps of the 3D medical imaging workflow, which can be challenging, error-prone, and time-consuming when done manually. We developed and validated an automatic tool that streamlines image orientation, segmentation of ROFs, registration, and generation of registered label maps and surface meshes. Although commercial companies such as 3Shape [1], and Dolphin [9] offer tools for orienting and registering CBCT, they are limited to a single-case use and are not open-source solutions.

Standardized image orientation is crucial to eliminate errors from variable patient positioning during image acquisition, which may lead to differences in the measurements of directional changes in 3D space, and assessments of growth and treatment [21]. Automatic orientation as the initialization of the registration process is crucial for achieving reproducibility.

Innovations in medical imaging workflow automation have increased in the last few decades, but they need to be validated to meet clinician standards for a precise automatic workflow. The significance thresholds are set to less than 2 mm (linear) and 4° (angular) differences to be clinically acceptable [25]. The average ± standard deviations of errors observed for both the ASO and AReg algorithms are within these acceptable ranges to assess the accuracy of craniofacial growth and treatment changes. Moreover, having tested the two AReg approaches separately and then combined, the slight differences observed between them also validated the fully-automated pipeline.

Although the proposed registration tool performs well for the angular and linear changes compared to the clinician expert that followed the manual registration steps [4], the mandibular registration of cases during the pubertal growth spurt presented approximately 2 mm of 3D differences at the mandibular condyles that are the major growth site in the mandible. However, AP, vertical, and transverse linear differences were all smaller than 2 mm.

The utilization of different frameworks, such as MONAI for training ML models and SimpleElastix for facilitating the registration process, has simplified the various steps of the novel tools.

While the proposed novel tool was tested and implemented as functionalities of a free open-source module (available at https://github.com/DCBIA-OrthoLab/SlicerAutomatedDentalTools) with a user-friendly interface in 3D Slicer (Fig. 5) for CBCT images, our pipeline is generalizable and can be extended to other imaging modalities such as spiral CT, Magnetic Resonance Imaging, micro-CT, and ultrasound. Adding supplemental scans to the training of the different ML tools used can also be beneficial to increase the quality of the overall image processing. Additionally, comprehensive evaluation within clinical settings can be conducted to thoroughly assess the advantages under real-world practical conditions.

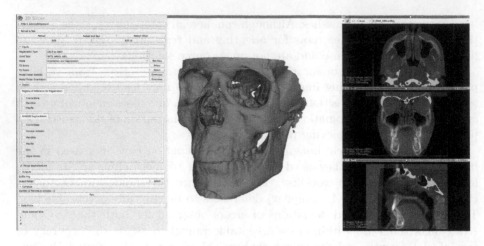

Fig. 5. AReg 3D Slicer module. On the left, the user interface which allows users to choose various parameters. On the right, the Slicer scene visualization of a cranial base registration between the pre (purple) and the post (green) treatment. (Color figure online)

6 Conclusion

The results obtained with the ASO and AReg tools represent a significant advancement in automating the CBCT image analysis digital workflow, as two major processing tasks are incorporated and validated. These study results are particularly relevant considering that machine learning techniques are increasingly employed to achieve cost-effective dental image analysis.

A Appendix

Table 4. SimpleElastix parameters for the different steps of the registration.

Step	Region of Reference	Registration Type	Number of Iterations	Number of Spatial Samples	Number of Resolutions
Approximation	Full Scan	Rigid	256	2048	4
Fine-Tuning	Masked Scan		10000	10000	1

Table 5. Description of the different landmarks used in this study.

Cranial Base	
Ba	Placed at the most posteroinferior point of the anterior margin of the foramen magnum in the midsagittal plane
S	Placed on the most central point of sella turcica from supero-inferior, antero-posterior, and transversal aspects
N	Placed at the most anterosuperior junction of the nasofrontal suture
RPo	Placed at the most superior point of the right external auditory canal
LPo	Placed at the most superior point of the left external auditory canal
Maxillary	
A	The most posterior point of the concavity of the anterior region of the maxilla
ANS	Placed at the anterior nasal spine
PNS	Placed at the posterior nasal spine
IF	Placed at the center of the incisive foramen
ROr	Placed at the most inferior point on the lower portion of the right orbit contour
LOr	Placed at the most inferior region of the left orbit contour
UR1O	Placed in the middle of the incisal edge of the maxillary right permanent central incisor
UR6O	Placed at the most central point in the crown occlusal surface of the maxillary right permanent first molar
UR6MB	Placed at the mesial buccal cusp of the maxillary right permanent first molars
UR6R	Placed at the center of the pulp chamber floor of the maxillary right permanent first molars
UL1O	Placed in the middle of the incisal edge of the maxillary left permanent central incisor
UL1R	Placed at the center portion of the root canal at the axial level of the cementoenamel junction of the maxillary left permanent central incisor
UL6O	Placed at the most central point in the crown occlusal surface of the maxillary left permanent first molar
UL6MB	Placed at the mesial buccal cusp of the maxillary left permanent first molars
UL6R	Placed at the center of the pulp chamber floor of the maxillary left permanent first molars
Mandibular	
B	Placed at the most posterior point of the concavity of the anterior region of the symphysis
Gn	Placed in the projection of a virtual bisector of a line adjacent to the Pog and Me landmarks
RGo	Placed in the projection of a virtual bisector of a line adjacent to the right mandibular base and right posterior border of mandible
LGo	Placed in the projection of a virtual bisector of a line adjacent to the left mandibular base and left posterior border of mandible
RCo	Placed at the most superior and central point of the right condyle
LCo	Placed at the most superior and central point of the left condyle
LR1O	Placed in the middle of the incisal edge of the mandibular right permanent central incisor
LR1R	Placed at the center portion of the root canal at the axial level of the cementoenamel junction of the mandibular right permanent central incisor
LR6O	Placed at the most central point in the crown occlusal surface of the mandibular right permanent first molar
LR6MB	Placed at the mesial buccal cusp of the mandibular right permanent first molars
LR6R	Placed at the center of the pulp chamber floor of the mandibular right permanent first molars
LL1O	Placed in the middle of the incisal edge of the mandibular left permanent central incisor
LL1R	Placed at the center portion of the root canal at the axial level of the cementoenamel junction of the maxillary left permanent central incisor
LL6O	Placed at the most central point in the crown occlusal surface of the mandibular left permanent first molar
LL6MB	Placed at the mesial buccal cusp of the mandibular left permanent first molars
LL6R	Placed at the center of the pulp chamber floor of the mandibular left permanent first molars

Table 6. Different anatomic lines used to compute the angular changes for the orientation.

Type	Yaw	Pitch	Roll
Cranial Base	S-N	S-N	RPo-LPo
Maxillary	UR6MB-UR1O	ANS-IF	UR6MB-UL6MB

Table 7. Different anatomic lines used to compute the angular errors for the registration.

Type	Yaw	Pitch	Roll
Cranial Base	RCo-LCo	Go-Gn	RCo-LCo
Maxillary	UR6MB-UL6MB	UR6MB-UR1O	UR6MB-UL6MB
Mandible	RCo-LCo	LR6MB-LR1O	RCo-LCo

Table 8. Linear differences for the cranial base registration for both approaches.

Type	R-L	A-P	S-I	3D Distance
N Displacement				
Approach 1	-0.06 ± 0.28	0.06 ± 0.19	0.01 ± 0.26	0.25 ± 0.35
Approach 2	0.08 ± 0.59	0.23 ± 0.42	0.24 ± 0.65	0.68 ± 0.76
A Displacement				
Approach 1	-0.05 ± 0.34	0.10 ± 0.40	0.01 ± 0.26	0.34 ± 0.48
Approach 2	0.11 ± 0.75	0.45 ± 1.04	0.25 ± 0.66	1.01 ± 1.15
B Displacement				
Approach 1	-0.04 ± 0.37	0.11 ± 0.56	0.00 ± 0.23	0.43 ± 0.57
Approach 2	0.12 ± 0.93	0.55 ± 1.50	0.23 ± 0.58	1.32 ± 1.43
Gn Displacement				
Approach 1	-0.03 ± 0.39	0.12 ± 0.62	0.00 ± 0.23	0.47 ± 0.61
Approach 2	0.13 ± 1.00	0.60 ± 1.73	0.24 ± 0.61	1.48 ± 1.60

Linear: Right (R) displacement (+), Left (L) displacement (−), Anterior (A) displacement (+), Posterior (P) displacement (−), Superior (S) displacement (+), Inferior (I) displacement (−); Approach 1: only the flexible VBR; Approach 2: fully automated registration

Table 9. Linear differences for the maxillary registration for both approaches.

Type	R-L	A-P	S-I	3D Distance
A Displacement				
Approach 1	-0.09 ± 0.34	-0.07 ± 0.43	0.76 ± 0.85	1.01 ± 0.76
Approach 2	-0.10 ± 0.30	-0.11 ± 0.52	0.51 ± 0.96	0.93 ± 0.81
UR1O Displacement				
Approach 1	-0.15 ± 0.40	-0.48 ± 0.75	0.65 ± 0.80	1.19 ± 0.77
Approach 2	-0.11 ± 0.40	-0.52 ± 0.93	0.43 ± 0.97	1.24 ± 0.92
UR6MB Displacement				
Approach 1	-0.07 ± 0.24	-0.34 ± 0.71	1.27 ± 1.15	1.50 ± 1.17
Approach 2	-0.02 ± 0.34	-0.37 ± 0.83	1.03 ± 1.12	1.41 ± 1.11
UL6MB Displacement				
Approach 1	-0.07 ± 0.24	-0.45 ± 0.77	1.35 ± 1.05	1.56 ± 1.15
Approach 2	$-0.01. \pm 0.35$	-0.52 ± 0.92	1.00 ± 0.98	1.42 ± 1.07

Linear: Right (R) displacement $(+)$, Left (L) displacement $(-)$, Anterior (A) displacement $(+)$, Posterior (P) displacement $(-)$, Superior (S) displacement $(+)$, Inferior (I) displacement $(-)$; Approach 1: only the flexible VBR; Approach 2: fully automated registration

Table 10. Linear differences for the mandibular registration for both approaches.

Type	R-L	A-P	S-I	3D Distance
B Displacement				
Approach 1	-0.07 ± 0.31	-0.07 ± 0.12	-0.20 ± 0.42	0.50 ± 0.29
Approach 2	-0.05 ± 0.33	-0.11 ± 0.18	-0.30 ± 0.80	0.60 ± 0.71
LR1O Displacement				
Approach 1	-0.04 ± 0.28	-0.29 ± 0.50	-0.17 ± 0.52	0.72 ± 0.42
Approach 2	-0.04 ± -0.42	-0.52 ± 0.51	-0.18 ± 0.74	0.88 ± 0.69
LR6MB Displacement				
Approach 1	0.04 ± 0.15	-0.27 ± 0.58	-0.56 ± 0.35	$0.81 + 0.43$
Approach 2	0.01 ± 0.25	-0.56 ± 0.63	-0.76 ± 1.36	1.17 ± 1.35
LL6MB Displacement				
Approach 1	0.04 ± 0.15	-0.38 ± 0.58	-0.39 ± 0.22	0.69 ± 0.47
Approach 2	0.01 ± 0.26	-0.61 ± 0.51	-0.64 ± 1.03	1.02 ± 1.05
RCo Displacement				
Approach 1	0.24 ± 0.45	-0.45 ± 1.30	-0.93 ± 0.89	1.62 ± 1.06
Approach 2	0.10 ± 0.42	-1.01 ± 1.54	-1.16 ± 1.26	2.07 ± 1.46
LCo Displacement				
Approach 1	0.36 ± 0.34	0.87 ± 0.81	0.90 ± 0.71	1.42 ± 0.96
Approach 2	0.12 ± 0.43	-1.14 ± 1.20	-1.19 ± 1.48	1.97 ± 1.62

Linear: Right (R) displacement $(+)$, Left (L) displacement $(-)$, Anterior (A) displacement $(+)$, Posterior (P) displacement $(-)$, Superior (S) displacement $(+)$, Inferior (I) displacement $(-)$; Approach 1: only the flexible VBR ; Approach 2: fully automated registration

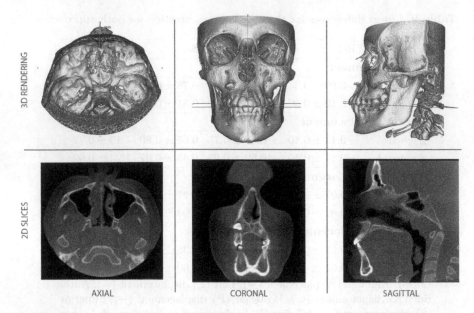

Fig. 6. Maxillary standardized orientation with Occlusal and Midsagittal plane. With both the 3D rendering and the different slice views.

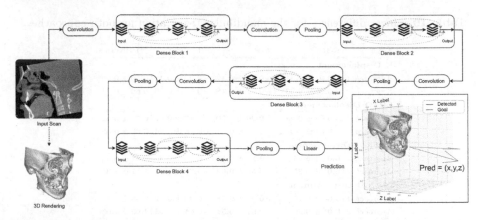

Fig. 7. Schematic representation of the MONAI Densenet used for detecting the orientation of CBCT scan. The directional vector prediction of the current input scan is in blue while the goal orientation is in red. (Color figure online)

References

1. Shape: https://www.3shape.com
2. Alam, F., Rahman, S.U., Hassan, M., Khalil, A.: An investigation towards issues and challenges in medical image registration. J. Postgrad. Med. Inst. **31**(3), 224–233 (2017)

3. Bagchi, P., Bhattacharjee, D., Nasipuri, M.: Reg3DFacePtCd: registration of 3D point clouds using a common set of landmarks for alignment of human face images. KI - Künstl. Intell. **33**(4), 369–387 (2019)
4. Bates, W.R., Cevidanes, L.S., Larson, B.E., Adams, D., De Oliveira Ruellas, A.C.: Three-dimensional cone-beam computed technology evaluation of skeletal and dental changes in growing patients with class II malocclusion treated with the cervical pull face-bow headgear appliance. Am. J. Orthod. Dentofacial Orthop. **162**(4), 491–501 (2022)
5. Cevidanes, L., Oliveira, A.E.F., Motta, A., Phillips, C., Burke, B., Tyndall, D.: Head orientation in CBCT-generated cephalograms. Angle Orthod. **79**(5), 971–977 (2009)
6. Cevidanes, L.H.C., Heymann, G., Cornelis, M.A., DeClerck, H.J., Tulloch, J.F.C.: Superimposition of 3-dimensional cone-beam computed tomography models of growing patients. Am. J. Orthod. Dentofacial Orthop. **136**(1), 94–99 (2009)
7. Chang, Y.J., Ruellas, A.C.O., Yatabe, M.S., Westgate, P.M., Cevidanes, L.H.S., Huja, S.S.: Soft tissue changes measured with three-dimensional software provides new insights for surgical predictions. J. Oral Maxillofac. Surg. **75**(10), 2191–2201 (2017)
8. CheibVilefort, P.L., et al.: Condyle-glenoid fossa relationship after Herbst appliance treatment during two stages of craniofacial skeletal maturation: a retrospective study. Orthod. Craniofac. Res. **22**(4), 345–353 (2019)
9. Dolphin: https://www.dolphinimaging.com
10. Fedorov, A., et al.: 3d slicer as an image computing platform for the quantitative imaging network. Magn. Reson. Imaging **30**(9), 1323–1341 (2012)
11. Gillot, M., et al.: Automatic multi-anatomical skull structure segmentation of cone-beam computed tomography scans using 3D UNETR. PLoS ONE **17**(10), e0275033 (2022)
12. Gillot, M., et al.: Automatic landmark identification in cone-beam computed tomography. Orthod. Craniofac. Res. (2023)
13. Huang, G., Liu, Z., Pleiss, G., van der Maaten, L., Weinberger, K.Q.: Convolutional networks with dense connectivity (2020)
14. Kumar, M., Shanavas, M., Sidappa, A., Kiran, M.: Cone beam computed tomography - know its secrets. J. Int. Oral Health **7**(2), 64–68 (2015)
15. Lowekamp, B.C., Chen, D.T., Ibáñez, L., Blezek, D.: The design of SimpleITK. Front. Neuroinform. **7**, 45 (2013)
16. Marstal, K., Berendsen, F., Staring, M., Klein, S.: Simpleelastix: a user-friendly, multi-lingual library for medical image registration. In: 2016 IEEE Conference on Computer Vision and Pattern Recognition Workshops (CVPRW), pp. 574–582 (2016)
17. Martinsson, T., Oda, T., Fernvik, E., Roempke, K., Dalsgaard, C.J., Svensjö, E.: Ropivacaine inhibits leukocyte rolling, adhesion and CD11b/CD18 expression. J. Pharmacol. Exp. Ther. **283**(1), 59–65 (1997)
18. Muenzing, S.E.A., van Ginneken, B., Murphy, K., Pluim, J.P.W.: Supervised quality assessment of medical image registration: application to intra-patient CT lung registration. Med. Image Anal. **16**(8), 1521–1531 (2012)
19. Nada, R.M., Maal, T.J.J., Breuning, K.H., Bergé, S.J., Mostafa, Y.A., Kuijpers-Jagtman, A.M.: Accuracy and reproducibility of voxel based superimposition of cone beam computed tomography models on the anterior cranial base and the zygomatic arches. PLoS ONE **6**(2), e16520 (2011)

20. Ruellas, A.C.D.O., et al.: Comparison and reproducibility of 2 regions of reference for maxillary regional registration with cone-beam computed tomography. Am. J. Orthod. Dentofacial Orthop. **149**(4), 533–542 (2016)
21. Ruellas, A.C.D.O., et al.: Common 3-dimensional coordinate system for assessment of directional changes. Am. J. Orthod. Dentofacial Orthop. **149**(5), 645–656 (2016)
22. Ruellas, A.C.D.O., et al.: 3D mandibular superimposition: Comparison of regions of reference for voxel-based registration. PLoS ONE **11**(6), e0157625 (2016)
23. Schroeder, W., Martin, K., Lorensen, B.: The visualization toolkit: an object-oriented approach to 3D graphics. Ingram (2006)
24. Stamatakis, H.C., Steegman, R., Dusseldorp, J., Ren, Y.: Head positioning in a cone beam computed tomography unit and the effect on accuracy of the three-dimensional surface mode. Eur. J. Oral Sci. **127**(1), 72–80 (2019)
25. Tonin, R.H., et al.: Accuracy of 3D virtual surgical planning for maxillary positioning and orientation in orthognathic surgery. Orthod. Craniofac. Res. **23**(2), 229–236 (2020)
26. Verhelst, P.J., et al.: Three-dimensional cone beam computed tomography analysis protocols for condylar remodelling following orthognathic surgery: a systematic review. Int. J. Oral Maxillofac. Surg. **49**(2), 207–217 (2020)
27. Yushkevich, P.A., et al.: User-guided 3D active contour segmentation of anatomical structures: significantly improved efficiency and reliability. Neuroimage **31**(3), 1116–1128 (2006)
28. Zhou, Y., Li, J.P., Lv, W.C., Ma, R.H., Li, G.: Three-dimensional CBCT images registration method for TMJ based on reconstructed condyle and skull base. Dentomaxillofac. Radiol. **47**(5), 20170421 (2018)

Deep Learning-Based Fast MRI Reconstruction: Improving Generalization for Clinical Translation

Nitzan Avidan⬤ and Moti Freiman$^{(\boxtimes)}$⬤

Faculty of Biomedical Engineering, Technion - IIT, Haifa, Israel
`moti.freiman@bm.technion.ac.il`

Abstract. Numerous deep neural network (DNN)-based methods have been proposed in recent years to tackle the challenging ill-posed inverse problem of MRI reconstruction from undersampled 'k-space' (Fourier domain) data. However, these methods have shown instability when faced with variations in the acquisition process and anatomical distribution. This instability indicates that DNN architectures have poorer generalization compared to their classical counterparts in capturing the relevant physical models. Consequently, the limited generalization hinders the applicability of DNNs for undersampled MRI reconstruction in the clinical setting, which is especially critical in detecting subtle pathological regions that play a crucial role in clinical diagnosis. We enhance the generalization capacity of deep neural network (DNN) methods for undersampled MRI reconstruction by introducing a physically-primed DNN architecture and training approach. Our architecture incorporates the undersampling mask into the model and utilizes a specialized training method that leverages data generated with various undersampling masks to encourage the model to generalize the undersampled MRI reconstruction problem. Through extensive experimentation on the publicly available Fast-MRI dataset, we demonstrate the added value of our approach. Our physically-primed approach exhibits significantly improved robustness against variations in the acquisition process and anatomical distribution, particularly in pathological regions, compared to both vanilla DNN methods and DNNs trained with undersampling mask augmentation.

Trained models and code for experiment replication are available at: https://github.com/nitzanavidan/PD_Recon.

Keywords: MRI reconstruction · Deep-learning · clinical translation

1 Introduction

Magnetic resonance imaging (MRI) is a non-invasive imaging modality widely used in clinical applications due to its ability to provide detailed soft-tissue images. The MRI signal is acquired in the Fourier space known as the 'k-space.'

© The Author(s), under exclusive license to Springer Nature Switzerland AG 2023
S. Wesarg et al. (Eds.): CLIP/FAIMI/EPIMI 2023, LNCS 14242, pp. 59–69, 2023.
https://doi.org/10.1007/978-3-031-45249-9_6

Subsequently, an inverse Fourier transform (IFT) is applied to the k-space data to generate meaningful MRI scans in the spatial domain [10]. However, the acquisition times required to sample the full k-space pose a significant limitation in achieving high spatial and temporal resolutions, reducing motion artifacts, improving patient experience, and reducing costs [17].

Partial sampling of the k-space offers a linear reduction in acquisition times. However, reconstructing an MRI image from undersampled k-space data poses a highly ill-posed inverse problem. Naïve reconstruction by zero-filling the missing k-space data and applying the inverse Fourier transform (IFT) leads to clinically meaningless images due to the presence of various artifacts [17]. Early research focused on k-space properties, such as partial Fourier imaging methods that leverage Hermitian symmetry to achieve reduced acquisition times [5].

Classical linear approaches for MRI reconstruction from undersampled data exploit advancements in parallel imaging, utilizing multiple receiver coils combined with linear reconstruction algorithms applied either in the k-space domain [6] or in the spatial domain [13]. However, the theoretical acceleration factor is constrained by the number of available coils [7]. Moreover, the practical acceleration factor is further limited due to noise amplification resulting from matrix inversion [13].

The non-linear compressed sensing (CS) approach [12] aims to reconstruct a high-quality image from undersampled k-space data by imposing sparsity constraints on the ill-posed inverse problem using a sparsifying linear transform. Although the CS objective function lacks a closed-form solution, it is a convex problem and can be addressed using various iterative algorithms [7].

In recent years, a multitude of deep neural network (DNN)-based methods have been proposed for undersampled MRI reconstruction, leading to significant improvements in both image quality and acceleration factors [3]. Similar to classical methods, DNN-based approaches can be applied in either the spatial domain or the k-space domain. For instance, the KIKI-net alternates between the image domain (I-CNN) and k-space domain (K-CNN) iteratively, enforcing a data consistency constraint in an interleaved manner [4]. A more recent approach, the End-to-End Variational Network (E2E-VarNet), simultaneously estimates coil-specific sensitivity maps and predicts the fully-sampled k-space from the undersampled data through a series of cascades [15].

Although currently available deep neural network (DNN) methods have shown promising performance, Antun et al. [2] and Jalal et al. [9] demonstrated that, unlike their classical counterparts, DNN-based methods exhibit instability in the presence of variations in the acquisition process and anatomical distribution. These variations include using different undersampling masks or acceleration factors during inference compared to those used in training, as well as the presence of small pathologies or different anatomies not encountered during training.

Early efforts have focused on addressing the stability gap in DNN-based MRI reconstruction through data augmentation techniques. For instance, Liu et al. enhanced overall reconstruction performance and robustness against sampling pattern discrepancies and images acquired at different contrast phases by augmenting the undersampled data with a wide range of varying undersampling patterns [11]. More recently, Jalal et al. combined a DNN-based generative prior with classical CS-based reconstruction and posterior sampling to mitigate the stability gap [9]. However, this approach does not directly target the stability gap in DNN-based MRI reconstruction methods.

Recently, the emergence of physics-driven deep learning methods has provided a powerful approach to enhance the generalization capacity of DNN-based undersampled MRI reconstruction. These methods encompass various techniques that incorporate the physics of MRI acquisitions, such as physics-driven loss functions, plug-and-play methods, generative models, and unrolled networks [7]. Notable examples include enforcing k-space consistency directly after image enhancement [8] and incorporating k-space consistency as an additional cost function term during training [16].

However, the stability of DNN-based methods against variations in the acquisition process and anatomical distribution remains an unresolved question [7]. Moreover, current DNN methods approach the ill-posed undersampled MRI reconstruction problem as a regression task, aiming to predict the fully-sampled k-space data or a high-quality image from the undersampled data. In doing so, they effectively eliminate the sampling mask used during acquisition from the regression process at inference time. This stands in contrast to classical counterparts, which incorporate the sampling mask as an integral part of the forward model during the optimization process.

In this study, our objective is to bridge the stability gap in DNN-based undersampled MRI reconstruction by introducing a physically-primed DNN architecture and training approach. In contrast to previous methods, our architecture incorporates the undersampling mask as an integral component in addition to the observed data, enhancing the model's ability to capture the undersampled MRI reconstruction problem. Moreover, we adopt a training approach that leverages data generated with diverse undersampling masks, promoting the model's generalization capability for undersampled scenarios.

Figure 1 succinctly presents our main result. The physically-primed approach (d, h) demonstrates improved generalization capacity and robustness against variations in the acquisition process (e.g., differences in undersampling masks) compared to baseline methods (b, f and c, g).

Trained models and code for experiment replication are available at: https://github.com/nitzanavidan/PD_Recon.

2 Methods

2.1 Background

The forward model of the undersampled single coil MRI acquisition process is given by:

$$k_{us} = M \circ \mathcal{F}x + n \tag{1}$$

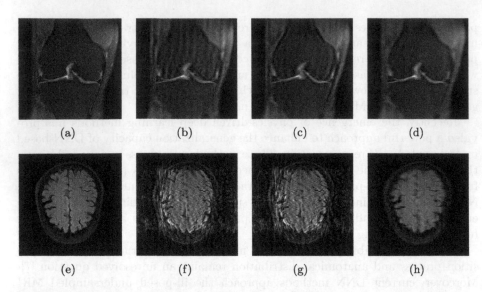

Fig. 1. The stability gap in DNN-based MRI reconstruction from undersampled data is demonstrated in the following scenarios. The first row illustrates variations in the acquisition process, specifically using different undersampling masks. The second row showcases variations in the anatomical distribution, where the model is trained on knee data and tested on brain data. The corresponding results are as follows: (a/e) the target image from fully-sampled k-space data, (b/f) reconstruction using a model trained with a fixed sampling mask (PSNR = 24.9/14.07, SSIM = 0.4684/0.1926), (c/g) reconstruction using a model trained with mask augmentations (PSNR = 25.35/16.56, SSIM = 0.4969/0.2812), and (d/h) reconstruction using our proposed physically-primed approach (PSNR = 26.4/26.05, SSIM = 0.5354/0.6359).

where $k_{us} \in \mathbb{C}^N$ are the observed measurements in the k-space, $x \in \mathbb{R}^N$ is the image representing the underlying anatomy, $\mathcal{F}$ is the Fourier operator, $M \in \mathbb{R}^N$ is a binary undersampling mask, $\circ$ is element-wise multiplication and n is an additive noise. For the sake of simplicity, we assume $n \sim \mathcal{N}\left(0, \sigma^2\right)$ [1]. Without loss of generality, deep-learning-based methods for MRI reconstruction define the reconstruction task as:

$$\widehat{k_{full}} = F_\theta\left(k_{us}\right) \tag{2}$$

where F_θ denotes the network function and θ represents the DNN weights. Taking a supervised learning approach, the DNN weights θ are estimated by minimizing the loss between the full k-space data predicted by the DNN, $\widehat{k_{full}}$, from the undersampled data, k_{us}, to the corresponding ground truth fully sampled k-space data, k_{full}, as follows:

$$\hat{\theta} = \arg\min_\theta \sum_{i=0}^{n_{data}} \left\| F_\theta\left(k_{us}^{(i)}\right) - k_{full}^{(i)} \right\|_1 \tag{3}$$

However, unlike the classical CS approaches [12], such methods are known to be unstable against the presence of variations in the acquisition process and the anatomical distribution [2,9].

A key observation is that while the CS approach [12] explicitly encodes the undersampling mask M as part of the system forward model, DNN-based approaches essentially ignore the undersampling mask during training an inference.

2.2 Physically-Primed DNN for MRI Reconstruction

Our main hypothesis is that by explicitly encoding the undersampling mask in the DNN architecture and leveraging this information during inference, the DNN will be more capable to accurately generalize the ill-posed inverse problem associated with MRI reconstruction from undersampled data. Thus, it will be more robust against variations in the acquisition process and anatomical distribution. We introduce a physically-primed, U-Net [14] based, DNN architecture operating on the k-space domain. We represent the complex-valued k-space data as a two-channel input, corresponding to the real and imaginary parts. We encode the undersampling mask M by adding a 3rd input channel to the DNN. The prediction of the full k-space data from the undersampled k-space data is defined as:

$$\widehat{k_{full}} = F_\theta \left(k_{us}, M\right) \tag{4}$$

The DNN weights θ are estimated by minimizing the loss between the full k-space data predicted by the DNN, $\widehat{k_{full}}$, from the undersampled data, k_{us} and the undersampling mask M, to the corresponding ground truth fully sampled k-space data, k_{full}, as follows:

$$\hat{\theta} = \arg \min_\theta \sum_{i=0}^{n_{data}} \left\| F_\theta \left(k_{us}^{(i)}, M^{(i)}\right) - k_{full}^{(i)} \right\|_1 \tag{5}$$

We encourage our physically-primed DNN model to generalize the ill-posed inverse problem of MRI reconstruction from undersampled k-space data by varying the undersampling mask M during training. We used the fastMRI U-Net [17] as our backbone model, while modifying the output to be the summation of the k-space input and the network output. Figure 2 illustrates our physically-primed approach for training.

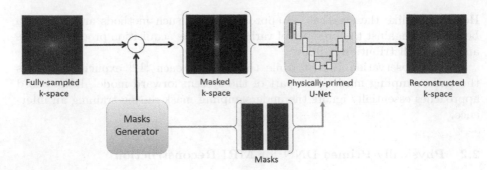

Fig. 2. The fully sampled k-space is multiplied by a binary mask to create the undersampled k-space. Then the undersampled k-space (blue line) is fed into the physically primed DNN model along with the mask (purple line) used for the sampling. (Color figure online)

3   Experiments

3.1   Dataset

We used the publicly available fastMRI dataset [17], consisting of raw k-space data of knee and brain volumes.

The training set consisted of 34742 knee slices. We split the fastMRI knee validation set into validation and test sets since the original fastMRI test set does not allow applying random undersampling. The splitting ratio was 2:1, yielding 5054 slices for validation and 2081 slices for testing. Brain images (1000 slices) were used for test purposes only.

To further evaluate the clinical impact of our approach, we also used the bounding box annotations generated by subspecialist experts on the fastMRI knee and brain dataset provided by the fastMRI+ dataset [18]. Each bounding box annotation includes its coordinates and the relevant label for a given pathology on a slice-by-slice level. Pathology annotations were marked in 39% of our training set, 28% of the validation set, and 51% of the test set. These annotations enabled us to examine the clinical relevance of our models' performance on pathological regions.

3.2   Experimental Methodology

We examined the generalization capability of the networks by testing their performances with varying acquisition conditions and anatomical distributions. Specifically, we evaluated reconstruction performance for: 1) undersampling patterns (equispaced for training and random for test) different from those used during training, but a similar acceleration factor (R = 4), 2) different acceleration factor (R = 4 for training and R = 8 for the test), and; 3) different anatomical distribution (knee for training and brain for the test).

We used standard evaluation metrics, including the average normalized mean square error (NMSE), Peak Signal-to-Noise Ratio (PSNR) and Structural Similarity (SSIM) to assess the different models' performances.

3.3   Results

Figure 3 presents representative reconstruction results in cases of similar anatomical distribution and variation in the acquisition process, i.e. knee images with R = 4 and different sampling masks in test. Figure 4 presents representative reconstruction results in cases of variation in the anatomical distribution, i.e. brain images in test.

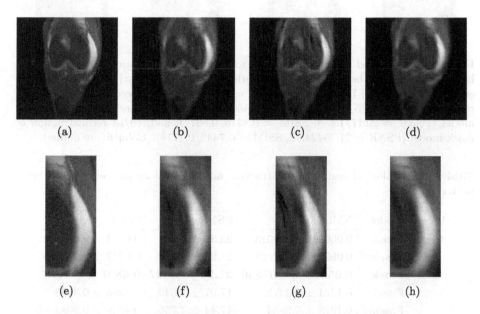

Fig. 3. Reconstructed images along with zoom-in clinical annotation (red bounding box) for variation in the acquisition process, i.e., knee data undersampled with a random mask and R = 4 in test. (a/e) target image, (b/f) reconstruction of fixed model (PSNR = 28.87/25.47, SSIM = 0.6043/0.6524), (c/g) reconstruction of baseline model (PSNR = 29.12/25.98, SSIM = 0.60122/0.663), and; (d/h) reconstruction of mask model (PSNR = 29.58/26.66, SSIM = 0.6325/0.7026). (Color figure online)

Table 1 summarizes model performance for the clinically relevant regions, for the same anatomical distribution (i.e. knee data) but with variations in the acquisition process (i.e. different equispaced sampling mask). Similarly, Table 2 summarizes model performance for random sampling masks in test. In all cases, the physically-primed model (Mask) has significantly better reconstruction accuracy (Paired student's t-test, $p \ll 0.01$). The improved accuracy

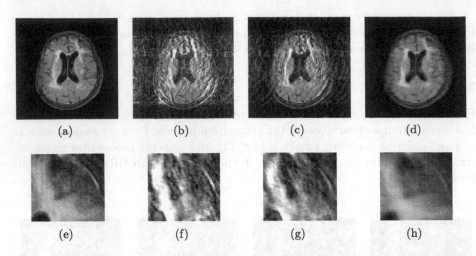

(a)   (b)   (c)   (d)

(e)   (f)   (g)   (h)

Fig. 4. Reconstructed images along with zoom-in clinical annotation (red bounding box) for variation in the anatomical distribution (brain data on test) undersampled with equispaced mask and R = 4 in test. (a/e) target image, (b/f) reconstruction of fixed model (PSNR = 14.85/10.92, SSIM = 0.2405/0.364), (c/g) reconstruction of baseline model (PSNR = 20.17/15.53, SSIM = 0.4315/0.5377), and; (d/h) reconstruction of mask model (PSNR = 27.75/24.67, SSIM = 0.7446/0.7779) (Color figure online)

Table 1. Pathological regions reconstruction accuracy for varying equispaced mask patterns in test.

R test	Model	NMSE	PSNR	SSIM
4	Fixed	0.06059 ± 0.09013	20.36 ± 5.133	0.6374 ± 0.266
	Baseline	0.05073 ± 0.07628	21.22 ± 5.332	0.6797 ± 0.2415
	Mask	**0.05013 ± 0.07356**	**21.26 ± 5.327**	**0.6807 ± 0.2408**
8	Fixed	0.1324 ± 0.3375	17.07 ± 4.649	0.4268 ± 0.3105
	Baseline	0.1223 ± 0.2854	17.44 ± 4.755	0.4528 ± 0.3083
	Mask	**0.1193 ± 0.3246**	**17.59 ± 4.659**	**0.4638 ± 0.3014**

Table 2. Pathological regions reconstruction accuracy for random mask in test.

R test	Model	NMSE	PSNR	SSIM
4	Fixed	0.07082 ± 0.1018	19.62 ± 4.839	0.5815 ± 0.2899
	Baseline	0.06401 ± 0.09236	20.12 ± 5.008	0.6091 ± 0.281
	Mask	**0.05946 ± 0.0852**	**20.43 ± 4.973**	**0.6206 ± 0.2713**
8	Fixed	0.1377 ± 0.2631	16.77 ± 4.577	0.391 ± 0.3098
	Baseline	0.1311 ± 0.2524	17.06 ± 4.713	0.4096 ± 0.3123
	Mask	**0.1228 ± 0.2367**	**17.31 ± 4.644**	**0.429 ± 0.3048**

indicates a better generalization and robustness against variations in the acquisition process compared to models trained with and without mask augmentation. It is important to note that the improved performance is evident mostly in the clinically-relevant regions which are critical for clinical diagnosis.

Table 3. Pathological regions reconstruction accuracy on brain dataset for fixed equispaced mask in test.

R test	Model	NMSE	PSNR	SSIM
4	Fixed	0.1542 ± 0.7087	16.41 ± 8.643	0.3731 ± 0.4088
	Baseline	0.105 ± 0.3544	17.65 ± 7.585	0.4185 ± 0.4031
	Mask	**0.02502 ± 0.06294**	**23.8 ± 7.615**	**0.6421 ± 0.2506**
8	Fixed	0.105 ± 0.235	16.87 ± 6.588	0.294 ± 0.356
	Baseline	0.09502 ± 0.2554	17.68 ± 7.031	0.3007 ± 0.3833
	Mask	**0.04952 ± 0.1054**	**20.78 ± 7.396**	**0.4544 ± 0.2891**

Table 3 demonstrates the generalization capacity of our physically-primed model in case of variation in the anatomical distribution, i.e. training on knee data while testing on brain data for the clinically-relevant regions. Our physically-primed model (mask) performed significantly better (Paired student's t-test, $p \ll 0.01$) than the models trained with and without mask augmentation.

The improved robustness of our physically-primed model against variations in both the acquisition process and the anatomical distribution suggests a better generalization capacity beyond more common data augmentation techniques.

4   Conclusions

We have introduced a physically-primed DNN architecture to address the stability gap in deep-learning-based MRI reconstruction from undersampled k-space data. The observed stability gap highlights the limited generalization of the ill-posed inverse problem associated with DNN methods compared to classical approaches. This stability gap poses practical challenges for DNN-based MRI reconstruction from undersampled data in clinical applications.

While previous approaches focused on data augmentation and physically motivated loss functions to mitigate the stability gap, our physically-primed DNN approach enhances generalization capacity by incorporating the forward model, including the undersampling mask, into the network architecture. Additionally, we introduce an appropriate training scheme using samples generated with various undersampling masks.

Our experimental results demonstrate that encoding the undersampling mask as part of the DNN architecture improves generalization capacity, particularly in clinically relevant regions, surpassing the performance of previously proposed data augmentation techniques across multiple scenarios. Consequently, our approach has the potential to facilitate the practical utilization of DNN-based MRI reconstruction methods in clinical settings.

Acknowledgements. This research was supported in part by a research grant from Microsoft Education and the Israel Inter-university computation center (IUCC).

References

1. Aja-Fernández, S., Vegas-Sánchez-Ferrero, G.: Statistical Analysis of Noise in MRI. Springer, Cham (2016). https://doi.org/10.1007/978-3-319-39934-8
2. Antun, V., Renna, F., Poon, C., Adcock, B., Hansen, A.C.: On instabilities of deep learning in image reconstruction and the potential costs of AI. Proc. Natl. Acad. Sci. **117**(48), 30088–30095 (2020)
3. Chen, Y., et al.: AI-based reconstruction for fast MRI-a systematic review and meta-analysis. Proc. IEEE **110**(2), 224–245 (2022)
4. Eo, T., Jun, Y., Kim, T., Jang, J., Lee, H.J., Hwang, D.: Kiki-net: cross-domain convolutional neural networks for reconstructing undersampled magnetic resonance images. Magn. Reson. Med. **80**(5), 2188–2201 (2018)
5. Feinberg, D.A., Hale, J.D., Watts, J.C., Kaufman, L., Mark, A.: Halving MR imaging time by conjugation: demonstration at 3.5 kg. Radiology **161**(2), 527–531 (1986)
6. Griswold, M.A., et al.: Generalized autocalibrating partially parallel acquisitions (GRAPPA). Magn. Reson. Med. Off. J. Int. Soc. Magn. Reson. Med. **47**(6), 1202–1210 (2002)
7. Hammernik, K., et al.: Physics-driven deep learning for computational magnetic resonance imaging. arXiv preprint arXiv:2203.12215 (2022)
8. Hyun, C.M., Kim, H.P., Lee, S.M., Lee, S., Seo, J.K.: Deep learning for undersampled MRI reconstruction. Phys. Med. Biol. **63**(13), 135007 (2018)
9. Jalal, A., Arvinte, M., Daras, G., Price, E., Dimakis, A.G., Tamir, J.: Robust compressed sensing MRI with deep generative priors. Adv. Neural. Inf. Process. Syst. **34**, 14938–14954 (2021)
10. Liang, Z.P., Lauterbur, P.C.: Principles of Magnetic Resonance Imaging. SPIE Optical Engineering Press, Bellingham (2000)
11. Liu, F., Samsonov, A., Chen, L., Kijowski, R., Feng, L.: SANTIS: sampling-augmented neural network with incoherent structure for MR image reconstruction. Magn. Reson. Med. **82**(5), 1890–1904 (2019)
12. Lustig, M., Donoho, D., Pauly, J.M.: Sparse MRI: the application of compressed sensing for rapid MR imaging. Magn. Reson. Med. Off. J. Int. Soc. Magn. Reson. Med. **58**(6), 1182–1195 (2007)
13. Pruessmann, K.P., Weiger, M., Scheidegger, M.B., Boesiger, P.: Sense: sensitivity encoding for fast MRI. Magn. Reson. Med. Off. J. Int. Soc. Magn. Reson. Med. **42**(5), 952–962 (1999)
14. Ronneberger, O., Fischer, P., Brox, T.: U-Net: convolutional networks for biomedical image segmentation. In: Navab, N., Hornegger, J., Wells, W.M., Frangi, A.F. (eds.) MICCAI 2015. LNCS, vol. 9351, pp. 234–241. Springer, Cham (2015). https://doi.org/10.1007/978-3-319-24574-4_28
15. Sriram, A., et al.: End-to-end variational networks for accelerated MRI reconstruction. In: Martel, A.L., et al. (eds.) MICCAI 2020. LNCS, vol. 12262, pp. 64–73. Springer, Cham (2020). https://doi.org/10.1007/978-3-030-59713-9_7

16. Yang, G., et al.: DAGAN: deep de-aliasing generative adversarial networks for fast compressed sensing MRI reconstruction. IEEE Trans. Med. Imaging **37**(6), 1310–1321 (2017)
17. Zbontar, J., et al.: fastMRI: an open dataset and benchmarks for accelerated MRI. arXiv preprint arXiv:1811.08839 (2018)
18. Zhao, R., et al.: fastMRI+: clinical pathology annotations for knee and brain fully sampled multi-coil MRI data. arXiv preprint arXiv:2109.03812 (2021)

Uncertainty Based Border-Aware Segmentation Network for Deep Caries

Gayeon Kim[1], Yufei Chen[1(✉)], Shuai Qi[2], Yujie Fu[2], and Qi Zhang[2]

[1] College of Electronics and Information Engineering, Tongji University, Shanghai, China
yufeichen@tongji.edu.cn

[2] Department of Endodontics, School and Hospital of Stomatology, Tongji University, Shanghai Engineering Research Center of Tooth Restoration and Regeneration, Shanghai, China

Abstract. Deep caries is a progressive and destructive disease affecting the hard surface of teeth. If left untreated, it can lead to serious health risks such as inflammation and apical periodontitis. According to clinical evidence, carious dentin can be categorized into soft, firm, and hard dentin based on its hardness. Precise assessment of carious lesions is critical for effective treatment; however, current methods rely on subjective judgments based on tactile feedback, leading to variability in dentin removal and potential treatment failure. To address this problem and provide accurate references to the dentist, we propose a border-aware network for deep caries segmentation using clinical Micro-computed tomography (MicroCT) data. The network performs segmentation of the three types of dentin within the cavity and incorporates the Signed Distance Field (SDF) method to enhance accuracy at the borders of caries. To evaluate the clinical feasibility, we simulate CBCT images by Gaussian-blurring MicroCT images and introduce evidence theory to estimate uncertainty in ambiguous border regions, ensuring robustness against low-quality inputs. We collect a real-world dental dataset which includes 25 MicroCT scans with deep caries. Through experiments, we demonstrate that our proposed method improves segmentation accuracy, especially in images with different degrees of blurriness. Moreover, by segmenting images with different levels of Gaussian blurring, we validate the robustness of our proposed method in handling low-quality inputs, thus showing its potential for future clinical applications.

Keywords: Dental Caries Segmentation · MicroCT · Uncertainty Estimation · Signed Distance Field

1 Introduction

Caries is a progressive and destructive disease that occurs in the crown of the teeth. As a severe type of caries, deep caries refers to the situation where a carious lesion reaches the inner quarter of the dentin, but with no apparent

S. Wesarg et al. (Eds.): CLIP/FAIMI/EPIMI 2023, LNCS 14242, pp. 70–80, 2023.
https://doi.org/10.1007/978-3-031-45249-9_7

pulp involvement. If left untreated, deep caries will continue to progress, leading to pulp inflammation and infection, posing a severe hazard to oral health [11]. From a clinical perspective, precise assessment of the carious lesion is crucial for the successful treatment of deep caries. According to existing clinical evidence, carious dentin can be categorized into soft, firm, and hard dentin based on its hardness, as shown in Fig. 1(a) [9,11]. During the treatment, it is not necessary to remove all the carious dentin, instead, dentists rely on tactile feedback from instruments to perceive changes in hardness, make real-time judgments, and determine whether to stop the removal operation based on their experience, as depicted in Fig. 1(b). However, this process is rather subjective, and individual variations in perception may lead to either conservative or aggressive carious dentin removal, increasing the risk of treatment failure. Therefore, it is of great significance to develop an objective reference for carious dentin differentiation.

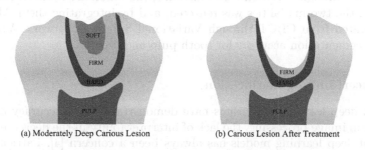

(a) Moderately Deep Carious Lesion   (b) Carious Lesion After Treatment

Fig. 1. Example for a moderately deep carious lesion.

To address the problem above, we propose a border-aware network for deep caries segmentation. The network initially performs segmentation on the three kinds of dentin constituting the cavity simultaneously. To enhance the segmentation accuracy at the borders, we incorporate the Signed Distance Field (SDF) method, which makes the model pay more attention to the borders, thereby providing more accurate clinical references. Subsequently, to further explore the clinical feasibility of the method, we simulate clinical CBCT data using Gaussian-blurred MicroCT and introduce evidence theory to estimate the uncertainty of ambiguous borders, ensuring the model's robustness against low-quality inputs. Through experimentation, we validate that the proposed method improves segmentation accuracy, especially in edges with different degrees of blurriness. Furthermore, by segmenting images subjected to different levels of Gaussian blurring, we verify the model's robustness in handling blurred images and demonstrate the feasibility of this method in future clinical applications.

2   Related Work

2.1   Dental Caries Image Segmentation

Particularly in the domain of dental caries segmentation, a few studies have been conducted. Zhu et al. [14] propose CariesNet, a novel deep-learning architecture incorporating a full-scale axial attention module into a U-shaped network. It aims to segment three types of caries lesions depending on the depth of caries: Shallow caries, Moderate caries, and Deep caries from panoramic radiographs images. Through similar research, Dayı et al. [1] propose Dental Caries Detection Network (DCDNet) with a Multi-Predicted Output (MPO) structure for caries segmentation. The network is designed to segment occlusal, proximal, and cervical caries lesions depending on their location from panoramic radiographs. Xiang, L et al. [7] proposed a new data pipeline to convert CBCT images to high resolution to approximate MicroCT to improve the segmentation performance of the pulp cavity and tooth in the CBCT image. In addition to dental caries, segmentation studies have also been conducted on other parts of the tooth, such as pulp and crown. Xiaoyu, Y et al. [13] proposed a new cross-modality method to generate MicroCT images from CBCT images, bridging the gap between these two modalities. Through variation-based distance, the difference between the distributions of the two modalities was reflected, and by integrating them, MicroCT was generated from CBCT through Variational Synthesis network (VSnet) to increase segmentation accuracy for tooth pulp and crown.

2.2   Uncertainty Quantification

Although deep learning algorithms have demonstrated high accuracy and performance in image processing, the lack of interpretability in the decision-making process of deep learning models has always been a concern [4]. Particularly in the medical domain, which is a safety-critical area, the need for not only accurate prediction results but also reliability and explainability of these results has emerged. Therefore, research on uncertainty quantification in medical image processing has gained prominence as a crucial step. Uncertainty can be broadly categorized into two types: Epistemic uncertainty and Aleatoric uncertainty, and methodologies for quantifying uncertainty can be divided into Bayesian-based approaches, ensemble-based approaches, evidential deep learning-based approaches, etc. Several studies have been proposed in the field of medical imaging for uncertainty quantification. Y. Kwon et al. [6] propose a method for quantifying uncertainty using Bayesian neural networks and apply it to Ischemic stroke lesion and Retinal blood vessel segmentation. Alireza et al. [8] propose an image segmentation model through confidence calibration and uncertainty estimation. By refining confidence scores and quantifying uncertainty using a deep ensemble approach, it enhances the reliability of segmentation results and provides valuable information for decision-making. Kohl et al. [5] propose a Probabilistic U-Net architecture that incorporates both a fully convolutional network (FCN) and a conditional variance autoencoder. This approach enables the generation of multiple segmentation hypotheses for images with ambiguous features. Zou, Ke et al. [15] apply subjective logic theory to model uncertainty and improve the reliability and robustness of brain tumor segmentation by measuring uncertainty for each segmented pixel.

3   Method

To address the aforementioned problems, in this section we propose an automatic multi-class segmentation network that is uncertainty-aware for carious dentin using labels for each carious tissue in MicroCT images. The architecture of the network for the proposed method is shown in Fig. 2. We adopt a U-Net [10] based network that shows good performance in medical image segmentation as a backbone network. We introduce a segmentation method that combines SDF to increase the segmentation accuracy of caries boundaries and introduces a segmentation method based on the theory of evidence by recognizing the uncertainty of the segmentation results for the ambiguity of the boundaries between different cavities.

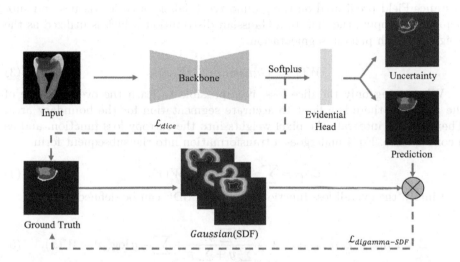

Fig. 2. Overview of Uncertainty Based Border-Aware Segmentation Network.

3.1   Border-Aware Network Using SDF

In our study, prior to incorporating the signed distance field (SDF) into tooth caries segmentation, we utilize two essential loss functions: dice loss and cross-entropy loss. These loss functions are employed to effectively address the imbalance between different caries classes and they are denoted as follows:

$$\mathcal{L} = \mathcal{L}_{Dice} + \mathcal{L}_{CE} = 1 - \frac{2 \sum yp}{\sum y + \sum p} + \sum -y \log(p), \tag{1}$$

where y and p are the label and prediction probability.

Signed Distance Field (SDF) has become an essential paradigm across multiple domains, most notably within computer graphics and implicit reconstruction. It presents an efficient and continuous framework for delineating the shape and spatial attributes of objects. It is characterized as the minimal distance from a point to the object's surface, with the convention that the distance is negative inside the object and positive outside. The surface of the object is a zero iso-surface δ. Formally, the signed distance field for a point x in the Euclidean space can be encapsulated by the ensuing mathematical representation:

$$SDF(x) = \text{sign}(x) \min_{g \in \delta} \|x - g\|, \quad \text{where } \text{sign}(x) = \begin{cases} -1 & \text{inside} \\ 1 & \text{outside} \end{cases} \quad (2)$$

Drawing inspiration from the SDF, this paper introduces the SDF loss function with the aim to enhance the segmentation accuracy in the boundary regions of caries tissues. In this regard, we put forward a novel loss function, meticulously crafted for this purpose. The methodology entails computing a 2D Signed Distance Field predicated on the ground truth labels of caries tissues and subsequently mapping the SDF to a Gaussian distribution, which is utilized as the weight for each pixel in segmentation.

$$W(x) = Gaussian(SDF(x)), \quad (3)$$

When using only the dice loss, it is possible to learn the overall shape of the area, but there is a limit to accurate segmentation for the boundary area. Therefore We integrate the pixel weights into the entropy loss function, and as a consequence, Eq. 4 undergoes a transformation into the subsequent form.

$$\mathcal{L}_{CE} = \sum -\boldsymbol{y} \log(\boldsymbol{p}) \otimes W(x), \quad (4)$$

Finally, the overall loss function combining SDF can be defined as:

$$\mathcal{L} = \mathcal{L}_{Dice} + \mathcal{L}_{CE-SDF} = 1 - \frac{2\sum \boldsymbol{y}\boldsymbol{p}}{\sum \boldsymbol{y} + \sum \boldsymbol{p}} + \sum -\boldsymbol{y} \log(\boldsymbol{p}) \otimes W(x), \quad (5)$$

In Fig. 3, a distinct observation can be made, where higher values are prominently observed at the edges of various tissues of caries. By incorporating the weights into the loss function, we are able to emphasize the significance of these edge regions, thereby improving the segmentation performance in these areas.

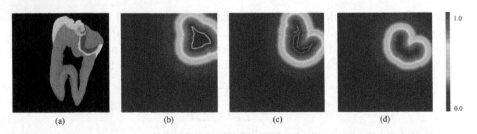

(a)   (b)   (c)   (d)

Fig. 3. Example of SDF weights for each category after mapping to a Gaussian distribution.

3.2 Uncertainty Based Caries Segmentation

In clinical practice, CBCT images are commonly used for dental caries treatment. However, there are practical challenges when it comes to segmenting carious tissues from tooth images obtained through CBCT. As a previous study on this problem, we add different sigma parameters of the Gaussian filter to the MicroCT image and convert the image to bring the MicroCT image closer to the CBCT image. However, this conversion process results in decreased image clarity, particularly affecting the clarity of caries boundaries. Due to the inherent image ambiguity, there will inevitably be uncertainty associated with the segmentation results (Fig. 4).

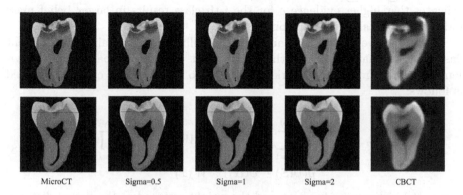

MicroCT   Sigma=0.5   Sigma=1   Sigma=2   CBCT

Fig. 4. Examples of MicroCT images using Gaussian blurriness with different parameters (sigma) and CBCT.

The Dempster-Shafer theory of evidence extends Bayesian reasoning to handle subjective probabilities [2]. In this theory, a "frame of discernment" represents a set of mutually exclusive possible states corresponding to a specific domain. Belief mass, assigned to subsets of the frame, expresses opinions about the truth of a particular state. Subjective logic formulates belief allocation in DST as a Dirichlet Distribution, enabling the quantification of belief mass and uncertainty [3].

Applying the theory to dental caries image segmentation, we establish a multi-class segmentation framework that calculates the probabilities and uncertainties for the caries tissue segmentation problem based on evidence. Given the segmentation results for image pixels, the sum of belief mass values for each class and the sum of uncertainties should equal 1. This relationship can be defined as follow:

$$\sum_{i=1}^{N} b_i + u = 1, \tag{6}$$

where belief mass and uncertainty are non-negative values representing the probability of a pixel belonging to the i-th class being true and the total uncertainty, respectively.

To obtain evidence e for the segmentation results, the output of the network passes through a softplus activation function layer, ensuring that $e \geq 0$. Subjective logic can relate the evidence e to the Dirichlet distribution, expressing it as $\alpha = e + 1$. Consequently, given derived evidence e the belief mass and uncertainty for each pixel can be denoted as:

$$b_i = \frac{e_i}{S} \quad \text{and} \quad u = \frac{K}{S}, \tag{7}$$

where $S = \sum \alpha = \sum(e + 1)$ represents the Dirichlet strength. This formulation indicates that higher evidence e for the i-th class results in larger belief mass and lower uncertainty for each pixel.

The subjective logic can obtain the probability and uncertainty for different classes by associating them with the Dirichlet distribution based on the evidence collected from the network. According to [12,15], Eq. 4 can be improved as follows:

$$\mathcal{L}_{digamma-SDF} = \int \left[\sum_{i=1}^{N} -y_i log(p_i) \otimes W(x) \right] \frac{1}{B(\alpha_i)} \prod_{i=1}^{N} p_i^{\alpha_i - 1} dp \tag{8}$$
$$= \boldsymbol{y}(\psi(\mathbf{S}) - \psi(\boldsymbol{\alpha})) \otimes W(x)$$

where $\psi(\cdot)$ is *digamma* function and the weight assigned to the signed distance field (SDF) in combination with cross-entropy ultimately becomes the weight of the digamma loss. Finally, the loss function we use in model training is described as:

$$\mathcal{L} = \mathcal{L}_{Dice} + \mathcal{L}_{digamma-SDF}$$
$$= 1 - \frac{2\sum \boldsymbol{yp}}{\sum \boldsymbol{y} + \sum \boldsymbol{p}} + \boldsymbol{y}(\psi(\mathbf{S}) - \psi(\boldsymbol{\alpha})) \otimes W(x) \tag{9}$$

4   Experiments and Discussion

4.1   Dataset and Settings

Dataset. In this study, we collect a dental dataset consisting of 25 tooth images with deep caries obtained through MicroCT scans. The dataset is collected between October 2022 and January 2023 from the Department of Oral and Maxillofacial Surgery, Affiliated Stomatology Hospital of Tongji University. The MicroCT images in the dataset are categorized into three labels: soft dentin, firm dentin, and hard dentin. All of the dentin regions are labeled by experienced dentists. To evaluate the performance of the proposed method, 70% of the samples and 20% of the samples from these dental images are respectively selected as the training dataset and validation set, while the remaining 10% of the samples are used as the test dataset.

Implementation. All experiments are conducted using the PyTorch platform and NVIDIA GeForce RTX 3090. The network is optimized during training using the Adam optimizer, with a maximum of 100 epochs. The initial learning rate is 0.0001. All images are transformed from 3D images to 2D slices and uniformly resized to 512×512 pixels. The output of the network consists of four classes: background, soft dentin, firm dentin, and hard dentin.

4.2 Verification of SDF Effectiveness

In Table 1, comparison experiments are conducted to evaluate the performance of different segmentation models: UNet, Attention UNet, and UNet++. The experiments involve comparing the baseline models combined with SDF loss. Two region-based evaluation metrics, Dice and IoU, are used to assess the segmentation quality. For boundary distance evaluation, Hausdorff (HD95) and Average Surface Distance (ASD) metrics are employed, comparing the boundaries predicted by the models to the ground truth label.

Table 1. Quantitative results with U-Net based methods.

Methods	Dice↑			IoU↑			HD5↓			SD95↓		
	soft	firm	hard	soft	firm	hard	soft	firm	hard	soft	firm	hard
UNet w/o SDF	0.800	0.802	0.496	0.716	0.678	0.330	14.090	**13.601**	**9.203**	4.797	13.601	3.006
Unet w/ SDF	**0.834**	**0.808**	**0.759**	**0.754**	**0.686**	**0.615**	**13.266**	13.672	10.383	**4.268**	**3.181**	**2.467**
Attention UNet w/o SDF	0.818	0.814	**0.773**	0.727	**0.691**	0.635	15.707	12.608	12.732	3.942	3.524	**2.683**
Attention UNet w/ SDF	**0.833**	**0.815**	0.770	**0.749**	0.688	**0.636**	**12.743**	**12.327**	**8.460**	**3.706**	**3.163**	2.799
UNet++ w/o SDF	0.775	0.773	0.717	0.675	0.639	0.562	14.701	14.977	10.122	5.521	3.973	2.641
UNet ++ w/ SDF	**0.826**	**0.831**	**0.805**	**0.743**	**0.715**	**0.676**	**13.558**	**6.697**	**7.239**	**4.396**	**3.011**	**2.313**

The results of all backbones indicate that combining SDF loss with the baseline models led to improved segmentation performance and the metrics with respect to boundaries decrease. In Fig. 5, baseline methods with and without SDF show good segmentation performance on the border between soft dentin and firm dentin. In contrast, the segmentation results of the model with SDF exhibit improved performance on the border. Notably, the SDF information is concentrated on each dental caries boundary, resulting in an enhanced segmentation accuracy around these edges.

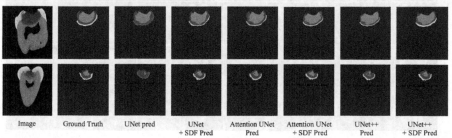

Image   Ground Truth   UNet pred   UNet + SDF Pred   Attention UNet Pred   Attention UNet + SDF Pred   UNet++ Pred   UNet++ + SDF Pred

Fig. 5. The comparisons of dental caries segmentation results on U-Net based Network with SDF loss.

4.3 Verification of Model Robustness

In Table 2, we conduct an experiment to verify the robustness of the model by reducing the image quality by adding a Gaussian blurriness of different degrees to the MicroCT images. The parameters (sigma) used are 0.5, 1, and 2. The results of U-Net combined with the SDF method for segmenting images with different sigma values show that as the image quality decreases, the segmentation performance significantly declines. On the other hand, the segmentation results of the method with uncertainty exhibit a relatively gradual decline as the image quality deteriorates. Remarkably, the segmentation performance for firm dentin and hard dentin remains stable. This experiment demonstrates that adding uncertainty improves the robustness of our proposed method.

Table 2. Quantitative results of the robustness of the proposed method on Gaussian blurriness images of different degrees.

Methods	sigma	Dice ↑			IoU↑			HD95↓			ASD↓		
		soft	firm	hard	soft	firm	hard	soft	firm	hard	soft	firm	hard
UNet +SDF	0	0.834	0.808	0.759	0.754	0.686	0.615	13.266	13.672	10.383	4.268	3.181	2.467
	0.5	0.720	0.719	0.566	0.622	0.566	0.398	17.432	16.260	58.062	4.645	6.39	95.39
	1	0.684	0.681	0.791	0.558	0.520	0.659	25.569	39.014	11.841	9.982	3.905	18.208
	2	0.628	0.429	0.423	0.501	0.275	0.269	26.725	49.988	39.813	10.200	7.118	7.685
UNet +SDF +UNC	0	0.811	0.820	0.779	0.722	0.698	0.642	13.534	18.773	24.860	4.894	4.615	4.082
	0.5	0.679	0.802	0.822	0.575	0.673	0.702	19.959	17.597	7.741	4.798	3.965	2.085
	1	0.737	0.761	0.668	0.618	0.627	0.513	19.817	19.817	12.073	9.094	4.835	3.139
	2	0.545	0.675	0.631	0.411	0.513	0.467	17.912	31.277	37.946	13.064	5.503	4.031

According to Fig. 6, it is shown that given a Gaussian blurriness applied to the image with different parameters (sigma), our proposed method using uncertainty obtains more accurate and robust segmentation results, especially on firm dentin and hard dentin. In addition, it is shown to be effective for segmentation performance and has practicable potential in clinical medicine.

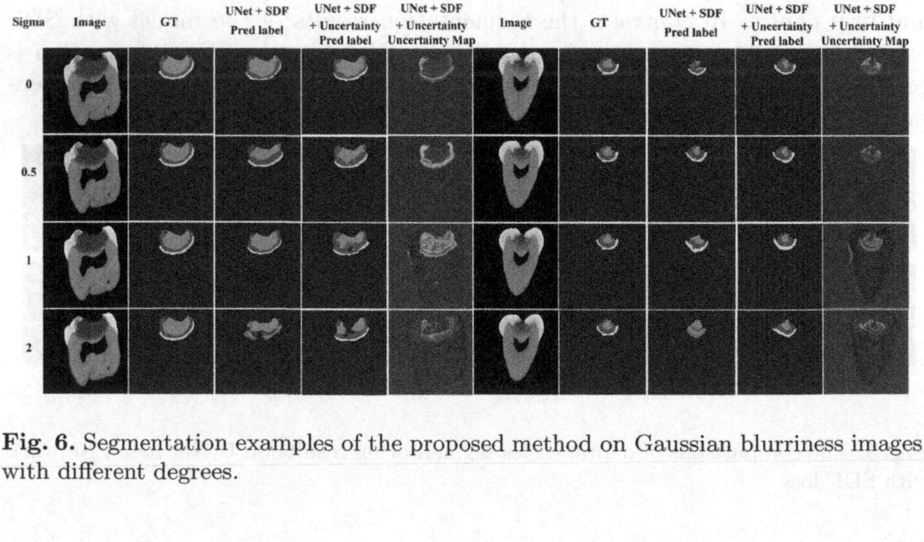

Fig. 6. Segmentation examples of the proposed method on Gaussian blurriness images with different degrees.

5   Conclusion

Compared with U-Net based methods, the proposed Uncertainty Based Border-Aware Network for deep caries segmentation can achieve more precise segmentation performance on caries borders by more concentrating border regions of caries tissue, providing uncertainty estimated based on evidence theory. It can provide information about the border of the dental caries, assisting clinicians in making precise decisions during treatment. This approach is a crucial technological foundation for the future development of clinically applicable deep caries assessment systems based on cone-beam computed tomography (CBCT).

Acknowledgements. This work was supported by the National Natural Science Foundation of China (No. 62173252).

References

1. Dayı, B., Üzen, H., Çiçek, İB., Duman, ŞB.: A novel deep learning-based approach for segmentation of different type caries lesions on panoramic radiographs. Diagnostics **13**(2), 202 (2023)
2. Dempster, A.P.: A generalization of Bayesian inference. J. Roy. Stat. Soc. Ser. B (Methodol.) **30**(2), 205–232 (1968)
3. Jsang, A.: Subjective Logic: A Formalism for Reasoning Under Uncertainty. Springer, Cham (2018). https://doi.org/10.1007/978-3-319-42337-1
4. Kendall, A., Gal, Y.: What uncertainties do we need in Bayesian deep learning for computer vision? In: Advances in Neural Information Processing Systems, vol. 30 (2017)
5. Kohl, S., et al.: A probabilistic u-net for segmentation of ambiguous images. In: Advances in Neural Information Processing Systems, vol. 31 (2018)
6. Kwon, Y., Won, J.-H., Kim, B.J., Paik, M.C.: Uncertainty quantification using Bayesian neural networks in classification: application to biomedical image segmentation. Comput. Stat. Data Anal. **142**, 106816 (2020)
7. Lin, X., et al.: Micro-computed tomography-guided artificial intelligence for pulp cavity and tooth segmentation on cone-beam computed tomography. J. Endod. **47**(12), 1933–1941 (2021)
8. Mehrtash, A., Wells, W.M., Tempany, C.M., Abolmaesumi, P., Kapur, T.: Confidence calibration and predictive uncertainty estimation for deep medical image segmentation. IEEE Trans. Med. Imaging **39**(12), 3868–3878 (2020)
9. Pitts, N.B., et al.: Dental caries. Nat. Rev. Dis. Primers **3**(1), 1–16 (2017)
10. Ronneberger, O., Fischer, P., Brox, T.: U-Net: convolutional networks for biomedical image segmentation. In: Navab, N., Hornegger, J., Wells, W.M., Frangi, A.F. (eds.) MICCAI 2015, Part III. LNCS, vol. 9351, pp. 234–241. Springer, Cham (2015). https://doi.org/10.1007/978-3-319-24574-4_28
11. Schwendicke, F., et al.: Managing carious lesions: consensus recommendations on carious tissue removal. Adv. Dent. Res. **28**(2), 58–67 (2016)
12. Sensoy, M., Kaplan, L., Kandemir, M.: Evidential deep learning to quantify classification uncertainty. In: Advances in Neural Information Processing Systems, vol. 31 (2018)

13. Yang, X., Chen, Y., Yue, X., Lin, X., Zhang, Q.: Variational synthesis network for generating micro computed tomography from cone beam computed tomography. In: 2021 IEEE International Conference on Bioinformatics and Biomedicine (BIBM), pp. 1611–1614. IEEE (2021)
14. Zhu, H., Cao, Z., Lian, L., Ye, G., Gao, H., Wu, J.: CariesNet: a deep learning approach for segmentation of multi-stage caries lesion from oral panoramic x-ray image. In: Neural Computing and Applications, pp. 1–9 (2022)
15. Zou, K., Yuan, X., Shen, X., Wang, M., Fu, H.: TBraTS: trusted brain tumor segmentation. In: Wang, L., Dou, Q., Fletcher, P.T., Speidel, S., Li, S. (eds.) MICCAI 2022. LNCS, vol. 13438, pp. 503–513. Springer, Cham (2022). https://doi.org/10.1007/978-3-031-16452-1_48

An Efficient and Accurate Neural Network Tool for Finding Correlation Between Gene Expression and Histological Images

Guy Shani[1]([✉]), Moti Freiman[2][iD], and Yosef E. Maruvka[1][iD]

[1] Faculty of Biotechnology and Food Engineering Technion, Haifa, Israel
guyshani3@gmail.com, yosi.maruvka@technion.ac.il
[2] Faculty of Biomedical Engineering, Technion, Haifa, Israel
moti.freiman@technion.ac.il

Abstract. Tumor development is clinically characterized through the manual review of histopathological Whole Slide Images (WSI). However, the molecular attributes influencing tumor morphology are not entirely comprehended. Here, we present RNALerner, an innovative tool designed to expedite the identification of correlations between gene expression and tumor morphology as presented in H&E WSI. RNALerner achieves its efficiency by transforming the problem from linear regression to binary classification of high versus low RNA levels, and the use of Resnet18, Convolutional Neural Network (CNN) model. Furthermore, the training phase of the model is halted after only 3 iterations. Upon comparing our results with previous work, we discovered a similar number of statistically significant correlated genes but with a reduction in the number of model parameters and processing time. Analysis of the significant pathways revealed both similarities to and deviations from earlier findings, bringing forth new pathways in the process. RNALerner represents an advancement toward the practical integration of machine learning in WSI analysis, which holds the potential to substantially improve disease diagnosis and guide more effective treatments.

Keywords: Convolutional neural network · H&E Whole slide image

1 Introduction

Histology involves the microscopic examination of cells and tissues to detect diseases and understand cellular structures. Analyzing histopathological images is crucial for diagnosing cancers and identifying therapeutic targets [4].

With the rise of digital pathology, especially H&E WSI, images can be stored digitally and analyzed using machine learning techniques. While creating WSI is efficient, manual analysis is slow and can be biased. This has led to an increased

Supported by ISF grant number 2070090.

emphasis on using advanced image processing and machine learning to detect cancerous regions in histology images. Yet, challenges like variable histological features and limited ground truth data persist, highlighting the need for continued research [6].

The introduction of Convolutional Neural Networks (CNN) has revolutionized medical image analysis. Recent efforts aim to extract molecular information from WSI, such as tumor mutations and markers [2]. Obtaining this data traditionally requires additional tests like DNA and RNA sequencing, emphasizing the potential of deep learning in deducing this information from standard H&E WSI. RNA expression levels, in particular, offer insights into tumor molecular mechanisms, with bulk RNA-seq providing a comprehensive perspective on transcript levels in cancer datasets.

Efforts using various CNN architectures to infer gene expression levels have been promising. Schmauch et al. [11] utilized a Resnet50 model with an MLP for predictions, while Wang et al. [13] and Fu et al. [3] chose Inception-V3/V4. Despite different training strategies and tumor types, these models showed comparable results regarding the correlation of gene expression levels and WSI.

Challenges include the unknown relationship between gene expression and histological features, leading to potential overfitting, and the significant computational power needed for model training.

We introduce RNALerner, a pipeline aiming to infer RNA levels from H&E WSI with fewer parameters and reduced run time. Using clustering for binary classification, we employed a pre-trained Resnet18 [5]. Our design minimizes overfitting, training time, and memory use. We utilized synthetic data to validate our model and conducted an enrichment analysis to pinpoint pathways linked with histological features.

2   Methodology

2.1   Data

Our data is compromised of Colorectal cancer (CRC) tumor samples, for each sample, we have the gene expression level of all the genes, and diagnostic H&E WSI from the GDC portal [1]. We selected 10,400 genes with the highest expression for analysis, which amounts to 35% of the expressed transcripts in the data.

We opted to use processed images created by Kather et al. [8] for our analysis, which are readily available for download at [7]. These images are diagnostic WSI from 356 tumor samples (one per patient), procured from the GDC portal, color-normalized, and segmented into 256 * 256 tiles as detailed in the aforementioned study. Each WSI is divided into several hundred tiles.

2.2   Label Generation

In order to simplify the analysis of the CNN model, we formulated the analysis of continuum expression levels as a binary classification, of high vs low expression levels. We clustered our samples using a Gaussian mixture model (GMM) [10] that adapts our data samples into two Gaussian densities in the following way:

$$p(x|\lambda) = \sum_{i=1}^{2} w_i g(x|\mu_i, \sigma_i)$$

where x is the continuous expression value, $w_i = w_1, w_2$ are the mixture weights and $g(x|\mu_i, \sigma_i), i = 1, 2$, are the component Gaussian densities. Each component density is a Gaussian function of the form,

$$g(x|\mu_i, \sigma_i) = \frac{1}{\sqrt{2\pi}\sigma_i} \exp\left(-\frac{(x - \mu_i)^2}{2\sigma_i^2}\right)$$

With mean μ_i and standard deviation σ_i. The mixture weights satisfy the constraint $\sum_{i=1}^{2} w_i = 1$. The complete GMM is parameterized by $\lambda = \{w_i, \mu_i, \sigma_i\}, i = 1, 2$, which are estimated using the maximum likelihood method and the expectation-maximization algorithm.

One of the powerful attributes of the GMM is its ability to form smooth approximations to arbitrarily shaped densities, which is crucial for our case where different genes have different expression density shapes. This technique is highly effective for accurately distinguishing samples in cases of bimodal expression distribution, and in other scenarios, it demarcates the distribution tail from the majority. Notably, most genes do not exhibit bimodal expression distribution. In such instances, the simplest classification methodologies typically establish an expression value threshold based on either mean or median expression. Our objective of ensuring maximal reflection of underlying biological differences in our classifications, combined with the inherent characteristics of RNA-seq distributions, makes it practical for us to identify the distribution's longer tail as a minority group. This group signifies genetic outliers, while the majority of the samples constitute the other group, which makes GMM a viable method of clustering the samples by expression levels for all genes. The label of each sample was assigned to each of its tiles. The average ratio of samples in the minority vs minority class was 0.25 for all the genes we tested.

2.3   CNN Training and Testing

In a study by Kather et al. [8], six state-of-the-art CNNs were used to classify CRC molecular subtypes from WSI. Resnet18 emerged as the top performer due to its minimal parameter count, which reduces the risk of overfitting the data and training time while achieving the same accuracy as the top performing models. We adopted this pre-trained model from the PyTorch library, standardizing image patches and substituting the final fully connected (FC) layer to produce two output classes. This FC layer, along with the fourth layer, were trained on the data while the remaining model parameters were kept constant, resulting in a total of 8,394,754 trainable parameters.

To manage class imbalance, we utilized a cross-entropy loss function with class ratio weighting and trained in batches of 16. The model gives a prediction for each image tile and the final probability score for the WSI is aggregated via the average across its tiles. We implemented a 4-fold cross-validation with a 75%-25% split for training and testing, resulting in four models per gene. Model training was halted after the third epoch to optimize resources. This decision was based on extensive testing on 10 selected genes, using different CNN models and hyperparameters, and observing that the models typically attain about 80% of peak performance after only three training epochs.

We chose a probability threshold of 0.5 for class predictions, and measured model performance with balanced accuracy, Cohen's κ coefficient, and normalized Precision-Recall AUC (PR-AUC), due to class imbalance considerations. PR-AUC was normalized by subtracting the ratio of class size from the value, which equals the no-skill classifier score. The whole process took 1.5 to 2 h per gene, including labeling, training, and testing with cross-validation to provide a prediction for each sample in our data.

2.4 Significance Testing and Gene Set Analysis

To ascertain significant model scores, we evaluated the random error magnitude of RNALerner using 500 "artificial genes." These were generated by randomly classifying each sample into binary classes, ensuring diverse class ratios similar to those produced by applying GMM to the data. These artificial genes underwent the same analysis as the experimental genes. We then used the distributions of balanced accuracy, Cohen's κ, and PR-AUC of this artificial data to estimate the distribution of the null hypothesis (H_0) - No correlation between model predictions and WSI features.

We first tested the artificial data distributions for normality using kolmogorov-smirnov test, and a quantile-quantile plot. We then calculated μ_0 and σ_0 for the artificial distributions and used them for calculating:

$$P[X > x_{gene}|x_{gene} \sim \mathcal{N}(\mu_0, \sigma_0^2)]$$

Where x_{gene} is the model performance score for the gene. P.values are then adjusted using the Benjamini-Hochberg (BH) procedure.

Upon determining the genes exhibiting correlation with tumor morphology, we undertook an evaluation of the biological pathways in which these genes are involved. For this purpose, we conducted a gene-set analysis using highly curated gene sets: KEGG, WikiPathways, Reactome, and Gene Ontology. Analysis was done using two approaches: Gene Set Enrichment Analysis (GSEA) and Over Representation Analysis (ORA), conducted with ClusterProfiler 4.0 package for R [14].

3   Results

3.1   Resnet Comparison

We compared Resnet18 to Resnet34 by using both models to predict the expression of a test set of known CRC associated genes: *KRAS, BRAF, DDX3X,*

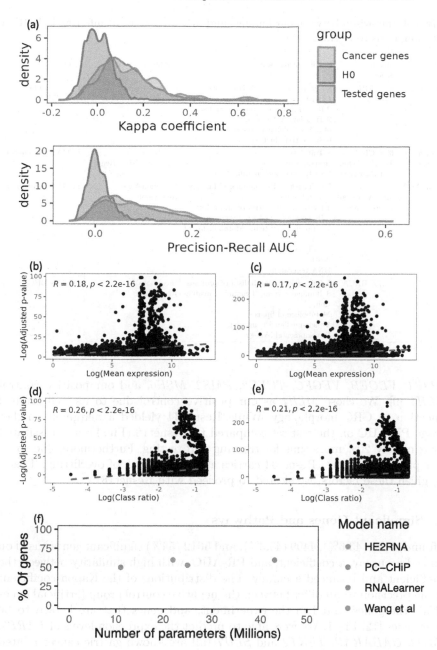

Fig. 1. (a) Density plots displaying the distributions of artificial genes, cancer genes and rest of the genes for: a) Top, κ coefficient, b) Bottom, PR-AUC. (b–e) Correlations between adjusted p.values, mean expression levels, and class ratio. b) Cohen's κ p.values VS Mean gene expression (counts), c) PR-AUC p.values VS mean gene expression (counts), d) Cohen's κ p.values VS class ratio (minority/majority class), e) PR-AUC p.values VS class ratio (minority/majority class). (f) Comparison between parameter number of the different methods VS percentage of significantly predicted genes.

Table 1. Comparison between our findings and other attempts at inferring bulk RNA expression levels from H&E WSI.

Study	Significant genes	Significant pathways	Method	Dataset
RNALearner	43%–54% CRC	6 Mitotic pathways, 5 Transport related pathways, 5 GTPases pathways 2 Post translational modifications, 2 Translation, Metabolism of lipids, Cellular response to stress, 6 Signal transduction.	Resnet18 - Binary classification	TCGA - COAD and READ
Schmauch et al. 2020	60% CRC 41.7% Across all tumor types	8 Translation, Response to damage, Post translational modification	HE2RNA - Custom CNN - Regression	TCGA - pan-cancer
Fu et al. 2020	60% CRC 42% Across all tumor types	5 Cell cycle, 3 Developmental biology, 8 Metabolism, 4 Extracellular matrix organisation, 8 Homeostasis, 11 Signal transduction, 11 Immune system, Metabolism of RNA, Muscle contraction, Programmed cell death, DNA replication	Inception-V4 - Regression	TCGA - pan-cancer
Wang et al. 2021	52.7% BRCA	5 Cell cycle, 2 Metabolism of proteins, 3 Immune system, 3 Signal transduction, Metabolism of lipids, Vesicle-mediated transport, Developmental biology	Inception-V3 - Regression	TCGA - BRCA

PPP3R1, *VEGFB*, *VEGFC*, *TGFB3*, *PMS2*, *MSH6*, and our positive control - *MLH1* [9]. We chose *MLH1* as our positive control due to its well studied connection to CRC morphology. While Resnet34 yielded a marginally higher average PR-AUC on the test set compared to Resnet18 (Fig. 1), it necessitated approximately 20% more time for training and testing. Furthermore, due to its larger parameter count, Resnet34 carries an elevated risk of overfitting. Therefore, given these factors, we elected to proceed with Resnet18.

3.2 Significant Genes and Pathways

We found, 4528 (43.6%), 4499 (43.3%), and 5612 (54%) significant genes based on balanced accuracy, κ coefficient, and PR-AUC, with high similarity between the κ coefficient and balanced accuracy. The distributions of the Kappa coefficient and the precision recall differ between the negative control group (artificial genes or H_0), the genes tested in the experiment, and genes that are known to be cancer genes [12] Fig. 1. We were able to predict the expression levels of *LARP7*, *SUMO2*, *GABARAP*, *TENT2* and *SLU7* (all are known gastric cancer related genes) with a balanced accuracy of over 90%.

In terms of the significant pathways, our analysis has produced similar pathways to other work in CRC and in other tumor types (Table 1).

3.3 Correlations Between Model Performance and Data Properties

To investigate the influence of mean gene expression levels and class imbalance on model performance, we computed the Pearson correlation between these parameters and adjusted p-values. There are substantial correlations between $-Log(adj.p-value)$ of both performance metrics and mean gene expression levels, with coefficients of 0.18 and 0.17 for κ and PR-AUC, respectively Fig. 1. Observing the point distributions in both plots, most high-performance models were trained on genes with $Log\left(\overline{expression}\right)$ exceeding 7, suggesting that most genes significantly impacting tumor morphology exhibit expression levels surpassing a certain threshold.

A higher correlation with class imbalance was noted, reflected by coefficients of 0.26 and 0.21 for κ and PR-AUC, respectively. This is anticipated given the small size of the dataset; with high class imbalance, fewer samples from the minority group are available for learning. As depicted in the figure, the best performing models demonstrate a class ratio of around 0.15, aligning with the biological context of the minority group with over/under gene expression while ensuring a sufficient sample size for model training.

3.4 Comparison of Findings with Other Methodologies

Our method yielded a similar proportion of significant genes (50%) as other work, using a smaller model in terms of parameters Fig. 1 and Table 1. Schmauch et al. [11] noted that their best predicted LIHC and BRCA datasets involved 765 and 786 genes, respectively, predicted with a correlation exceeding 0.4, equating to 2.6% of the tested genes. Our method identified 296 genes with a κ value surpassing 0.4, representing 2.8% of tested genes. This comparison indicates that our method yields comparable results in terms of high accuracy predictions, utilizing fewer trainable parameters and minimal optimization.

Comparatively, Fu et al. [3] found that 6% of gene predictions had a correlation exceeding 0.5 with the actual values, and 0.2% surpassed 0.75. Our method yielded corresponding gene predictions for 1.7% and 0.1% respectively.

All other methodologies trained one model to predict all genes with a much higher training time in total except for Wang et al. that trained a model for each gene similar to our method but still required more training time for each model compare to our method.

4 Conclusions

Here, we introduce RNALearner, a novel CNN-based methodology aimed at predicting RNA expression levels from WSI. With a primary focus on mitigating model overfitting, our strategy involved utilizing a Gaussian mixture model for data clustering, thereby converting the task into binary classification. As a negative control group, we generated synthetic data and found that approximately 50% of the examined genes demonstrated statistically significant predictions, consistent with previous studies. Our method successfully predicted the expression levels of 100 genes with a balanced accuracy surpassing 80%.

Our findings demonstrate that RNALearner provides comparative performance to state-of-the-art RNA expression level prediction models despite requiring fewer trainable parameters and less training data, hinting at the prominence of overfitting in current deep learning models applied to H&E WSI.

The results underscore the significant promise held by machine learning, particularly CNN, for inferring gene expression levels directly from H&E WSI, an established tool in cancer diagnosis and prognosis. This methodology has the potential to lower the financial burden of cancer treatment, enhance patient outcomes, and facilitate further cancer research, all achievable with compact, easily-trainable machine learning models.

References

1. GDC data portal. https://portal.gdc.cancer.gov/
2. Davri, A., et al.: Deep learning on histopathological images for colorectal cancer diagnosis: a systematic review. Diagnostics **12**(4), 837 (2022)
3. Fu, Y., et al.: Pan-cancer computational histopathology reveals mutations, tumor composition and prognosis. Nat. Can. **1**(8), 800–810 (2020)
4. Gurcan, M.N., Boucheron, L.E., Can, A., Madabhushi, A., Rajpoot, N.M., Yener, B.: Histopathological image analysis: a review. IEEE Rev. Biomed. Eng. **2**, 147–171 (2009). https://doi.org/10.1109/RBME.2009.2034865
5. He, K., Zhang, X., Ren, S., Sun, J.: Deep residual learning for image recognition. In: Proceedings of the IEEE Conference on Computer Vision and Pattern Recognition, pp. 770–778 (2016)
6. He, L., Long, L.R., Antani, S., Thoma, G.R.: Histology image analysis for carcinoma detection and grading. Comput. Methods Programs Biomed. **107**(3), 538–556 (2012). https://doi.org/10.1016/j.cmpb.2011.12.007, https://www.sciencedirect.com/science/article/pii/S0169260711003245
7. Kather, J.N.: Histological images for tumor detection in gastrointestinal cancer (2019). https://doi.org/10.5281/zenodo.2530789
8. Kather, J.N., et al.: Deep learning can predict microsatellite instability directly from histology in gastrointestinal cancer. Nat. Med. **25**(7), 1054–1056 (2019)
9. Lawrence, M.S., et al.: Discovery and saturation analysis of cancer genes across 21 tumour types. Nature **505**(7484), 495–501 (2014)
10. Reynolds, D.A., et al.: Gaussian mixture models. Encyclopedia Biomet. **741**, 659–663 (2009)
11. Schmauch, B., et al.: A deep learning model to predict RNA-SEQ expression of tumours from whole slide images. Nat. Commun. **11**(1), 3877 (2020)
12. Sondka, Z., Bamford, S., Cole, C.G., Ward, S.A., Dunham, I., Forbes, S.A.: The cosmic cancer gene census: describing genetic dysfunction across all human cancers. Nat. Rev. Cancer **18**(11), 696–705 (2018)
13. Wang, X., Price, S., Li, C.: Multi-task learning of histology and molecular markers for classifying diffuse glioma. arXiv preprint arXiv:2303.14845 (2023)
14. Wu, T., et al.: clusterprofiler 4.0: a universal enrichment tool for interpreting omics data. Innovation **2**(3) (2021)

FAIMI

De-identification and Obfuscation of Gender Attributes from Retinal Scans

Chenwei Wu[1]([✉]), Xiyu Yang[1], Emil Ghitman Gilkes[1], Hanwen Cui[1],
Jiheon Choi[1], Na Sun[2], Ziqian Liao[1], Bo Fan[2], Mauricio Santillana[1,3],
Leo Celi[2], Paolo Silva[4], and Luis Nakayama[2]

[1] Harvard University, Cambridge, MA 02138, USA
chenweiwu@g.harvard.edu
[2] Massachusetts Institute of Technology, Cambridge, MA 02139, USA
[3] Northeastern University, Boston, MA 02115, USA
[4] Joslin Diabetes Center, Boston, MA 02215, USA

Abstract. Retina images are considered to be important biomarkers and have been used as clinical diagnostic tools to detect multiple diseases. We examine multiple techniques for de-identifying retina images while maintaining their clinical ability for detecting diabetic retinopathy (DR), using gender as a proxy for identifiability. We apply two differential privacy algorithms, Snow and VS-Snow, on the entire image (globally) and on blood vessels only (locally) to obfuscate important image features that can predict a patient's sex. We evaluate the level of privacy and retained clinical predictive power of these de-identified images by using attacking gender classifier models and downstream disease classifiers. We show empirically that our proposed VS-Snow framework achieves strong privacy while preserving a meaningful clinical predictive power across different patient populations.

Keywords: Fundus images · Data Privacy · De-identification

1 Introduction

Automated machine learning (ML) systems have shown great potential to enhance medical decisions and improve healthcare access. However, public medical datasets sharing has always been a bottleneck for the development of equitable clinical AI [19]. The release of such datasets is strictly regulated by the Health Insurance Portability and Accountability Act (HIPAA) to control the risk of leakage of private attributes. This risk is particularly notable in ophthalmology, as retinal fundus photos are considered accurate and identifiable biomarkers.

C. Wu and X. Yang—These authors contributed equally.

The original version of the chapter has been revised. A correction to this chapter can be found at https://doi.org/10.1007/978-3-031-45249-9_30

Supplementary Information The online version contains supplementary material available at https://doi.org/10.1007/978-3-031-45249-9_9.

S. Wesarg et al. (Eds.): CLIP/FAIMI/EPIMI 2023, LNCS 14242, pp. 91–101, 2023.
https://doi.org/10.1007/978-3-031-45249-9_9

Rapid advancements in artificial intelligence also introduce novel privacy risks. Previous research has demonstrated that deep learning models could accurately predict private attributes like gender from retinal fundus images [8], raising concerns for potential malicious uses of this information.

More importantly, datasets that contain biases in private attributes like gender tend to produce biased AI algorithms [9]. Consequently, biased algorithms tend to produce stereotypical diagnoses and under-perform on minority patient groups, which is extremely dangerous in the context of healthcare as this can yield discriminatory outcomes.

We propose a framework to mitigate the privacy and fairness risks from the root: datasets. We aim not only at promoting AI fairness, but we envision making private data sharing a viable future. Specifically, our work investigates privacy concerns arising from the public release of two retina fundus image datasets, using gender as a proxy and a starting point for privacy. We show that our clinically-inspired de-identification algorithms significantly reduce the ability of an adversary to distinguish a patient's gender, while retaining most utility for downstream tasks like the identification of diabetic retinopathy (DR).

1.1 Differential Privacy for Image Obfuscation

Originally designed for statistical databases (i.e., the US Census), differential privacy algorithms have also been reinvented for medical image obfuscation. The purpose of image obfuscation is to modify an image such that sensitive information is no longer discernible in the image. Recently, differential privacy has been applied to iris images, which are also human biomarkers. For example, the Snow method [6] employs pixel-level noise by arbitrarily re-assigning pixel intensities to a constant value, i.e., 127 for grayscale images, based on parameter p, which determines the proportion of modified pixels. Snow achieves $(0, \delta)$-differential privacy with $\delta = 1 - p$ and protects individual pixels in the input image. More formally, $P[snow(x) \in S] \leq P[snow(x') \in S] + \delta$, where x are an arbitrary set of pixels and x' are the neighboring pixels of x ($snow(x)$ is defined in **De-identification Framework** section).

1.2 Deep Learning for Diabetic Retinopathy and Sex Classification

In recent years, researchers have applied various computer vision models for diabetic retinopathy detection on diverse data and population [16,18]. Some models achieve high performance in detecting referable diabetic retinopathy, which is comparable to the performance of ophthalmologists [4,12].

We are also interested in using gender classifiers to simulate real-life privacy attacks. Several recent studies have attempted to perform sex prediction on color fundus images and achieved reliable results [7,8,13,15]: Munk et al. [13] built a ResNet-152-based model in predicting patient information such as age and sex from three distinct retinal imaging modalities with accuracy of 0.73; Korot et al. [8] achieved around 0.84 accuracy using retinal fundus images. The aforementioned studies generally identify regions within images that are important for sex classification. Kim et al. [7] conducted an experiment to examine the sex prediction results after erasing fovea and blood vessels from the fundus images. The results show that erasing both anatomical regions decreases the model performance, indicating that both regions are helpful for sex prediction. Despite these results, retina specialists have not reached a consensus on fundus structures that are distinct for different sexes [8] (Table 1).

2 Materials and Methods

2.1 Dataset

Table 1. Demographic characteristics

	BRSET (Train-val)	BRSET (Test)	D-C (Train-val)	D-C (Test)
# Patients	7153	627	15379	3845
# Images	13401	1182	15379	3845
Mean Age (St. Dev.)	57.53 (18.1)	55.57 (19)	54.47 (17.18)	54.43 (17.29)
Gender (% female)	62.14	55.75	48.08	46.68
Normal (%)	93.14	92.98	79.08	78.67
NPDR (%)	4.28	4.4	13.92	14.07
PDR (%)	2.57	2.62	7.1	7.26

Note: (1) D-C is Diabetes Center. (2) Numbers reported are post-removal of images with N/A ICDR rating.

This work explores two distinct datasets, one publicly available and one private. The first one is the Brazilian Open-Access Ophthalmological Dataset (BRSET) [3,14]. The BRSET consists of retinal fundus images, clinical diagnosis and sensitive demographic variables like patient sex and age, from three Brazilian ophthalmological centers in Sao Paulo with a total of 16,266 images from 8,108 patients seen from 2010 to 2020. In terms of label distribution, the dataset is highly imbalanced: Only 4.2% of the images have positive DR. The images were captured by a Nikon NF505 and a Canon CR-2 professional retinal camera, resulting in images with a resolution of about 900 × 1000 pixels and a centered composition.

The second dataset is a private Diabetes Center Dataset with 19,224 ultrawide field retinal patient images collected through a DR screening program in the US. The dataset includes information about the presence and severity of diabetic retinopathy in each patient's eye. This dataset is more balanced, as 21% of the images have positive DR. The images were captured with a high-resolution ultra-wide camera of 4000 × 4000 pixels resolution. The Diabetes Center images do not all have a centered composition, and the orientation of the images varies significantly. The Diabetes Center dataset images also contain some extraneous noise, such as eyelashes and fingers obscuring parts of the retina.

2.2 Pre-processing

Our images were normalized based on the dataset statistics (mean, std) and resized to 256 × 256. In training, we employed a combination of horizontal flip and Shift-Scale-Rotate both with probabilities of 0.25 as data augmentations to add model generalizability.

The diabetic retinopathy severity in both datasets is rated following the International Clinical Diabetic Retinopathy (ICDR) Severity Scale. Because we have highly-skewed data, where most patients do not have diabetic retinopathy, we regrouped patients with mild, moderate, and severe non-proliferative diabetic retinopathy (NPDR) into one diabetic retinopathy group. We kept patients with proliferative diabetic retinopathy (PDR) as the other group, leaving us three final disease categories – healthy, NPDR, PDR. In this way, we have sufficient samples for both NPDR and PDR groups. The images from the BRSET and Diabetes Center data are split into a training set, a validation set, and a test set of ratios 70%, 10%, and 20%, respectively, by patient stratified sampling to prevent data leakage.

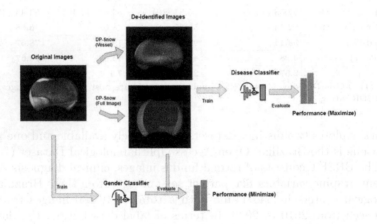

Fig. 1. De-identification & evaluation workflow

2.3   De-identification Framework

Figure 1 shows our general de-identification and evaluation workflow with some example retinal scans and de-identified images.

Full Image Obfuscation: Snow. To determine the effectiveness of differentially private image obfuscation techniques on retinal fundus images, we first applied the Snow algorithm to the full image region. Snow is a differentially private de-identification technique first introduced by John et al. [6]. It randomly adds noise to the image to obfuscate sensitive demographic information. Here we modified it to fit the RGB retina images de-identification task. Specifically, the Snow method first computes the whole dataset pixel intensity average for each RGB channel. Next, for each RGB channel, the method randomly selects a proportion, p, of image pixels. It assigns the corresponding channel average pixel intensity to the selected pixels. Consider a retinal fundus image as $I(x)$ where x refers to the index of each pixel in the image. For a 256×256 image, $x \in [1, 2, ..., 65536]$. The intensity of the pixel x is represented by $I(x) \in [0, 255]$. A subset S of size $p * I_{rows} * I_{cols}$ indices are randomly selected, and a new image $I'(x)$ is created such that

$$I'(x) = \begin{cases} \mu_c & x \in S \\ I(x) & x \notin S \end{cases}, \tag{1}$$

where μ_c is the average pixel intensity of channel c of all images in the given dataset.

Vascular Region Obfuscation: VS-Snow. Our observation of the Snow method (in results section) is that a full image de-identification technique may be sacrificing too much utility for privacy, and the whole image pixel noising approach would lose local details that are clinically useful. Inspired by studies showing that blood vessels are related to sex identifications [7], we propose VS-Snow, a local de-identification method to only apply obfuscation to the vessel regions. VS-Snow is composed of two parts: First, the blood vessel region mask was segmented using a modified Full Resolution network (FR-UNet) [11], demonstrating state-of-the-art retinal vessel segmentation performance. Next, we randomly selects a proportion, p, of the vessel mask pixels, (as opposed to full image in Snow) and change their values to the average of their neighboring vessel pixels (as opposed to the average of full pixels in Snow). To better fit our use case, we changed the aggregation module of the U-Net to perform depth-wise convolution across RGB channels and incorporated a random sliding window sampler in the dataloading module. Consider a retinal fundus image as $I(x)$ where x refers to the index of each pixel in the image. For a 256×256 image, $x \in [1, 2, ..., 65536]$. Denote the vessel mask as $m \in [1, 2, ..., 100]$, a subset of $I(x)$, and the $n_i \in m$ as the set of neighboring vessel pixels for each given x_i Then a subset S of size $p * Size(m)$ indices are randomly selected, and a new image $I'(x)$ is created such that

$$I'(x_i) = \begin{cases} \mu_{ni} & x_i \in S \\ I(x_i) & x_i \notin S \end{cases} \tag{2}$$

For both Snow and VS-Snow we generated images with $p = 0.1$, $p = 0.3$, $p = 0.5$ on both BR-Set and private Diabetes Center dataset.

2.4   Evaluation Framework

To evaluate how robust our de-identified images are against attacking models and how much clinical utility these images retain compared to the original image, we designed the two following pipelines.

Gender Classifier (Attacker). The gender classifier's (Attacker) objective is to re-identify patients' identity, in our settings, to recognize the gender of a patient. In a realistic scenario, a hacker would train the gender classifier on available datasets with private information and test on our released dataset to infer gender. Thus we trained our attacker models on the original images with gender labels and tested on 1) original images, 2) Snow de-identified images and 3) VS-Snow de-identified images, to remove gender-related information as much as possible via de-identification. We quantify the level of privacy gain by computing the performance drop of the attacker model successfully recognizing gender on de-identified images compared to original images.

Diabetic Retinopathy (DR) Classifier (Downstream Task). The diabetic retinopathy classifier's (Downstream Task) objective is to classify the disease categories from the images. In a realistic setting, a researcher would take our de-identified images and train models to predict diabetic retinopathy. Thus we trained our downstream models on all three types of images with DR labels, (original, Snow de-identified images, and VS-Snow images) and tested on respective test sets. We quantify the level of clinical utility by computing the performance drop of the downstream model on original images compared to de-identified images.

Implementation Details. Separate models were trained for sex and DR classification using PyTorch 2.0.1 as the deep-learning library and Nvidia V100 GPUs. Our backbone network is ResNet-200D [5], a modification of the ResNet architecture that utilizes an average pooling tweak for downsampling. We chose ResNet-200D after comparing its performance with various other architectures like EfficientNet [17] and XceptionNet [2]. Furthermore, in the unmodified ResNet, the 1×1 convolution for the downsampling block ignores 3/4 of the input feature maps, whereas ResNet-D takes the whole feature maps as inputs, and no information will be ignored. Our ResNet-200d was pre-trained in general image classification from the ImageNet database, as transfer learning is generally faster with better performance than training from scratch [20]. We modified the fully connected last layer (changed output dimension from 1000 to 2 for sex classification or 3 for DR classification) to tailor the CNN to desired outputs. The cross-entropy loss function was adopted in the sex prediction model for binary classification; the class-weighted focal loss (gamma = 0.6) [10] was adopted for the 3-class DR classification because it has the appropriate properties to handle the class imbalance and overfitting issues. We used Adam as the optimization scheme (learning rate = 1e−4, weight decay = 1e−5) and trained through 50 epochs with early stopping mechanism. The CNNs were validated for each epoch, and the model with highest F1, accounting for data imbalance, were selected as the final predictor.

3   Results

3.1   Full Image Snow Results

We tested our attacker and downstream models on the original and de-identified sets of images (Table 2). As expected, the model performance decreases as we increase the Snow method's parameter, p, since more information is obfuscated. The gender classifier performed as high as 0.75 and 0.83 on the two datasets, proving itself to be a good attacker on par with other gender prediction work [8]. To balance patient privacy and the images' clinical utility, we obfuscated the images with the goal of reducing the sex classification accuracy to approximately 60% for BRSET and 50% for Diabetes Center data. We selected 60% as the random sex classifier for BRSET because of the imbalance in the sex class. At the same time, we maintained an F1 score for DR classification as close to the original images as possible. For the Brazilian dataset (BRSET), the sex classification accuracy dropped to 61.35% when we set the parameter p to 0.3. In this case, the F1 score of the DR classification is 69.6%, which reflects a 13% drop concerning the baseline (82.6%). On the Diabetes Center dataset, we reduced the sex classifier's accuracy to 51% by only setting parameter p to 0.3. Yet this led to a significant decrease in the DR classifier's performance from 79.5% on the original images to 58.4% on obfuscated images, with p equal to 0.3. These results show that the Snow approach successfully reduces privacy but may obfuscate images too much at the expense of losing clinical utility. Moreover, the F1 scores for the male group are 5% or 11.5% higher than those for the female group when we set p to 0.3 or 0.5 in the BRSET. The result indicates that the Snow method might still be vulnerable to gender bias during diabetes retinopathy prediction.

Table 2. Snow De-identification

Data	Classification	Original	$p = .1$	$p = .3$	$p = .5$
BRSET	Sex (Acc)	75.12	64.99	61.35	52.15
	DR (F1)	82.6	72.3	69.6	64.5
	DR-Female (F1)	83.4	72.8	66.9	60.4
	DR-Male (F1)	81.8	70.3	71.9	71.9
Diabetes Center	Sex (Acc)	83.33	57.43	51.00	42.59
	DR (F1)	79.5	65.8	58.4	55.9
	DR-Female (F1)	79.2	65.3	57.8	55.4
	DR-Male (F1)	79.7	66.1	58.9	56.2

3.2   VS-Snow Results

A more targeted de-identification approach –only applied to the vessel regions and neighboring vessel pixels– improves the clinical usefulness of the fundus images. Similar to Snow experiments, the classifiers' performance decreases as we increase the value of the p parameter. For the BRSET, at p parameter of 0.3, the sex classifier's attacking ability was reduced to 60.1% accuracy, close to a random sex classifier for BRSET, while the DR classifier experienced a 7.6% F1 score drop from training on the original images (75% vs. 82.6%). For the diabetes center dataset, we successfully reduced the sex classification accuracy to 50.7% with only 0.1 p parameter. We maintained a DR classification F1 score very close to the training results on the original images (75.9% vs. 79.5%). These results show that VS-Snow can preserve privacy and utility at a good standing across patient populations. Furthermore, when we compared DR prediction between males and females, we found the model performances between sex groups are closer to each other, showing that VS-Snow better alleviates sex bias than Snow (Table 3).

Table 3. VS-Snow De-identification

Data	Classification	Original	$p = .1$	$p = .3$	$p = .5$
BRSET	Sex (Acc)	75.12	67.7	60.1	58.1
	DR (F1)	82.6	78.1	75	72
	DR-Female (F1)	83.4	78.0	74.6	73.8
	DR-Male (F1)	81.8	77.0	74.4	70.3
Diabetes Center	Sex (Acc)	83.33	50.7	47.9	47.5
	DR (F1)	79.5	75.9	73.8	74.6
	DR-Female (F1)	79.2	75.5	73.6	74.2
	DR-Male (F1)	79.7	76.1	74.0	75.0

4 Discussion

4.1 Privacy-Utility Tradeoff

Our results show a clear privacy-utility tradeoff, indicating an inverse relationship between privacy protection (obfuscating the patient sex) and statistical utility (identifying DR) of the images. In order to reach almost perfect privacy (indicated when our sex classifier has a similar performance to a random classifier), the F1 score of the DR classifier reduces at least 7.6 and 3.6 points on the BRSET and Diabetes Center datasets, respectively. Still, our de-identification framework is able to strike a reasonable balance between privacy and utility. For both datasets, we are able to achieve almost perfect privacy while maintaining an F1 score ≥ 70. The only experiment that yielded unsatisfactory results was applying Snow to the entire image on the Diabetes Center dataset, implying that local obfuscation may be better than global obfuscation.

4.2 Importance of Vasculature

Comparing the results from applying Snow to the entire image and VS-Snow to only the blood vessels demonstrates the importance of the vasculature for both sex and DR classification. However, it appears that the vasculature is much more important for sex classification than DR classification. This is because, compared to sex classifier's performance, the DR classifier performance is significantly better when applying Snow to only the vessels ($p = 0.5$) than when it's applied to the entire image. This is contrary to the performance of the sex classifier, as the performance is comparable regardless of the de-identification method. In Fig. 2 we shows an example of saliency maps and it is clear that gender information is much more prevalent in the vessel regions than the disease information, supporting our hypothesis that a local obfuscation method might be more optimal for retinal de-identification.

(a) Original Images   (b) Gender Classification   (c) DR Classification

Fig. 2. Example of original retina image vs saliency maps

4.3 Limitations and Future Work

In the future, we would like to extend this work to examine and improve VS-Snow's ability to obfuscate other sensitive demographic features, such as income, race and age on retinal fundus data. Moreover, other advanced image models, such as large pretrained vision transformers, would be tested to determine whether our de-identification method is robust against different attackers. VS-Snow's two-step framework could be potentially adapted to other scenarios. I. Atas [1] was able to predict gender from panoramic dental X-Rays with high accuracy and found the saliency map focused on the mandible and teeth. Using a forensic medicine-inspired approach, we could mask out and obfuscate the mandible area for privacy.

References

1. Ataş, I.: Human gender prediction based on deep transfer learning from panoramic dental radiograph images. Traitement du Signal **39**(5), 1585 (2022)
2. Chollet, F.: Xception: deep learning with depthwise separable convolutions. In: Proceedings of the IEEE Conference on Computer Vision and Pattern Recognition, pp. 1251–1258 (2017)

3. Goldberger, A., et al.: Physiobank, physiotoolkit, and physionet: components of a new research resource for complex physiologic signals. Circulation [Online] **101**(23), E215–E220 (2000). https://doi.org/10.1161/01.cir.101.23.e215

4. Gulshan, V., et al.: Development and validation of a deep learning algorithm for detection of diabetic retinopathy in retinal fundus photographs. JAMA **316**(22), 2402–2410 (2016). https://doi.org/10.1001/jama.2016.17216

5. He, T., Zhang, Z., Zhang, H., Zhang, Z., Xie, J., Li, M.: Bag of tricks for image classification with convolutional neural networks (2018)

6. John, B., Liu, A., Xia, L., Koppal, S., Jain, E.: Let it snow: adding pixel noise to protect the user's identity. In: ACM Symposium on Eye Tracking Research and Applications. ETRA 2020 Adjunct, New York, NY, USA. Association for Computing Machinery (2020). https://doi.org/10.1145/3379157.3390512

7. Kim, Y.D., et al.: Effects of hypertension, diabetes, and smoking on age and sex prediction from retinal fundus images. Sci. Rep. **10**, 4623 (2020). https://doi.org/10.1038/s41598-020-61519-9

8. Korot, E., et al.: Predicting sex from retinal fundus photographs using automated deep learning. Sci. Rep. **11**, 10286 (2021). https://doi.org/10.1038/s41598-021-89743-x

9. Larrazabal, A.J., Nieto, N., Peterson, V., Milone, D.H., Ferrante, E.: Gender imbalance in medical imaging datasets produces biased classifiers for computer-aided diagnosis. Proc. Natl. Acad. Sci. **117**(23), 12592–12594 (2020)

10. Lin, T.Y., Goyal, P., Girshick, R., He, K., Dollár, P.: Focal loss for dense object detection (2018)

11. Liu, W., et al.: Full-resolution network and dual-threshold iteration for retinal vessel and coronary angiograph segmentation. IEEE J. Biomed. Health Inform. **26**(9), 4623–4634 (2022). https://doi.org/10.1109/JBHI.2022.3188710

12. Liu, X., et al.: Deep learning to detect oct-derived diabetic macular edema from color retinal photographs: a multicenter validation study. Ophthalmol. Retina **6**(5), 398–410 (2022). https://doi.org/10.1016/j.oret.2021.12.021

13. Munk, M.R., Kurmann, T., Márquez-Neila, P., Zinkernagel, M.S., Wolf, S., Sznitman, R.: Assessment of patient specific information in the wild on fundus photography and optical coherence tomography. Sci. Rep. **11**, 8621 (2021). https://doi.org/10.1038/s41598-021-86577-5

14. Nakayama, L.F., et al.: A Brazilian multilabel ophthalmological dataset (brset) (2023). https://doi.org/10.13026/xcxw-8198

15. Poplin, R., et al.: Prediction of cardiovascular risk factors from retinal fundus photographs via deep learning. Nat. Biomed. Eng. **2**, 158–164 (2018). https://doi.org/10.1038/s41551-018-0195-0

16. Ruamviboonsuk, P., et al.: Deep learning versus human graders for classifying diabetic retinopathy severity in a nationwide screening program. NPJ Dig. Med. **2**, 25 (2019). https://doi.org/10.1038/s41746-019-0099-8

17. Tan, M., Le, Q.: Efficientnet: rethinking model scaling for convolutional neural networks. In: International Conference on Machine Learning, pp. 6105–6114. PMLR (2019)

18. Ting, D.S.W., et al.: Development and validation of a deep learning system for diabetic retinopathy and related eye diseases using retinal images from multiethnic populations with diabetes. JAMA **318**(22), 2211–2223 (2017). https://doi.org/10.1001/jama.2017.18152

19. Yala, A., et al.: Syfer: neural obfuscation for private data release. arXiv preprint arXiv:2201.12406 (2022)
20. Yu, Y., Lin, H., Meng, J., Wei, X., Guo, H., Zhao, Z.: Deep transfer learning for modality classification of medical images. Information **8**(3) (2017). https://doi.org/10.3390/info8030091, https://www.mdpi.com/2078-2489/8/3/91

Unveiling Fairness Biases in Deep Learning-Based Brain MRI Reconstruction

Yuning Du[1]($\boxtimes$), Yuyang Xue[1], Rohan Dharmakumar[2],
and Sotirios A. Tsaftaris[1,3]

[1] School of Engineering, The University of Edinburgh, Edinburgh EH9 3FG, UK
yuning.du@ed.ac.uk
[2] Krannert Cardiovascular Research Center, Indiana University School of Medicine,
Indianapolis, IN, USA
[3] The Alan Turing Institute, London NW1 2DB, UK

Abstract. Deep learning (DL) reconstruction particularly of MRI has led to improvements in image fidelity and reduction of acquisition time. In neuroimaging, DL methods can reconstruct high-quality images from undersampled data. However, it is essential to consider fairness in DL algorithms, particularly in terms of demographic characteristics. This study presents the first fairness analysis in a DL-based brain MRI reconstruction model. The model utilises the U-Net architecture for image reconstruction and explores the presence and sources of unfairness by implementing baseline Empirical Risk Minimisation (ERM) and rebalancing strategies. Model performance is evaluated using image reconstruction metrics. Our findings reveal statistically significant performance biases between the gender and age subgroups. Surprisingly, data imbalance and training discrimination are not the main sources of bias. This analysis provides insights of fairness in DL-based image reconstruction and aims to improve equity in medical AI applications.

Keywords: Fairness · Image Reconstruction · Algorithm Bias · Neuroimaging

1 Introduction

Magnetic resonance imaging (MRI) is routinely used to help diagnose or ascertain the pathophysiological state in a noninvasive and harmless manner. However, MRI is characterised by long acquisition times. There is an interest in improving imaging fidelity whilst reducing acquisition time. A solution is to subsample the frequency domain (k-space). This introduces aliasing artefacts in the image domain due to the violation of the Nyquist sampling theorem, causing difficulties such as biomarkers extraction and interpretation in neuroimaging.

Recently, deep learning (DL) methods based on convolutional neural networks (CNNs) have been proposed to reconstruct high-quality images from the undersampled k-space data [5]. By learning complex patterns from large amounts

S. Wesarg et al. (Eds.): CLIP/FAIMI/EPIMI 2023, LNCS 14242, pp. 102–111, 2023.
https://doi.org/10.1007/978-3-031-45249-9_10

of training data and filling in missing k-space data, these DL models successfully reconstruct images that closely resemble those obtained through fully sampled acquisitions. Advances in DL-based image reconstruction enable both accelerated acquisition and high-quality imaging, providing significant benefits.

Deep learning methods may be subject to biases (e.g., from the training dataset) which can lead to fairness and lack of equity. For example, recent studies have shown that image segmentation algorithms can be unfair: Puyol-Antón et al. [8] found racial bias can exist in DL-based cine CMR segmentation models when training with a race-imbalanced dataset. This leads us to ask: *Could DL-based image reconstruction algorithms be also unfair?*

To date, such a, at least empirical, study is lacking, and this article precisely addresses this gap. Our primary objective is to investigate the biases in the algorithm resulting from demographic information present in the training data. To the best of our knowledge, this is the first fairness analysis in a DL-based image reconstruction model. We make the following contributions:

- We identify existing bias in performance between gender and age groups using the publicly available OASIS dataset [6].
- We investigate the origin of these biases by mitigating imbalances in the training set and training paradigm with different bias mitigation strategies.
- We discuss the factors that may impact the fairness of the algorithm, including inherent characteristics and spurious correlations.

2 Background

2.1 Fairness Definitions

Amongst the various definitions of fairness, since we study the fairness for different demographic subgroups, we consider only group fairness in our analysis.

Group Fairness: Group fairness aims to ensure equitable treatment and outcomes for different demographic or subpopulation groups. It recognises the potential for biases and disparities in healthcare delivery and seeks to address them to promote fairness and equity [3]. To ensure fairness, equalised odds [2] is used as a criterion that focuses on mitigating bias, stating as "The predictor $\hat{Y}$ satisfies equalised odds with respect to protected attribute A and outcome Y, if $\hat{Y}$ and A are independent conditional on Y." The criterion can be formulated as

$$\forall y \in \{0,1\} : P(\hat{Y} = 1 | A = 0, Y = y) = P(\hat{Y} = 1 | A = 1, Y = y). \tag{1}$$

Fairness in Image Reconstruction: It requires the reconstructed image to faithfully represent the original one without distorting or altering its content based on certain attributes such as race, gender, or other protected attributes.

When applying equalised odds as the fairness criterion, while the original equation focuses on fairness in predictive labels, image reconstruction tasks typically involve matching pixel values or image representations. Thus, we reformulate the problem based on probabilistic equalised odds, as proposed by [7]. We let $P \subset \mathbb{R}^k$ be the input space of an image reconstruction task, $(\mathbf{x}, \mathbf{y}) \sim P$ represent a patient, with $\mathbf{x}$ representing the undersampled image, and y representing the fully sampled image or ground truth image. Also, we assume the presence of two groups $g_1, g_2 \subset P$, which represent the subsets defined by the protected attribute $\mathbf{A}$. Fairness using probabilistic equalised odds is formulated as:

$$\forall \mathbf{y} \in \mathcal{Y} : \mathbb{E}(\mathbf{x}, \mathbf{y}) \sim g_1[f(\mathbf{x}) \mid \mathbf{Y} = \mathbf{y}] = \mathbb{E}(\mathbf{x}, \mathbf{y}) \sim g_2[f(\mathbf{x}) \mid \mathbf{Y} = \mathbf{y}]. \quad (2)$$

Here, f represents the DL-based reconstruction network. With this formulation, we aim to achieve fairness by ensuring that the quality or fidelity of the reconstructed image is consistent across different data distributions irrespective of different demographic characteristics.

2.2   Source of Bias

Data imbalance can be a significant source of bias in medical scenarios [17]. It can refer to the imbalanced distribution of demographic characteristics, such as gender and ethnicity, within the dataset. For example, the cardiac magnetic resonance imaging dataset provided by the UK Biobank [10] is unbalanced with respect to race: $> 80\%$ of the subjects are of white ethnicity, resulting in unequal representation of features correlated to ethnicity. This imbalance can introduce bias in the analysis and interpretation of the data.

Training discrimination is another source of bias, possibly occurring concurrently with data imbalance [4]. An imbalanced dataset can lead to imbalanced minibatches drawn for training. Hence, the model mainly learns features from the dominant subgroup in each batch, perpetuating bias in the training process.

Spurious correlations can also contribute to bias [17]. This refers to the presence of misleading or incorrect correlations between the training data and the features learned by the model. For instance, a model can learn how to classify skin diseases by observing markings made by dermatologists near lesions, rather than fully learning the diseases [15]. This is particularly likely to happen in the minority subgroup due to limited presence in training dataset, leading to overfitting during the training process and further exacerbating bias.

Inherent characteristics can also play a role in bias, even when the model is trained with a balanced dataset [17]. Certain characteristics may inherently affect the performance of different subgroups. For instance, in skin dermatology images, lesions are often more challenging to recognise in darker skin due to lower contrast compared to lighter skin. As a result, bias based on ethnicity can still exist even if the dataset is well-balanced in terms of proportions.

3   Methods

Our Main Goal: Our goal is to identify the bias in image reconstruction models and any potential sources of bias related to demographic characteristics. To investigate fairness in image reconstruction tasks, we systematically design and conduct experiments that eliminate potential origins of bias w.r.t. various demographic characteristics. We start by establishing a baseline model using Empirical Risk Minimisation (ERM) to assess the presence of bias in relation to diverse demographic subgroups. Then, we employ a subgroup rebalancing strategy with a balanced dataset in terms of demographic attributes, to test the hypothesis that bias is caused by data imbalance. Then, we use the minibatch rebalancing strategy to evaluate the effects of training discrimination for each subgroup.

Reconstruction Networks: We use a U-Net [12] as the backbone for the reconstruction network. The reconstruction network is trained using undersampled MRI brain scans, which are simulated by applying a random Cartesian mask to the fully sampled k-space data. Details of the data and the experimental setup of the reconstruction network are provided in Sect. 4.

Baseline Network: We follow the principle of Empirical Risk Minimisation (ERM) [14]. ERM seeks to minimise the overall risk of a model by considering the entire population, instead of the composition of specific groups and hence without controlling for the distribution of protected attributes.

Subgroup Rebalancing Strategy: This strategy aims to examine the performance when a perfectly balanced dataset of the protected attributes is used. Instead of randomly selecting data from the entire dataset to define a training set, the training set consists of an equal number of subjects from different subgroups according to demographic characteristics. This approach ensures that all subgroups have equal chances during the training phase, helping us identify if data imbalance is the source of bias.

Minibatch Rebalancing Strategy: This strategy examines the performance when balanced minibatches in terms of protected attributes are used to eliminate discrepancy before training [9]. Hence, each minibatch has an equal presence of subjects with different demographic characteristics and all subgroups have an equal opportunity during each iteration to influence the model weights.

Evaluation Metrics: Although several fairness metrics have been proposed, most of the current work is focused on image classification and segmentation tasks, which may not be directly applicable to image reconstruction tasks. Therefore, we analyse the fairness of image reconstruction using image reconstruction metrics and statistical analysis. The performance of the reconstruction is evaluated using Structural Similarity Index (SSIM, higher is better), Peak Signal-to-Noise Ratio (PSNR, higher is better) on the patient level.

To investigate bias between subgroups with different demographic characteristics, we performed the non-parametric Kruskal-Wallis ANOVA test (as available within OriginPro 2023) to test the omnibus hypothesis that there are differences in subgroups with $p < 0.05$ as the threshold for statistical significance. The test will provide Chi-Square value and p-value as results. Higher Chi-Square values indicate the presence of more significant differences between subgroups. This approach allows us to assess the potential bias in the image reconstruction process specifically instead of relying on fairness metrics designed for other tasks.

Table 1. Statistics of demographic subgroups in OASIS. Patients are categorised into young adult (below 40), middle-aged adult (40 to 65) and older adult (above 65).

Category	All	Female	Male	Young	Middle-aged	Older
Count	416	256	160	156	82	178
Proportion (%)	100.0	61.5	38.5	37.5	19.7	42.8

4   Experimental Analysis

4.1   Dataset and Pre-processing

Dataset: We select the publicly available Open Access Series of Imaging Studies (OASIS) dataset [6] to evaluate the fairness of the image reconstruction task. The initial data set consists of a cross-sectional collection of 416 subjects(316 subjects is healthy and 100 subjects is clinically diagnosed with very mild to moderate Alzheimer's disease)and for each subject, three or four individual T1-weighted MRI scans obtained in single imaging sessions are included. To simulate clinical practice with uncertainty about patients' conditions, we used an entire dataset consisting of a mix of patients, including both healthy subjects and patients with Alzheimer's disease (AD), without providing explicit labels for their conditions. To study the fairness regarding inherent demographic information, we choose gender and age information provided in the dataset as the protected attributes. Since the patients are aged 18 to 96, we categorise the patients into young adult (age below 40), middle-aged adult (ages 40 to 65), and older adult (age above 65) according to the criteria proposed by [13]. The statistics of the subgroups are summarised in Table 1.

According to Table 1, there is a clear imbalance in the distribution of demographic characteristics in the OASIS dataset. In the protected attribute gender, the female is the dominant group with 256 subjects, while the male is the disadvantaged group with only 160 subjects. In terms of age, compared to the middle-aged adults group, the young and older adults groups are dominant groups.

Data Pre-processing: To ensure the equal size of dataset for methods in Sect. 3, the dataset is firstly categorised into six age-gender subgroups (e.g., middle-aged female adults) and sampled according to the size of minority subgroups, which is 27 from middle-aged male adults, to maintain the balance for both age and gender distribution among sampled dataset (162 subjects in total). Then, we sampled 5 subjects from six age-gender subgroups to form test set, which is 30 subjects in total. For the training and validation set, we sampled the rest 22 subjects from each age-gender subgroups for the rebalancing and minibatch rebalancing strategies, which is 132 subjects in total. While for the baseline network, the training and validation set are randomly sampled with a size of 132 subjects. The train-validation-test splits all follow the proportions of 20 : 2 : 5. For each patient, we select 122 central slices out of 208 slices in one volume.

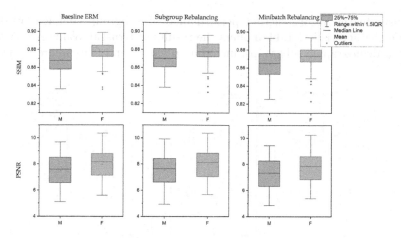

Fig. 1. Image Reconstruction Performance for Gender Subgroups. In the figure, 'F' represents 'Female' and 'M' represents 'Male'. This figure indicates performance gap between two gender subgroups in image reconstruction task under different strategies.

Table 2. Statistics of Image Reconstruction Performance under three strategies.

	Baseline ERM		Subgroup Rebalancing		Minibatch Rebalancing	
	SSIM	PSNR	SSIM	PSNR	SSIM	PSNR
Whole	0.872 (0.012)	7.742 (0.112)	0.867 (0.011)	7.529 (0.109)	0.867 (0.011)	7.529 (0.109)
Female	0.876 (0.010)	7.999 (0.099)	0.876 (0.010)	8.006 (0.095)	0.871 (0.010)	7.767 (0.095)
Male	0.868 (0.013)	7.485 (0.118)	0.870 (0.013)	7.509 (0.117)	0.864 (0.013)	7.292 (0.117)
Young Adults	0.876 (0.010)	8.690 (0.092)	0.877 (0.009)	8.729 (0.090)	0.872 (0.009)	8.496 (0.090)
Middle-aged Adults	0.874 (0.011)	7.859 (0.108)	0.875 (0.011)	7.877 (0.106)	0.869 (0.011)	7.645 (0.106)
Older Adults	0.867 (0.010)	6.676 (0.102)	0.867 (0.010)	6.666 (0.099)	0.861 (0.010)	6.448 (0.099)

4.2 Implementation Details

We employ a U-Net as backbone. Its first feature map size is 32, with 4 pooling cascades, resulting in a total of 7.8M parameters. We employ the Adam optimiser with a learning rate of 10^{-4} with a step-based scheduler with a decay gamma of 0.1. Both the ℓ_1 loss and the SSIM loss were incorporated into our experiments. Models were trained for 40 epochs with batch size 6.

5-fold cross validation is used to mitigate sample bias. Our experimental setup uses the PyTorch Lightning framework and we trained on an NVIDIA A100 Tensor Core GPUs. The implementation of our code is inspired by the fastMRI repository.[1] Our code is publicly available at: https://github.com/ydu0117/ReconFairness.

4.3 Results

Table 2 reports the SSIM and PSNR results (mean and standard deviation) from 5-fold cross-validation under three different strategies. Figures 1 and 2

---

[1] https://github.com/facebookresearch/fastMRI.

demonstrate the reconstruction performance of subgroups defined by demographic characteristics. Table 3 offers the results of Kruskal-Wallis ANOVA test between demographic subgroups, including p-values and Chi-Square values.

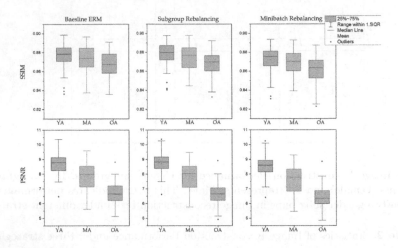

Fig. 2. Image Reconstruction Performance for Age Subgroups under Three strategies. In the figure, 'YA' represents 'Young Adults', 'MA' represents 'Middle-aged Adults' and 'OA' represents 'Older Adults'. This figure indicates performance gaps between age subgroups in image reconstruction task under different strategies.

Table 3. Kruskal-Wallis ANOVA results for the Three Strategies Testing for Influence of "Gender" and "Age Group". The results include Chi-Square values with statistical significance (p-value) indicated as *** $p < 0.001$, ** $p < 0.01$, * $p < 0.05$.

	Gender		Age Group	
	SSIM	PSNR	SSIM	PSNR
Baseline ERM	13.44***	6.45*	11.64**	78.91***
Subgroup Rebalancing	10.90**	6.01*	14.64**	81.08***
Minibatch Rebalancing	10.30**	5.44*	13.70**	78.62***

Presence of Bias: Focusing on the baseline ERM model, our results show that there is a significant performance difference between subgroups categorised by gender and age. Among the gender subgroups, the female group outperforms the male group. This difference is obvious in Fig. 1, showing that the baseline model provides better performance for female subjects compared to male subjects. This difference is statistically significant in Table 3.

Among the three age groups, the results demonstrate an obvious performance gap. Referring to Table 2, the young adults group provides a better performance in all three metrics. Furthermore, the results indicate that as age increases, the

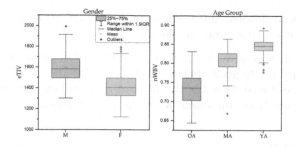

Fig. 3. Relations between Demographic Features and Neuroanatomy Metrics.

reconstruction performance worsens (both metrics). The trend is visually evident in Fig. 2 and is statistically significant in Table 3.

Is Dataset Imbalance the Source of Unfairness? The performance under rebalancing strategies shows that the imbalance of data and the discrimination of training are not the major cause of bias. Specifically, in Table 2, when comparing the performance of the subgroups under different training strategies, the performance gaps evidenced before still exist. These biases are also visually illustrated in Figs. 1 and 2 and are again statistically significant in Table 3.

However, the Chi-square values under rebalancing strategies in Table 3, is reduced compared to the baseline ERM network among gender subgroups. This reduction in Chi-Square values indicates that the rebalancing either of the training set or the minibatch may mitigate partial bias, illustrating that dataset imbalance and training discrimination towards gender may be sources of bias, but not the main source. However, it is noticeable that the balancing strategies result in performance reduction of the dominant subgroup.

5   Discussion

What is the Source of Unfairness: We find that data imbalance and training discrimination do not significantly contribute to bias. Instead, the bias may stem from spurious correlations and inherent characteristics. Specifically, the model may focus on neuroanatomical features that are associated with demographic factors [1,11]. In Fig. 3, the relations between demographic features and neuroanatomy metrics including estimated Total Intracranial Volume (eTIV) as well as normalised Whole Brain Volume (nWBV) are analysed. Our results show that women tend to have smaller eTIV compared to men, and young adults have the highest nWBV among age subgroups. Thus, these differences in eTIV between gender and nWBV between age may result in spurious correlations that lead to bias, which requires further investigation in future work.

Clinical Relevance: It is noticeable that the difference in SSIM among subgroups is in the second or third decimal place in some cases. Although the small difference may not be clinically meaningful in practice, it can lead to additional errors and bias in downstream tasks such as segmentation and classification, ultimately leading to inaccurate diagnoses.

Limitations: Previous studies [16] have reported data imbalances among different racial groups due to geographic limitations of the datasets. In our analysis, due to the lack of racial data, the training set may still exhibit an imbalance in terms of race, even if we implement a rebalancing strategy.

6   Conclusion

In this study, we conducted an initial analysis of fairness in DL-based image reconstruction tasks with respect to demographic characteristics, specifically gender and age. We employed three strategies to investigate the bias caused by these characteristics. Through the use of rebalancing strategies, we found that imbalanced training sets and training discrimination were not the major contributors to bias. However, further investigation is needed to identify the sources of bias in image reconstruction tasks. Correspondingly, we need to propose bias mitigation strategies to ensure fairness in DL-based image reconstruction applications.

Acknowledgements. This work was supported in part by National Institutes of Health (NIH) grant 7R01HL148788-03. Y. Du and Y. Xue thank additional financial support from the School of Engineering, the University of Edinburgh. S.A. Tsaftaris also acknowledges the support of Canon Medical and the Royal Academy of Engineering and the Research Chairs and Senior Research Fellowships scheme (grant RCSRF1819n 8n 25), and the UK's Engineering and Physical Sciences Research Council (EPSRC) support via grant EP/X017680/1. The authors would like to thank Dr. Chen and K. Vilouras for inspirational discussions and assistance. Data used in Sect. 4.1 were provided by OASIS-1: Cross-Sectional: Principal Investigators: D. Marcus, R, Buckner, J, Csernansky J. Morris; P50 AG05681, P01 AG03991, P01 AG026276, R01 AG021910, P20 MH071616, U24 RR021382.

References

1. Gunning-Dixon, F.M., Brickman, A.M., Cheng, J.C., Alexopoulos, G.S.: Aging of cerebral white matter: a review of MRI findings. Int. J. Geriatric Psychiatry: J. Psychiatry Late Life Allied Sci. **24**(2), 109–117 (2009)
2. Hardt, M., Price, E., Srebro, N.: Equality of opportunity in supervised learning. In: Advances in Neural Information Processing Systems, vol. 29 (2016)
3. Hellman, D.: When is Discrimination Wrong? Harvard University Press, Cambridge (2008)
4. Kamiran, F., Calders, T.: Data preprocessing techniques for classification without discrimination. Knowl. Inf. Syst. **33**(1), 1–33 (2012)
5. Lin, D.J., Johnson, P.M., Knoll, F., Lui, Y.W.: Artificial intelligence for MR image reconstruction: an overview for clinicians. J. Magn. Reson. Imaging **53**(4), 1015–1028 (2021)

6. Marcus, D.S., Wang, T.H., Parker, J., Csernansky, J.G., Morris, J.C., Buckner, R.L.: Open access series of imaging studies (oasis): cross-sectional MRI data in young, middle aged, nondemented, and demented older adults. J. Cogn. Neurosci. **19**(9), 1498–1507 (2007)

7. Pleiss, G., Raghavan, M., Wu, F., Kleinberg, J., Weinberger, K.Q.: On fairness and calibration. In: Advances in Neural Information Processing Systems, vol. 30 (2017)

8. Puyol-Antón, E., et al.: Fairness in cardiac magnetic resonance imaging: assessing sex and racial bias in deep learning-based segmentation. Front. Cardiovasc. Med. **9**, 859310 (2022)

9. Puyol-Antón, E., et al.: Fairness in Cardiac MR image analysis: an investigation of bias due to data imbalance in deep learning based segmentation. In: de Bruijne, M., et al. (eds.) MICCAI 2021. LNCS, vol. 12903, pp. 413–423. Springer, Cham (2021). https://doi.org/10.1007/978-3-030-87199-4_39

10. Raisi-Estabragh, Z., Harvey, N.C., Neubauer, S., Petersen, S.E.: Cardiovascular magnetic resonance imaging in the UK biobank: a major international health research resource. Eur. Heart J.-Cardiovasc. Imaging **22**(3), 251–258 (2021)

11. Ritchie, S.J., et al.: Sex differences in the adult human brain: evidence from 5216 UK biobank participants. Cereb. Cortex **28**(8), 2959–2975 (2018)

12. Ronneberger, O., Fischer, P., Brox, T.: U-net: convolutional networks for biomedical image segmentation. In: Navab, N., Hornegger, J., Wells, W.M., Frangi, A.F. (eds.) MICCAI 2015. LNCS, vol. 9351, pp. 234–241. Springer, Cham (2015). https://doi.org/10.1007/978-3-319-24574-4_28

13. Slijepcevic, D., et al.: Explaining machine learning models for age classification in human gait analysis. Gait Posture **97**, S252–S253 (2022)

14. Vapnik, V.: Principles of risk minimization for learning theory. In: Advances in Neural Information Processing Systems, vol. 4 (1991)

15. Winkler, J.K., et al.: Association between surgical skin markings in dermoscopic images and diagnostic performance of a deep learning convolutional neural network for melanoma recognition. JAMA Dermatol. **155**(10), 1135–1141 (2019)

16. Zhang, H., Dullerud, N., Roth, K., Oakden-Rayner, L., Pfohl, S., Ghassemi, M.: Improving the fairness of chest x-ray classifiers. In: Conference on Health, Inference, and Learning, pp. 204–233. PMLR (2022)

17. Zong, Y., Yang, Y., Hospedales, T.: Medfair: benchmarking fairness for medical imaging. arXiv preprint arXiv:2210.01725 (2022)

Brain Matters: Exploring Bias in AI for Neuroimaging Research

Sophie A. Martin[1](✉), Francesca Biondo[1], James H. Cole[1,2], and Beatrice Taylor[1]

[1] Centre for Medical Image Computing, UCL, London, UK
s.martin.20@ucl.ac.uk
[2] Dementia Research Centre, UCL, London, UK

Abstract. Developing fair and unbiased models is important for good scientific practice and clinical utility. This paper delves into the specific biases associated with artificial intelligence (AI) in neuroimaging research, and highlights the structural issues that underpin them. We propose a range of mitigation strategies, encompassing both behavioural and technical considerations. By recognising these challenges, we can encourage more accurate and equitable insights into neuroimaging research.

Keywords: Ethics · Fairness · Neuroimaging · Machine Learning

1 Introduction

With the growing potential of artificial intelligence (AI) to increase our understanding of neurological conditions, conversations about model fairness are becoming increasingly important. By 2022, there were 517 published studies that used AI models for understanding psychiatric disorders based on neuroimaging alone [8]. Fairness in AI encompasses a variety of topics such as bias, transparency, and privacy [6]. Bias is especially crucial in healthcare, where AI decisions can have significant impact on patient care and outcomes. Failing to account for this bias can lead to problems around justice, autonomy, beneficence and nonmaleficence [23].

Existing literature has highlighted potential biases in AI models applied to healthcare [32] and medical imaging [10,23]. In the context of neuroimaging, we define a set of desiderata for unbiased AI models: i) representation in the dataset, ii) appropriateness of the model to the research question and iii) transparency of research choices. In this paper we frame this in terms of both protected characteristics (e.g. ethnicity, race, gender, religious and political beliefs) and other characteristics (e.g. having a particular health condition) that can lead to the systematic exclusion of certain groups. We consider the predominant sources of bias inherent in neuroimaging research and propose strategies to mitigate them.

S. A. Martin, F. Biondo and B. Taylor—These authors contributed equally.

© The Author(s), under exclusive license to Springer Nature Switzerland AG 2023
S. Wesarg et al. (Eds.): CLIP/FAIMI/EPIMI 2023, LNCS 14242, pp. 112–121, 2023.
https://doi.org/10.1007/978-3-031-45249-9_11

2   Current Problems

2.1   Structural Problems

Before delving into specific examples of bias in neuroimaging, we first discuss some broader structural problems. We conceptualise structural problems as those rooted in politics and society which are difficult to solve through the individual efforts of researchers, including access to healthcare, education, and the dynamics of research funding. These problems require long-term, systemic changes beyond the academic community, but play a key role in understanding how specific biases in neuroimaging arise. Similar frameworks for understanding bias in AI have been formalised in existing literature [33] (Fig. 1).

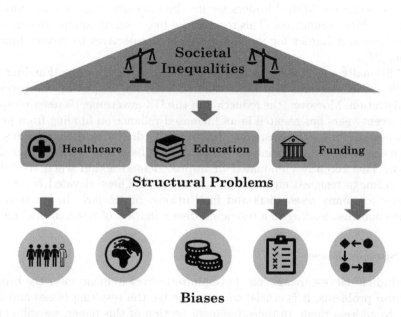

Fig. 1. Societal inequalities lead to structural problems. These structural problems underpin the biases that arise in neuroimaging research.

2.1.1   Disparities in Access to Healthcare and Education

Healthcare disparities are exhibited on both international and national scales. There are substantial differences in the availability, affordability and quality of healthcare services between countries and within the same country. Geographical, infrastructural and socioeconomic barriers lead to non-representative samples and consequently, the findings derived from such data may not accurately reflect the whole population [7].

Another structural problem is inequity in education, which has downstream effects on research outputs. Certain demographics face barriers that limit their participation and progression in higher education [3]. This is illustrated in the underrepresentation of Black female professors in UK universities (41 out of 22,000 as of May 2022 [31]). Overall, a researcher's background and experiences will influence their choice of research topics - whether consciously or not [24]. This may exacerbate research biases.

2.1.2   Issues Within Research and Academia

Academic and research institutions are another important area in which structural problems are apparent. Funding allocation tends to perpetuate a cycle where the same established leaders secure the majority of grants and form the decision-making committees. This results in a bias towards similar research topics and makes it harder for potentially novel perspectives to receive financial support [27].

Additionally, grant schemes may reflect diverse motivations. Funding bodies such as UK Research and Innovation (UKRI) are shaped by the governing administration. Moreover, the reduction in the UK government's research spending in recent years has resulted in an increased reliance on funding from private companies [35], whose interests may be primarily driven by profit margins [12].

The excessive costs associated with publishing have also become a significant concern. This recently culminated in unprecedented action when the editors of NeuroImage resigned en masse in protest [34]. These elevated costs create a barrier for many researchers and institutions, particularly in low-to-middle income countries, leading to a restricted dissemination of research findings.

2.2   Specific Biases

While individual researchers may have limited direct influence over the broader, structural problems, it is crucial to acknowledge the resulting biases and make efforts to address them. In the subsequent section of this paper, we will explore specific biases encountered in neuroimaging research in greater detail.

2.2.1   Bias in Curated Datasets

The majority of neuroimaging datasets are curated as a result of intentional research efforts. This can lead to selection bias, whereby participants with specific characteristics are more likely to be included. For example, it has been shown that white populations with a higher level of education are more likely to participate in neuroimaging research [14]. Another example is the UK Biobank dataset, where participants tended to be older, female and live in less socioeconomically deprived areas as shown by Fry and colleagues [13]. This bias is closely intertwined with the structural problems of access to health services and education (see Sect. 2.1.1).

2.2.2   Bias in Routinely-Collected Healthcare Data

In contrast to curated research studies, routinely-collected healthcare data has a higher likelihood of representing diverse populations. For example, Biondo and colleagues investigated dementia risk in memory clinic data, which exhibited higher ethnic diversity compared to a typical dementia research dataset [4]. However, routinely-collected healthcare datasets can still underrepresent specific groups. This includes individuals who are not served by national healthcare providers such as: undocumented migrants; individuals who may refrain for religious, cultural or health reasons; and individuals who face barriers due to geographic or socioeconomic factors [7,19].

Moreover, certain populations are underreported due to the nature of discretising demographic categories [26]. For instance, the binary classification of gender in data collection can obscure a spectrum of identities. Additionally, this inaccuracy could discourage an individuals' participation [5].

2.2.3   Bias Resulting from Geographic and Economic Differences

Many large neuroimaging datasets predominantly originate from high-income regions such as Europe and North America where there is a greater concentration of research funding, excluding certain groups on a global scale [15].

The prohibitive cost of certain imaging modalities introduces an inherent bias into who has the financial means to collect this data [11]. For example, magnetic resonance imaging (MRI) scanners, particularly those with high field strengths are expensive to set up and run (the UK's National Health Service spent in excess of 1 million US dollars per 1.5T scanner between 2001 and 2011 [36]). Consequently, these scanners are rarely accessible in low-to-middle-income countries where healthcare resources may be limited. Furthermore, even in countries with MRI scanners, there may be a lack of regional availability [25]. As a result individuals residing in more rural areas may not be able to travel the necessary distance to scanners due to time or money constraints.

2.2.4   Bias Due to Barriers in Data Acquisition

Certain data collection methodologies may disproportionally exclude specific populations due to inherent constraints. For instance, the design of electroencephalogram (EEG) sensors can lead to challenges in recruiting participants from certain populations by inadequately accommodating diverse hairstyles [29]. Furthermore, MRI necessitates a participants' ability to enter and comfortably remain in the scanner, which can be difficult for those with mobility, auditory, claustrophobia or other issues. This could have downstream effects on the model in limiting discovery of findings specific to these groups.

2.2.5   Bias in Preprocessing Stages

Preprocessing pipelines are frequently employed as off-the-shelf solutions, without necessarily verifying whether the algorithms are trained on a broadly representative population or customised for a specific dataset. For example, MNI152 is a widely used brain template for registration and segmentation, despite being constructed from a relatively small sample of 152 Canadian participants [9]. Unrepresentative pipelines can lead to problems if there is a population-specific pathology that is confounded with brain signals. This is evident in recent work by Holla and colleagues who developed a set of population-specific Indian brain atlases and highlighted significant structural differences compared to MNI152 [18].

When dealing with data with missing variables, it is common practice to remove participants or utilise augmentation techniques. However, if the missingness is not at random and is intrinsically linked to a specific group, improper handling could introduce bias. For instance, in a multi-modal, multi-centre, study it is possible that a subset of the cohort might be missing a particular imaging modality [1]. This missingness could be linked to a specific health centre with limited resources, and hence certain characteristics present in the local population.

2.2.6   Bias in Model Development

A key decision in model development is how to split training and test datasets. Often, researchers aim to balance maximising the amount of training data with ensuring that the test set can provide an accurate estimation of model performance. However, it is also important to consider the representation of data characteristics that are relevant to the outcome of the model, across each split. Overlooking this consideration may result in an inaccurate assessment of model performance and generalisability. This was demonstrated by Wen and colleagues [38], who compared several models for dementia prediction and found that classification performance generalised poorly across datasets with different inclusion criteria and group characteristics.

3   Mitigation Strategies

Here we present several suggestions to mitigate these problems and encourage representative, appropriate and transparent neuroimaging research in line with our desiderata for fairness.

3.1   Collect More Representative Data

One key mitigation strategy is to collect more representative data [29]. Whilst the concept of representativeness is closely tied to the study's objectives, it is essential for data collection efforts to consider underrepresented groups. This can be achieved in two primary ways.

Firstly, targeted community engagement and advertisement can be used to directly recruit underrepresented individuals. Secondly, we can support the integration of pre-existing electronic health records (EHRs), within the confines of privacy and ethical considerations. Despite the caveat that EHRs are more likely to have missing or mislabelled data, they have a distinct advantage over curated ones. This is due to the nature of healthcare settings which typically encompass a broader cross-section of the general population compared to research studies. Nonetheless, both strategies may still face the challenges of structural access, as discussed in Sect. 2.1.1.

3.2 Share and Collaborate

Whilst collecting new data is an important endeavour, it is a time-intensive and costly pursuit. An alternative approach to addressing data scarcity for underrepresented groups is through cross-institutional data sharing. Encouraging such collaborations can be facilitated by open research initiatives [39]. For instance, the National Alzheimer's Coordinating Center initiative (NACC) leverages pre-existing data and provides a rich representation of different ethnic groups compared to typical research datasets.

However, data sharing is fraught with technological and legal complexities [16]. Legal challenges may arise from potential violations of data privacy laws, such as the General Data Protection Regulation (GDPR) in Europe and the UK. These risks are amplified when dealing with sensitive information such as protected characteristics (e.g. ethnicity, political affiliations, religious beliefs, gender identity) and health data. If misappropriated, such information could be exploited for discriminatory purposes.

One potential solution to this issue is to fully anonymise the data, although this may not always be feasible due to inherent properties (e.g. medical histories may serve as potential identifiers). An alternate strategy is to utilise a federated learning approach. In this framework, a global model is trained using data that remains within the local environment (e.g. within an institution's firewall). This approach circumvents the need for centralised data sharing, thus minimising the risk of data leakage and exposure of sensitive information while optimising collaborative data efforts [30].

3.3 Reduce Reliance on Inaccessible Data Collection Methods

Although high-resolution data can be advantageous to enhance model performance, it can also impact research fairness. For instance, high-resolution neuroimaging data acquisition is not globally accessible (see Sect. 2.2.3). Instead, sharing coarser parcellations of volumetric brain data [2] or using more accessible and affordable neuroimaging techniques (e.g. EEG, CT and ultra-low field MRI [25]) could foster larger and more representative datasets for model development. A further advancement in this area is the utilisation of image enhancement approaches (e.g. SynthSR [20]) to translate low-resolution data to high-resolution.

3.4   Develop Both Generic and Specific Models and Employ Transfer Learning

It can be difficult to decide whether to prioritise generic models that enable enhanced comparability - a "one-model-fits-all" approach - or lean towards specific models optimised for a particular group. Given that utility is largely contextual, a compelling case emerges for maintaining both generic and specific models. Rather than being mutually exclusive, they can offer complementary information.

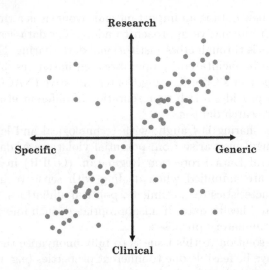

Fig. 2. Specific models may have higher clinical value by optimising output based on the local group characteristics, whilst generic models may have higher comparability and thus be more useful for research purposes.

Borrowing from concepts of global and local norms employed in various fields (e.g. human growth charts [28]) it may be beneficial to adopt a dual testing approach: one that concurrently applies a universally agreed-upon generic model along with specific ones. Results from generic models, referencing a predefined benchmark, add value to worldwide research. In contrast, specific models, aligning with group characteristics, aim to maximise clinical benefits such as diagnostic precision and prognostic value (Fig. 2).

The interplay between generic and specific models is relevant in transfer learning, where a pre-trained model is used to fine-tune a different one. Transfer learning is becoming popular in neuroimaging research as it requires less data and time, and often outperforms models trained from scratch. This is particularly advantageous in the context of small sample sizes (e.g. patient populations, underrepresented groups). For example, Wee and colleagues used a small Asian sample to fine-tune a neural network originally trained on a large, predominantly Caucasian dataset. This led to improved model performance in predicting dementia in the Asian dataset [37].

3.5   Consider the Use of Data Augmentation

Data augmentation strategies such as bootstrapping or the use of general adversarial networks, can synthetically reduce the disparity between under- and overrepresented groups [23]. Alternatively, Kenfack and colleagues used bias-aware objective functions to encourage models to learn fairer representations [21]. Here it is important to note, that in conjunction with describing the data, it is also crucial to transparently document any augmentation strategies used.

3.6   Raise Awareness of Bias and Engage in PPI

As researchers, we can flag potential sources of bias and foster open dialogues. These conversations can take place by engaging in thoughtful self-critique, having discussions with colleagues and organising formal training and workshops [22].

We should also seek out perspectives beyond our immediate research setting. When implemented effectively, public and patient involvement (PPI) increases communication between researchers and patients, facilitating a richer understanding and a shared sense of value [17]. It is also a proactive way to include more diverse voices within the research process.

4   Limitations

Whilst we have made an effort to acknowledge and discuss various biases, our list is not exhaustive. It is important to acknowledge that these biases are often nuanced, and the appropriate mitigation strategies may vary depending on the context. Intervention can inadvertently lead to undesirable effects, warranting careful consideration. We also recognise that our discussion is influenced by our individual backgrounds and experiences. In particular, we are all UK-based medical imaging researchers, so our discussion is limited in its perspective. We welcome discussions with others to broaden our understanding.

5   Conclusion

We believe this paper can serve as a starting point for researchers to reflect on what constitutes best practice in the field and how to achieve it. By continually examining and addressing the problems and biases we can work towards a more equitable and fair approach in neuroimaging research. Our key message is to continue identifying existing biases and transparently report them. Many biases are deeply rooted in societal issues, making them challenging to address. Fairness will require time and a concerted effort.

Acknowledgements. We would like to thank members of the Centre for Medical Image Computing, UCL and the Dementia Research Centre, UCL, for interesting and insightful discussions which helped shape this work.

References

1. Aghili, M., Tabarestani, S., Adjouadi, M.: Addressing the missing data challenge in multi-modal datasets for the diagnosis of Alzheimer's disease. J. Neurosci. Methods **375**, 109582 (2022)
2. Bethlehem, R.A.I., et al.: Brain charts for the human lifespan. Nature **604**(7906), 525–533 (2022)
3. Bhopal, K.: Gender, ethnicity and career progression in UK higher education: a case study analysis. Res. Pap. Educ. **35**(6), 706–721 (2020)
4. Biondo, F., et al.: Brain-age is associated with progression to dementia in memory clinic patients. NeuroImage Clin. **36**, 103175 (2022)
5. Cameron, J.J., Stinson, D.A.: Gender (mis)measurement: Guidelines for respecting gender diversity in psychological research. Soc. Pers. Psychol. Compass **13**(11), e12506 (2019)
6. Carneiro, D., Veloso, P.: Ethics, transparency, fairness and the responsibility of artificial intelligence. In: de Paz Santana, J.F., de la Iglesia, D.H., López Rivero, A.J. (eds.) DiTTEt 2021. AISC, vol. 1410, pp. 109–120. Springer, Cham (2022). https://doi.org/10.1007/978-3-030-87687-6_12
7. Chen, I.Y., et al.: Ethical machine learning in healthcare. Ann. Rev. Biomed. Data Sci. **4**(1), 123–144 (2021)
8. Chen, Z., et al.: Evaluation of risk of bias in neuroimaging-based artificial intelligence models for psychiatric diagnosis: a systematic review. JAMA Netw. Open **6**(3), e231671–e231671 (2023)
9. Diedrichsen, J., et al.: A probabilistic MR atlas of the human cerebellum. Neuroimage **46**(1), 39–46 (2009)
10. Drukker, K., et al.: Toward fairness in artificial intelligence for medical image analysis: identification and mitigation of potential biases in the roadmap from data collection to model deployment. J. Med. Imaging **10**(6), 061104 (2023)
11. Duncan, N.W.: Geographical and economic influences on neuroimaging modality choice. Center for Open Science (2023)
12. Fabbri, A., et al.: The influence of industry sponsorship on the research agenda: a scoping review. Am. J. Public Health **108**(11), e9–e16 (2018)
13. Fry, A., et al.: Comparison of sociodemographic and health-related characteristics of UK Biobank participants with those of the general population. Am. J. Epidemiol. **186**(9), 1026–1034 (2017)
14. Ganguli, M., et al.: Who wants a free brain scan? Assessing and correcting for recruitment biases in a population-based sMRI pilot study. Brain Imaging Behav. **9**(2), 204–212 (2015)
15. Henrich, J., Heine, S.J., Norenzayan, A.: The weirdest people in the world? Behav. Brain Sci. **33**(2–3), 61–83 (2010)
16. Hlávka, J.P.: Chapter 10 - Security, Privacy, and Information-Sharing Aspects of Healthcare Artificial Intelligence, pp. 235–270. Academic Press (2020)
17. Hoddinott, P., et al.: How to incorporate patient and public perspectives into the design and conduct of research. F1000Res **7**, 752 (2018)
18. Holla, B., et al.: A series of five population-specific Indian brain templates and atlases spanning ages 6–60 years. Hum. Brain Mapp. **41**(18), 5164–5175 (2020)
19. Hui, A., et al.: Exploring the impacts of organisational structure, policy and practice on the health inequalities of marginalised communities: Illustrative cases from the UK healthcare system. Health Policy **124**(3), 298–302 (2020)

20. Iglesias, J.E., et al.: SynthSR: a public AI tool to turn heterogeneous clinical brain scans into high-resolution T1-weighted images for 3D morphometry. Sci. Adv. **9**(5), eadd3607 (2023)
21. Kenfack, P.J., et al.: Learning fair representations through uniformly distributed sensitive attributes. In: 2023 IEEE Conference on Secure and Trustworthy Machine Learning (SaTML), pp. 58–67 (2023)
22. Kusnoor, S.V., et al.: Design and implementation of a massive open online course on enhancing the recruitment of minorities in clinical trials - faster together. BMC Med. Res. Methodol. **21**(1), 1–11 (2021)
23. Lara, M.A.R., et al.: Addressing fairness in artificial intelligence for medical imaging. Nat. Commun. **13**(1), 4581 (2022)
24. Longino, H.E.: The Fate of Knowledge. The Fate of Knowledge. Princeton University Press, Princeton (2002)
25. McLane, H.C., et al.: Availability, accessibility, and affordability of neurodiagnostic tests in 37 countries. Neurology **85**(18), 1614–22 (2015)
26. Moseson, H., et al.: The imperative for transgender and gender nonbinary inclusion: beyond women's health. Obstet. Gynecol. **135**(5), 1059–1068 (2020)
27. Murray, D.L., et al.: Bias in research grant evaluation has dire consequences for small universities. PLoS ONE **11**(6), e0155876 (2016)
28. Natale, V., Rajagopalan, A.: Worldwide variation in human growth and the World Health Organization growth standards: a systematic review. BMJ Open **4**(1), e003735 (2014)
29. Ricard, J.A., et al.: Confronting racially exclusionary practices in the acquisition and analyses of neuroimaging data. Nat. Neurosci. **26**(1), 4–11 (2023)
30. Rieke, N., et al.: The future of digital health with federated learning. NPJ Dig. Med. **3**(1), 119 (2020)
31. Showunmi, V.: Visible, invisible: Black women in higher education. Front. Sociol. **8**, 974617 (2023)
32. Starke, G., De Clercq, E., Elger, B.S.: Towards a pragmatist dealing with algorithmic bias in medical machine learning. Med. Health Care Philos. **24**(3), 341–349 (2021). https://doi.org/10.1007/s11019-021-10008-5
33. Suresh, H., Guttag, J.: A framework for understanding sources of harm throughout the machine learning life cycle. Association for Computing Machinery (2021)
34. The MIT Press: A conversation with Dr. Stephen M. Smith, editor-in-chief of imaging neuroscience (2023). https://mitpress.mit.edu/a-conversation-with-dr-stephen-m-smith-editor-in-chief-of-imaging-neuroscience/
35. UKRI: Consequences of the 2021 ODA Budget Cuts: Key Findings. UKRI ODA Review (2022). https://www.ukri.org/publications/consequences-of-the-2021-oda-budget-cuts-key-findings-report/
36. Wald, L.L., et al.: Low-cost and portable MRI. J. Magn. Reson. Imaging **52**(3), 686–696 (2020)
37. Wee, C.Y., et al.: Cortical graph neural network for AD and MCI diagnosis and transfer learning across populations. NeuroImage Clin. **23**, 101929 (2019)
38. Wen, J., et al.: Convolutional neural networks for classification of Alzheimer's disease: overview and reproducible evaluation. Med. Image Anal. **63**, 101694 (2020)
39. Wiener, M., Sommer, F.T., Ives, Z.G., Poldrack, R.A., Litt, B.: Enabling an open data ecosystem for the neurosciences. Neuron **92**(3), 617–621 (2016)

Bias in Unsupervised Anomaly Detection in Brain MRI

Cosmin I. Bercea[1,2,5]($\boxtimes$), Esther Puyol-Antón[5,6], Benedikt Wiestler[3], Daniel Rueckert[1,3,4], Julia A. Schnabel[1,2,5], and Andrew P. King[5]

[1] Technical University of Munich, Munich, Germany
cosmin.bercea@tum.de
[2] Helmholtz AI and Helmholtz Center Munich, Munich, Germany
[3] Klinikum Rechts der Isar, Munich, Germany
[4] Imperial College London, London, UK
[5] King's College London, London, UK
[6] HeartFlow Inc, London, UK

Abstract. Unsupervised anomaly detection methods offer a promising and flexible alternative to supervised approaches, holding the potential to revolutionize medical scan analysis and enhance diagnostic performance.

In the current landscape, it is commonly assumed that differences between a test case and the training distribution are attributed solely to pathological conditions, implying that any disparity indicates an anomaly. However, the presence of other potential sources of distributional shift, including scanner, age, sex, or race, is frequently overlooked. These shifts can significantly impact the accuracy of the anomaly detection task. Prominent instances of such failures have sparked concerns regarding the bias, credibility, and fairness of anomaly detection.

This work presents a novel analysis of biases in unsupervised anomaly detection. By examining potential non-pathological distributional shifts between the training and testing distributions, we shed light on the extent of these biases and their influence on anomaly detection results. Moreover, this study examines the algorithmic limitations that arise due to biases, providing valuable insights into the challenges encountered by anomaly detection algorithms in accurately capturing the variability in the normative distribution. Here, we specifically investigate Alzheimer's disease detection from brain MR imaging as a case study, revealing significant biases related to sex, race, and scanner variations that substantially impact the results. These findings align with the broader goal of improving the reliability, fairness, and effectiveness of anomaly detection.

Keywords: Unsupervised Anomaly Detection · Bias · Fairness

1 Introduction

Unsupervised anomaly detection (UAD) methods have gained significant attention in the medical image analysis research literature due to their potential to

S. Wesarg et al. (Eds.): CLIP/FAIMI/EPIMI 2023, LNCS 14242, pp. 122–131, 2023.
https://doi.org/10.1007/978-3-031-45249-9_12

identify anomalies without the need for labeled training data. However, recent literature has shown that UAD methods are vulnerable to non-pathological out-of-distribution (OoD) data [6]. As a result, notable failures in such approaches have raised concerns regarding bias and fairness in their evaluation. For example, Meissen et al. [9] presented cases where polyp detection algorithms achieved excellent performance even when the actual polyps were removed from the error maps. Similarly, Bercea et al. [1] demonstrated nearly perfect OoD detection using popular reconstruction-based methods that solely relied on analyzing background pixels. Moreover, a predominant focus in the recent literature on UAD has been on the detection of hyper-intense lesions in brain MRIs [3,7,11,17]. A recent study demonstrated that many reconstruction-based methods struggled to generalize to other types of anomalies, indicating a bias in the algorithmic performance [2]. In light of these notable failures, it is essential to thoroughly investigate these concerns to enable the development of more robust and reliable models that exhibit fair and unbiased behavior.

There has been relatively little research into bias in UAD techniques. This is in contrast to other medical imaging applications, where in recent years there has been an increasing focus on bias and fairness. For example, Gichoya et al. [4] demonstrated the presence of race-based distributional shifts across several imaging modalities, highlighting the potential for bias when training models with imbalanced data. Additionally, studies such as Larrazabal et al. [8] and Seyyed et al. [13] have examined bias in chest X-ray classification, Guo et al. [5] reviewed work on biases in skin cancer detection algorithms, and Puyol-Anton et al. [12] have identified race bias in cardiac MR segmentation. In recent years, there have been several studies that have examined biases in neuroimaging data [14–16], including the task of Alzheimer's disease (AD) detection [10].

However, these studies have all focused on supervised approaches. In contrast, our study specifically investigates unsupervised models, which are underpinned by the need to learn the normative training distribution and thus could be more susceptible to biases. These studies underscore the importance of addressing bias in medical applications and sand pave the way for further exploration in UAD. The motivation for studying bias in UAD is twofold. First, given that most experimental setups in the research literature have involved distinct data sources for healthy and pathological distributions, it is essential to analyze potential shifts to ensure fair evaluations and prevent correlations that can significantly skew the performance of these methods. Second, as UAD models strive to represent the entire variability of the normative distribution and effectively identify anomalies by isolating pathological shifts, it becomes increasingly more important to identify their algorithmic limitations.

In this work, we investigate both types of biases in anomaly detection, aiming to fill the gap in this important research area. Our focus is on a case study of AD detection from brain MR images. In summary, our main contributions are:

- To the best of our knowledge, this work represents the first comprehensive investigation into the biases present in UAD.

- Through rigorous analysis, we have uncovered evidence of scanner, sex, race, and metrics biases that significantly impact the performance of UAD.
- We examined other factors like age and brain volume but found no additional correlations for the observed performance drops.

Table 1. Datasets. We present the data splits utilized in our experiments. The abbreviations (Abv.) are linked to the experiments in Table 2. We refer to the healthy training distribution as "Control", the healthy cohort from a different distribution as "Healthy". "Alzheimer's (AD)" represent the pathology set. We mark in blue the shifts in distribution compared to the control set.

Abv.	Dataset	#Scans	Group	Race	Sex	Scanner
T	Training (Control)	434	Control	White	Female	Siemens
V	Validation (Control)	54	Control	White	Female	Siemens
C	Test (Control)	122	Control	White	Female	Siemens
AD	Test (Baseline)	131	AD	White	Female	Siemens
H1	Test (Healthy, Scanner Shift 1)	171	Healthy	White	Female	Philips
AD1	Test (Scanner Shift 1)	73	AD	White	Female	Philips
H2	Test (Healthy, Scanner Shift 2)	70	Healthy	White	Female	GE
AD2	Test (Scanner Shift 2)	36	AD	White	Female	GE
H3	Test (Healthy, Sex Shift)	480	Healthy	White	Male	Siemens
AD3	Test (Sex Shift)	188	AD	White	Male	Siemens
H4	Test (Healthy, Race Shift 1)	103	Healthy	Black	Female	Siemens
AD4	Test (Race Shift 1)	16	AD	Black	Female	Siemens
H5	Test (Healthy, Race Shift 2)	18	Healthy	Asian	Female	Siemens
AD5	Test (Race Shift 2)	8	AD	Asian	Female	Siemens

By shedding light on these biases, we strive to enhance the reliability, fairness, and effectiveness of anomaly detection methods in medical imaging, ultimately benefiting both healthcare providers and patients.

2 Materials and Methods

Dataset. Data used in this study were obtained from the Alzheimer's Disease Neuroimaging Initiative (ADNI) database[1]. ADNI offers a rich collection of magnetic resonance imaging (MRI) scans, accompanied by comprehensive metadata including MR scanner information, and demographic factors such as age,

---

[1] https://adni.loni.usc.edu.

sex, and race. This dataset offers an ideal opportunity to isolate specific factors and evaluate their impact on anomaly detection. In Table 1, we present an overview of the data utilized in this paper, specifically detailing the partitioning of the ADNI dataset into different training, validation and test subsets. As can be seen, our training dataset was acquired from white females using Siemens scanners with a field strength of 3T. This choice was made to maximize the availability of training images and facilitates a more comprehensive assessment of the model's performance.

UAD Method. We utilized a state-of-the-art variational auto-encoder architecture as our UAD method[2]. This recent model incorporates advanced techniques, including perceptual and adversarial loss functions, to enhance the accuracy of image reconstructions, while constraining the latent distribution using the Kullback-Leibler divergence.

Table 2. Bias in UAD. The conventional approach to evaluating UAD methods involves using the control set from the training distribution (denoted as 'C' in Table 1) as the healthy subjects during testing. We present these results as the Naive AD Detection. A relative increase ▲ x% and decrease ▼ x% in performance compared to the baseline (computed as (b-a)/a*100) signifies the presence of bias, while ► x% suggests no bias. To focus solely on the methodological bias, we also report the True AD Detection, which involves using both healthy and pathological subjects from the same source at test time, such as H1/AD1. ↑ x% demonstrates improved performance. We show the distributions of the residual errors and visualize the bias shifts in Fig. 2.

Test set	Naive AD Detection Evaluation & Methodological Bias			True AD Detection Methodological Bias		
	Data	AUROC ↑	AUPRC ↑	Data	AUROC ↑	AUPRC ↑
Baseline	C/AD	64.60	64.82	C/AD	64.60	64.82
Scanner (Philips)	C/AD1	50.30 ▼ 28%	41.71 ▼ 55%	H1/AD1	54.72 ▼ 18%	35.23 ▼ 84%
Scanner (GE)	C/AD2	60.22 ▼ 7%	30.73 ▼ 111%	H2/AD2	64.64 ► 0%	59.80 ▼ 8%
Sex (Male)	C/AD3	86.68 ▲ 25%	89.77 ▲ 28%	H3/AD3	61.69 ▼ 5%	38.39 ▼ 69%
Race (Black)	C/AD4	55.53 ▼ 16%	12.77 ▼ 408%	H4/AD4	56.43 ▼ 14%	15.07 ▼ 330%
Race (Asian)	C/AD5	65.06 ► 1%	9.65 ▼ 572%	H5/AD5	73.61 ↑ 12%	67.59 ↑ 4%

Metrics. We use a range of metrics to evaluate the performance of our method from different perspectives. To assess the reconstruction quality of the methods, we use the mean absolute error (MAE). To assess the anomaly detection ability of our method, we use area under the receiver operator curve (AUROC) and area under the precision-recall curve (AUPRC). Additionally, we include the subjective assessment of a clinician expert to evaluate the quality of reconstructions and the localization of anomalies. Finally, for assessing statistical significance,

---

[2] https://github.com/Project-MONAI/GenerativeModels.

we used Pearson's correlation to identify the impact of potential confounders on the residual errors, and the Kolomogorov-Smirnov test to identify distributional shifts between the AD sets of the training distribution and AD sets of the target distributions. We considered results with p-values lower than 0.05 significant.

3 Experiments and Results

In Subsect. 3.1, we conduct a comprehensive evaluation of the proposed method under ideal conditions, where no distributional shifts other than the pathological one are expected. This evaluation provides insight into the ability of the model to accurately detect AD. Subsequently, in Subsect. 3.2, we systematically introduce changes in the data distribution by modifying a single factor other than pathology, such as MRI scanner manufacturer, sex, or race and evaluate the impact of biases on the performance. Finally, in Subsect. 3.3 we perform further analysis to uncover potential causes of the performance drops.

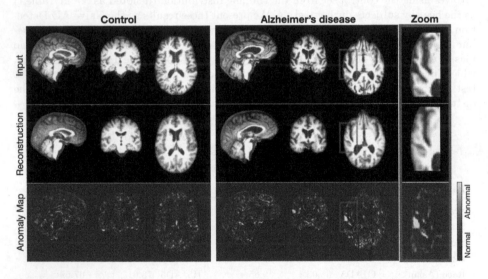

Fig. 1. Qualitative results of the baseline experiment. In the case of AD (right), the atrophy in the Sylvian fissure, which is a typical feature of the disease, is reduced in the reconstruction. This leads to a clear highlight in the anomaly map.

3.1 Baseline Performance

We first evaluated the baseline performance under ideal conditions, where the only factor of change was the presence of AD pathology. See Table 2 for quantitative results and Fig. 1 for a visual example. A clinician assessed that the VAE effectively reverses AD-related pathological changes, such as ventricle dilation and Sylvian fissure abnormalities. Consequently, such areas are highlighted in the residual anomaly map and thus can be readily interpreted for their plausibility. The distributions plot in Fig. 2 (Baseline) shows increased reconstruction errors for AD compared to the control set. Therefore, the method achieved moderate discriminative performance in detecting Alzheimer's pathology with AUROC and AUPRC scores of approximately 65%.

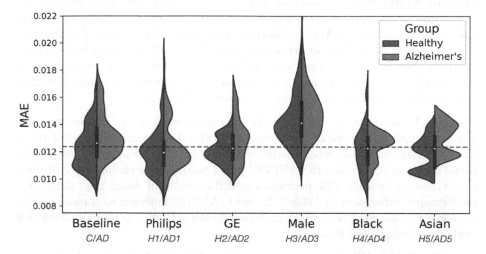

Fig. 2. The distributions of residual errors for different distributional shifts demonstrates the impact of biases. Evaluation bias is characterized by a shift in the overall mean of the residual error, either above or below the mean of the training distribution (shown as a dotted line). Methodological biases are observed when comparing the distributions of healthy and AD groups within a specific shift (violin), revealing a lack of clear distinction between the two distributions. See Table 2 for numerical results.

3.2   Impact of Bias

In this section, we conducted controlled experiments to systematically investigate the impacts of distributional shifts caused by various variables, such as scanner type, sex, and race, aiming to identify potential sources of bias that might impact the performance of UAD. We summarize the results quantitatively in Table 2, and visualize the distributions of the residual errors in Fig. 2.

We considered two distinct scenarios in our analysis. The first scenario, which we call the "naïve" approach, is commonly used in the research literature. Since there is a lack of healthy data within many publicly available datasets, most UAD use alternative sources of healthy data as controls and evaluate their performance on pathology data from a different distribution. However, this approach introduces additional distributional shifts into the evaluation process beyond the presence of pathology itself. Consequently, the evaluation is not clinically realistic and many cases of failure reported in the literature can be attributed to these confounders. In the second scenario, we aimed to address these confounders and isolate the impact of pathology by using an evaluation set that includes both healthy and pathological subjects from the same source. In doing so, we sought to assess any methodological shortcomings and evaluate the effects of bias on UAD. In both scenarios, we observed indications of significant bias stemming from factors such as scanner type, sex and race within the target groups.

Domain Shifts. First, we observed a significant domain shift in the distribution of residual errors, as depicted in Fig. 2. This shift manifests as a uniform vertical shift of both healthy and AD distributions compared to the baseline distribution. Note that this would lead to artificially high UAD performance when controls from the baseline distribution are employed. It is essential to recognize and address these evaluation biases, as they have the potential to strongly influence and distort the UAD results, as demonstrated in Table 2.

Metrics Bias. Next, we examined the bias introduced by the choice of evaluation parameters. Specifically, we found that the AUROC metric tended to be too optimistic, especially when the healthy and pathological samples were highly imbalanced. In some cases the AUROC failed to recognize the presence of bias, e.g., GE scanner shifts (0% performance difference) and Asian race shifts (1% performance difference) in Table 2. Instead, AUPRC emerged as a more robust measure for assessing performance, demonstrating that it is more suitable for evaluating imbalanced datasets.

Scanner Shifts. When using the conventional approach of using the training distribution control set as a reference (Naive AD Detection), we observed a performance decrease of 55% in AURPC for the Philips scanner (C/AD1) compared to the baseline. Similarly, the GE scanner (C/AD2) showed a substantial 111% decrease in AUPRC. To isolate methodological bias, we used both healthy and pathological distributions from the same source (True AD Detection). Interestingly, the performance on Philips (H1/AD1) still showed a considerable drop of 84%, while for GE (H2/AD2) we only observed a minor 8% performance drop.

Sex Shifts. A notable distinction between Naive and True AD detection performance occurs when examining the presence of sex bias. In the naive approach, AD detection performance for the male group increased significantly to an AUPRC of 89.77. However, a closer examination of the distribution plot shown in Fig. 2 reveals that both the healthy and pathology distributions show higher residual errors. This finding suggests a pronounced evaluation bias associated with the naive approach. In contrast, when considering the True AD scenario, which takes into account both healthy and pathological distributions from the same source, the performance dropped dramatically to only 38.39%.

Race Shifts. The race shift analysis revealed notable biases in anomaly detection performance. Specifically, we observed a considerable decrease in performance of 330% when evaluating samples from the black race. Conversely, there was a slight improvement in the True AD Detection performance for subjects from the Asian race, albeit with a limited number of samples.

Table 3. Sources of Bias. Statistical correlations among age, ventricular volume (VV), Hippocampal volume (HV), and whole brain volume (WBV) are examined to explore potential underlying causes for performance drops in the presence of domain shifts. Significant correlations between the analyzed confounds and increased/decreased residual errors in AD samples are denoted by a checkmark (✓), while no correlation is indicated by a cross (✗). Significant shifts in the AD distributions are highlighted in **bold** and the combination of both (✓ and **bold**) is shown in red .

Shift →	Philips (AD1)		GE (AD2)		Male (AD3)		Black (AD4)		Asian (AD5)	
Age	✗	(0.224, 0.015)	✗	(0.255, 0.041)	✓	**(0.346, 0.000)**	✗	(0.321, 0.082)	✓	(0.381, 0.172)
VV	✓	(0.216, 0.089)	✓	(0.265, 0.073)	✓	**(0.373, 0.000)**	✓	**(0.643, 0.002)**	✗	(0.429, 0.105)
HV	✓	(0.119, 0.728)	✗	(0.403, 0.002)	✓	(0.195, 0.063)	✓	**(0.551, 0.015)**	✗	(0.277, 0.543)
WBV	✓	(0.230, 0.065)	✓	(0.207, 0.242)	✓	**(0.494, 0.000)**	✓	**(0.581, 0.008)**	✓	(0.476, 0.051)

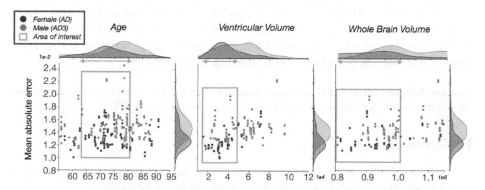

Fig. 3. Visual analysis of the correlations for the sex shift reveals distribution differences in various factors compared to the AD training distribution (distribution plots on top). However, these factors do not seem to be the cause of the performance drop. A closer examination of the target area (highlighted by an orange rectangle) indicates that males have larger overall residual errors than females, suggesting an unaccounted underlying cause for the performance drops. (Color figure online)

3.3   Sources of Bias

In this section, our objective is to explore potential causes for the observed performance drops under different shifts. Table 3 presents the results of our statistical analysis, focusing on distributional shifts resulting from potential confounders, including age, ventricular volume (VV), Hippocampal volume (HV), and whole brain volume (WBV). We mark in red identified significant correlations, where a confounder significantly (according to Pearson's correlation p-values) impacts the residual errors (marked with a checkmark) and there is a significant (according to a Kolmogorov-Smirnov test) distributional shift between the training and target AD distributions (indicated in bold). To summarise, we identified significant correlations for the male and black female distributions. We further inspect the sex shift visually in Fig. 3. The analysis demonstrates elevated residual errors for males, even within the shared ranges of the analyzed confounding factors between males and females. This suggests the presence of another underlying cause beyond the factors evaluated. Further investigations, exploring additional potential confounders are necessary to uncover potential explanations and causal factors contributing to the observed performance drops in the presence of bias.

4   Conclusion

In conclusion, our study highlights the presence of bias, including bias due to scanner, sex and race, in the performance of UAD algorithms. The results indicate that non-pathological distributional shifts can introduce significant distortions in UAD performance. These biases not only impact the overall error distribution, i.e., evaluation bias, but also affect the ability of the methods to accurately detect AD disease. It is essential to understand and address these biases in order to develop robust and reliable UAD algorithms. Future research should prioritize efforts to mitigate these biases and ensure accurate and precise detection in diverse populations and imaging environments.

Acknowledgements. C.I.B. is in part supported by the Helmholtz Association under the joint research school "Munich School for Data Science - MUDS".

References

1. Bercea, C.I., Rueckert, D., Schnabel, J.A.: What do we learn? Debunking the myth of unsupervised outlier detection. arXiv preprint arXiv:2206.03698 (2022)
2. Bercea, C.I., Wiestler, B., Rueckert, D., Schnabel, J.A.: Generalizing unsupervised anomaly detection: towards unbiased pathology screening. In: International Conference on Medical Imaging with Deep Learning (2023)
3. Chen, X., You, S., Tezcan, K.C., Konukoglu, E.: Unsupervised lesion detection via image restoration with a normative prior. Med. Image Anal. **64**, 101713 (2020)
4. Gichoya, J.W., B., et al.: AI recognition of patient race in medical imaging: a modelling study. Lancet. Digit. Health **7500**(22), e406–e414 (2022)
5. Guo, L.N., Lee, M.S., Kassamali, B., Mita, C., Nambudiri, V.E.: Bias in, bias out: underreporting and underrepresentation of diverse skin types in machine learning research for skin cancer detection - a scoping review. J. Am. Acad. Dermatol. **87**(1), 157–159 (2021)
6. Heer, M., Postels, J., Chen, X., Konukoglu, E., Albarqouni, S.: The OOD blind spot of unsupervised anomaly detection. In: Medical Imaging with Deep Learning (2021). https://openreview.net/forum?id=ZDD2TbZn7X1
7. Kascenas, A., Pugeault, N., O'Neil, A.Q.: Denoising autoencoders for unsupervised anomaly detection in brain MRI. In: International Conference on Medical Imaging with Deep Learning (2022)
8. Larrazabal, A.J., Nieto, N., Peterson, V., Milone, D., Ferrante, E.: Gender imbalance in medical imaging datasets produces biased classifiers for computer-aided diagnosis. Proc. Natl. Acad. Sci. U S A **117**(23), 12592–12594 (2020)
9. Meissen, F., Lagogiannis, I., Kaissis, G., Rueckert, D.: Domain shift as a confounding variable in unsupervised pathology detection. In: Medical Imaging with Deep Learning (2022). https://openreview.net/forum?id=6tsAzh_tnyF

10. Petersen, E., et al.: Feature robustness and sex differences in medical imaging: a case study in MRI-based Alzheimer's disease detection. In: Wang, L., Dou, Q., Fletcher, P.T., Speidel, S., Li, S. (eds.) MICCAI 2022. LNCS, vol. 13431, pp. 88–98. Springer, Cham (2022). https://doi.org/10.1007/978-3-031-16431-6_9
11. Pinaya, W.H.L., et al.: Unsupervised brain anomaly detection and segmentation with transformers. arXiv preprint arXiv:2102.11650 (2021)
12. Puyol-Antón, E., et al.: Fairness in cardiac magnetic resonance imaging: assessing sex and racial bias in deep learning-based segmentation. Front. Cardiovasc. Med. **9**, 859310 (2022)
13. Seyyed-Kalantari, L., Zhang, H., McDermott, M., et al.: Underdiagnosis bias of artificial intelligence algorithms applied to chest radiographs in under-served patient populations. Nat. Med. **27**, 2176–2182 (2021)
14. Stanley, E.A.M., Wilms, M., Forkert, N.D.: Disproportionate subgroup impacts and other challenges of fairness in artificial intelligence for medical image analysis. In: Baxter, J.S.H., et al. (eds.) EPIMI ML-CDS TDA4BiomedicalImaging 2022. LNCS, vol. 13755, pp. 14–25. Springer, Cham (2022). https://doi.org/10.1007/978-3-031-23223-7_2
15. Stanley, E.A.M., Wilms, M., Mouches, P., Forkert, N.D.: Fairness-related performance and explainability effects in deep learning models for brain image analysis. J. Med. Imaging **9**(6), 061102 (2022)
16. Wang, R., Chaudhari, P., Davatzikos, C.: Bias in machine learning models can be significantly mitigated by careful training: evidence from neuroimaging studies. Proc. Natl. Acad. Sci. U S A **120**(6), e2211613120 (2023)
17. Zimmerer, D., Isensee, F., Petersen, J., Kohl, S., Maier-Hein, K.: Unsupervised anomaly localization using variational auto-encoders. In: Shen, D., et al. (eds.) MICCAI 2019. LNCS, vol. 11767, pp. 289–297. Springer, Cham (2019). https://doi.org/10.1007/978-3-030-32251-9_32

Towards Unraveling Calibration Biases in Medical Image Analysis

María Agustina Ricci Lara[1,2]([✉]) [ID], Candelaria Mosquera[1,2] [ID],
Enzo Ferrante[3] [ID], and Rodrigo Echeveste[3] [ID]

[1] Health Informatics Department, Hospital Italiano de Buenos Aires, Buenos Aires,
Argentina
[2] Universidad Tecnológica Nacional, Buenos Aires, Argentina
{maria.ricci,candelaria.mosquera}@hospitalitaliano.org.ar
[3] Research Institute for Signals, Systems and Computational Intelligence sinc(i)
(FICH-UNL/CONICET), Santa Fe, Argentina
{eferrante,recheveste}@sinc.unl.edu.ar

Abstract. In recent years the development of artificial intelligence (AI)
for medical image analysis has gained enormous momentum. At the same
time, a large body of work has shown that AI systems can systematically
and unfairly discriminate against certain populations in various application scenarios, motivating the emergence of algorithmic fairness studies.
Most research on healthcare algorithmic fairness to date has focused on
the assessment of biases in terms of classical discrimination metrics such
as AUC and accuracy. Potential biases in terms of model calibration,
however, have only recently begun to be evaluated. This is especially
important when working with clinical decision support systems, as predictive uncertainty is key to optimally evaluate and combine multiple
sources of information. Here we study discrimination and calibration
biases in models trained for automatic detection of malignant dermatological conditions from skin lesions images. Importantly, we show how
several typically employed calibration metrics are systematically biased
with respect to sample sizes, and how this can lead to erroneous conclusions if not taken into consideration. This is of particular relevance
to fairness studies, where data imbalance results in drastic sample size
differences between demographic sub-groups, which could act as confounders.

Keywords: Fairness · Bias · Calibration · Skin Lesion Analysis

1 Introduction

Different studies have shown that data-driven algorithms can perform differently
across sub-populations, and even amplify historically observed biases against

---

Supplementary Information The online version contains supplementary material
available at https://doi.org/10.1007/978-3-031-45249-9_13.

S. Wesarg et al. (Eds.): CLIP/FAIMI/EPIMI 2023, LNCS 14242, pp. 132–141, 2023.
https://doi.org/10.1007/978-3-031-45249-9_13

particular groups [5,27,30]. Here biased algorithms are taken as those that perform unevenly when evaluated in sub-groups defined in terms of a protected attribute, such as sex, gender, age, skin tone, among others. This behaviour has raised an alarm in the scientific community and the evaluation of algorithmic fairness [20] has gained increasing attention. Multiple studies examined this phenomenon in the domain of Medical Imaging Computing (MIC) [7,27], considering a variety of demographic sub-groups and image modalities, like gender and ethnicity in x-ray images [18,28] or skin-tone in the detection of diabetic retinopathy from fundus images [6]. While the number of studies of fairness in MIC has started to increase, there are still significant areas of vacancy to be explored [27]. Here we are interested in better understanding the properties of certain evaluation metrics in the context of healthcare. In particular, considering that in fairness studies there is usually a majority group and an underrepresented group, if the metrics selected for the assessment are systematically affected by the sample size, conclusions might be misleading. Moreover, few cases of potential bias assessment in terms of calibration have been observed in this domain [27].

Classification performance results from a combination of discrimination and calibration capabilities [3]. A system is said to be well-calibrated when its output reflects the uncertainty about a sample's class given the input information [8,23]; for example, when the output can be interpreted as the probability that the sample belongs to the class of interest given the input [22]. A distinctive characteristic of the medical imaging domain is the need for interpretable outputs to assist in clinical decision-making. In this regard, quantifying model uncertainty becomes crucial [17]. As with discrimination metrics, the selection of calibration metrics for auditing fairness in MIC must also take into account sample-size differences usually seen between sub-groups, to ensure their effectiveness in detecting and mitigating biases that may disproportionately affect underrepresented groups. For instance, one of the most widely used calibration estimators in the scientific community, known as Expected Calibration Error (ECE), is highly sensitive and heavily biased with respect to sample size, and becomes monotonically worse with fewer samples in the evaluation set [13].

Although recently the relation between discrimination and calibration biases has started to be explored [25], to our knowledge no studies have been carried out so far in the field of medical imaging analyzing the impact of sub-group underrepresentation in the outcome of algorithmic fairness audits in terms of both discrimination and calibration metrics. Here we focused on skin lesion classification and studied a set of commonly used metrics, analyzing their behavior in the context of a specific data imbalance problem: light-skinned vs. dark-skinned cases in the Fitzpatrick scale [11]. This problem is of particular significance, since skin cancer prevalence is reported to be lower on dark-skinned individuals [1] and is usually detected at a more advanced stage leading to a worse prognosis [14]. These factors, combined with historical biases and insufficient representation of dark-skinned individuals in medical studies, may be some of the underlying causes for their frequent underrepresentation in the databases used to train machine learning models [16]. Whether AI models applied to the classification of skin lesions tend to be biased in terms of skin-tone is still up for debate. While some studies found that vanilla AI models had a lower classification performance in dark-skinned individuals [19], other studies found no systematic pattern in the accuracy of the classifiers between sub-groups [12,15].

Here we studied performance gaps both in terms of discrimination and calibration between light-skinned and dark-skinned individuals in the detection of malignant lesions in dermatological images. We used an available open database of skin clinical images that include skin tone information of subjects [24]. We observed that, although discrimination performance can show no significant difference between protected groups, calibration metrics naively used may appear to show that the model outputs do not represent uncertainty fairly between them. However, by matching sample sizes, we demonstrated that the differences observed from the naive estimators of calibration between sub-groups were actually spurious. In order to understand the mechanistic relationship between sample sizes and calibration estimators we additionally performed a series of synthetic experiments in which we emulated different de-calibration scenarios. Importantly, we show that, as models are overall better calibrated, finite size effects become more noticeable, which is relevant in scenarios where models are post-hoc re-calibrated after training. Overall, our study sheds light on the importance of considering differences in sample sizes among population sub-groups when performing fairness studies.

2   Numerical Experiments on Real Data

In this work we address the problem of binary classification of benign and malignant skin lesions in clinical (i.e., not dermoscopic) images. We study algorithmic fairness in terms of discrimination and calibration considering the binarized patient's skin tone (dark/light) as the protected attribute.

2.1   Data

The analysis was carried out employing the open access database PAD-UFES-20 [24] with 2,298 clinical images from 1,641 skin lesions belonging to 1,373 patients. Lesion are classified into six categories: three benign conditions (actinic keratosis, seborrheic keratosis and nevus) and three malignant conditions (basal cell carcinoma, squamous cell carcinoma and melanoma). Images come accompanied by metadata regarding lesions details and patient information. Only a subset of 1,494 images from the dataset include the skin tone on the Fitzpatrick scale.

Regarding class balance with respect to the classification target (malignancy), PAD-UFES-20 includes 1,089 malignant lesions (47.38% of the cases). Regarding class balance with respect to the protected attribute, the number of light-skinned subjects greatly exceeds the number of dark-skinned subjects in the dataset with approximately 20 times more light-skinned cases.

We randomly divided PAD-UFES-20 into training, validation and test subsets with a patient-level split. To stratify the split both by classification target (benign or malignant) and protected attribute (light, dark or unknown), we applied a two stratified K-Folds strategy of five iterations, totalling 25 runs. This ensures that the distribution of lesion malignancy (Fig. S1**A**) and patient skin tone (Fig. S1**B**) are preserved across subsets. As cases of unknown skin tone were later removed from the test set, the relative frequency of benign and malignant cases in the target variable in this subset was slightly affected (see Supplementary Material).

2.2 Model Training

We used a VGG-19 architecture [29] pretrained on ImageNet, performing image normalization with the channel's mean and standard deviation on each dataset, and applying horizontal flips, random rotations and zoom as data augmentation. Models were trained using all skin tone cases combined, with weighted binary cross-entropy as loss function and Adam as optimizer for 100 epochs using Keras. Validation loss was monitored with a patience of 20 epochs to avoid overfitting. We used a learning rate of 0.0005 and 32 images per batch.

2.3 Platt Scaling

We evaluated the effect of applying a monotonic transformation to the models outputs (without changing discrimination performance). We fitted Platt Scaling as calibrator (logistic regression) [26] using the log likelihood ratios (LLRs) of the validation subset on each run. All available samples in every validation set were employed, regardless of the value of the skin tone variable. The fitted parameters were then used to transform the scores of the corresponding test sets. The performance of these calibrated scores was compared between groups, and they were especially used to calculate the calibration loss (Sect. 2.4).

2.4 Performance Evaluation

Classification performance metrics were computed on the test set of each split separately for each protected group (Fig. 1). Regarding discrimination metrics, we report the area under the receiving operating characteristic curve (AUC_{ROC}) and the area under the precision-recall curve (AUC_{PR}). For comparison with prior works on skin imaging fairness [15], we also present the balanced accuracy computed using a threshold of 0.5. In addition, we report a modified version of the AUC_{PR} that normalizes this metric considering the area under the curve for a random guess classifier (AUC_{PRG}) [4]. To assess the calibration performance solely, we performed Platt Scaling on every validation subset and the performance metrics of the transformed scores on each corresponding test set were used as reference to decompose proper scoring rules (PSR) into discrimination and calibration components [10]. We calculated the difference in cross-entropy

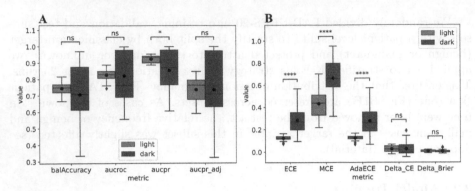

Fig. 1. Comparison of metrics between sub-groups. Discrimination metrics (**A**) and calibration metrics (**B**) were computed over 25 runs. Intentionally, calibration metrics were computed naively without taking sub-group sample sizes into account, resulting in *spurious* significant differences. Boxplots show the mean, median and interquartile range.

(CE) between model outputs before calibration and their recalibrated versions to obtain the calibration component (a.k.a. calibration loss or $DeltaCE$) of the metric, and applied the same procedure with the Brier score ($DeltaBrier$). We also calculated bin-based metrics that are usually reported in computer vision studies as calibration metrics: the expected calibration error (ECE) and the maximum calibration error (MCE) [22]. In addition, we calculated the adaptive ECE ($AdaECE$) [21], which fits bin boundaries so that the number of samples is equal across bins. Wilcoxon Signed Rank tests for two related samples were used to evaluate significant differences (at a 0.05 level) between groups ($^*p < 0.05,^{**}p < 0.005,^{***}p < 0.0005,^{****}p < 0.00005$).

2.5   Results

Discrimination and Calibration Performance Across Protected Attributes on the Original Test Set. Firstly we compared the performance between sub-groups with respect to discrimination metrics, which showed no statistical significance (Fig. 1**A**). This is in line with previous studies like Ref. [15]. Next we compared sub-groups using calibration metrics naively without taking sample-sizes into account (Fig. 1**B**), and found a trend favoring light-skinned individuals in many of them. Critically, as we will show in the next sub-section, these differences were actually spurious, and a result of the different sample sizes between sub-groups.

Sample Size Analysis on Performance Metrics. As shown in Fig. S1, there are very few cases belonging to the minority class (dark-skinned individuals) in the test set. We further analyzed whether this imbalance can influence the differences observed in the calibration metrics. By subsampling the larger group so that the number of samples matches that of the minority group, we find that previously observed differences are eliminated (Fig. 2). In addition, we show how randomly subsampling the entire data set with different sampling ratios affects the value of most calibration metrics, which, if not carefully taken into account, may give rise to misleading findings (Fig. 3**A**).

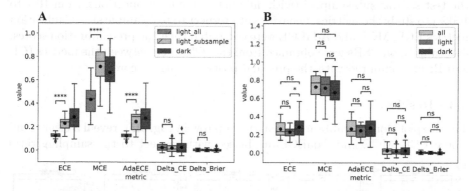

Fig. 2. Effect of sample sizes on calibration metric fairness audits. Significant differences on (**A**) show the impact of sample size in most of the metrics when comparing the original light-skinned test set and a sub-sampled version, matching the number of samples in the dark-skinned test set. When metrics are computed with equal sample sizes, they do not exhibit such differences (**B**).

3   Synthetic Experiments

3.1   Data

When subsampling, previously significant results may vanish either because a metric is biased with respect to sample size, or simply because the variances of the estimators increase and one has lost statistical power. The second component will necessarily always be present in part, but our hypothesis is that the first reason is playing a key role in spurious differences. In order to test this hypothesis we constructed a synthetic dataset, where these different factors can be independently controlled. We constructed a binary classification problem where we know the true posterior probabilities for each sample generated at random (n = 10e6), and can manipulate calibration and sample size at will. Random values associated with these samples were also generated to emulate the output scores of a model. Different scenarios of model de-calibration were simulated by the use of a cumulative beta-distribution of varying α and β parameters (see Fig. S2 from Supplementary Material for more details), without affecting discrimination metrics since the threshold was kept fixed at 0.5.

3.2   Performance Evaluation

In each synthetic scenario the emulated scores were modified by each of the beta functions, and in each case 100 random splits were performed to obtain train, validation and test sets. Each validation set was used to fit a Platt Scaling recalibrator, which was then applied to the corresponding test set data. Later, the test set was sub-sampled (with different sampling proportions, from 10% to 100%) to study the metrics' behavior with respect to test sample size. Calibration metrics ECE, MCE and AdaECE were computed with the pre-calibration scores, and calibration PSRs were obtained from the difference between the metrics (CE and Brier) computed with the pre- and post-calibration scores.

3.3   Results

The boxplots for the calibrated scenario (top row in Fig. 3B), reveal a clear trend where ECE, MCE and AdaECE are biased with respect to the sampling ratio

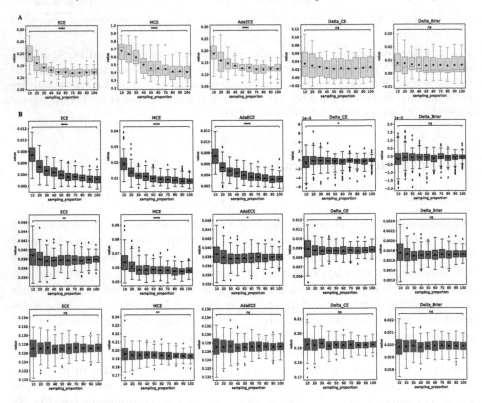

Fig. 3. Calibration metrics calculated over the runs with different sampling ratios of the original test sets. Results of the numerical experiments on real data **(A)** show the trend on ECE, MCE and AdaECE values with respect to the sampling proportion. The results of the synthetic experiments **(B)** are arranged according to the different de-calibration scenarios: perfectly calibrated ($\alpha = \beta = 1$) in the top row, slightly out of calibration ($\alpha = \beta = 1.5$) in the middle row and highly miscalibrated ($\alpha = \beta = 5$) in the bottom row.

(i.e. test set size), supported by the significant differences found between the value of these metrics computed on 100% and 10% of the data (Wilcoxon Signed Rank test at a 0.05 level).

We note however that differences become less pronounced as de-calibration becomes more prominent (middle and bottom rows in Fig. 3B). Comparing the scales on the y-axis we note that the finite size effect component in the metrics is eventually dominated by the miscalibration component itself. On the other hand, it can be observed that sample size does not seem to have a major effect on the PSRs.

4 Discussion

Algorithmic fairness analysis can have important implications for defining the best way to present model results when this may impact clinical decision making [9]. Obtaining a fair calibration performance across demographic groups allows to use a common fixed threshold to obtain binary predictions that will imply the same error costs for all groups. On the contrary, if a common threshold is applied when the relationship between model outputs and uncertainty is different across protected groups, the majority group will likely have a real diagnostic performance that is closer to the expected one while the minority group will have unexpected false positive and negative rates. Moreover, when multiple sources of information need to be combined (where the uncertainty of the combined prediction is dominated by the source of lowest uncertainty [2]), miscalibrated models will lead to higher error rates.

Here we observed that, even in the absence of biases in discrimination metrics, differences in terms of calibration may appear when these metrics were used naively, without taking relative sample sizes into account. However, on closer inspection, taking into account the extreme imbalance in the data of the different sub-groups, as often is the case in fairness studies, we find that these differences may actually be spurious, serving as a cautionary tale for researchers conducting these types of audits. The impact of sample size on ECE has been demonstrated in other scenarios [13, 25], presenting systematic biases in line with results shown here. Moreover we found that some of the metrics used in the literature appear to present higher susceptibility to sample size imbalance than others. We also found that sample size dependence seems to be stronger for better calibrated models, which raises a second warning, since models are these days usually not left uncalibrated, but re-calibrated. Despite the importance of calibration and fairness in the context of healthcare, to our knowledge our work is the first to evaluate fairness in dermatological imaging AI using calibration metrics. Overall, our study emphasizes the importance of careful and comprehensive assessments of model fairness, including calibration, to ensure accurate and equitable machine learning outcomes.

Acknowledgments. We thank the Artificial Intelligence and Data Science in Health Program at Hospital Italiano de Buenos Aires for providing the space to discuss and work on these issues. This work was supported by Argentina's National Scientific and Technical Research Council (CONICET), which covered the salaries of R.E. and E.F. The work of E.F. was partially supported by the ARPH.AI project funded by a grant (Number 109584) from the International Development Research Centre (IDRC) and the Swedish International Development Cooperation Agency (SIDA). We also acknowledge the support of Universidad Nacional del Litoral (Grants CAID-PIC-50220140100084LI, 50620190100145LI), Agencia Nacional de Promoción de la Investigación, el Desarrollo Tecnológico y la Innovación (Grants PICT 2018-3907, PRH 2017-0003, PICT-2020-SERIEA-01765, PRH 2022-00002) and Santa Fe Agency for Science, Technology and Innovation (Award ID: IO-138-19). This work is supported by the Google Award for Inclusion Research (AIR) Program. We also thank Luciana Ferrer and Celia Cintas for the fruitful discussions.

References

1. Agbai, O.N., et al.: Skin cancer and photoprotection in people of color: a review and recommendations for physicians and the public. J. Am. Acad. Dermatol. **70**(4), 748–762 (2014)
2. Bejjanki, V.R., Clayards, M., Knill, D.C., Aslin, R.N.: Cue integration in categorical tasks: insights from audio-visual speech perception. PLoS ONE **6**(5), e19812 (2011)
3. Blattenberger, G., Lad, F.: Separating the brier score into calibration and refinement components: a graphical exposition. Am. Stat. **39**(1), 26–32 (1985)
4. Bugnon, L.A., Yones, C., Milone, D.H., Stegmayer, G.: Deep neural architectures for highly imbalanced data in bioinformatics. IEEE Trans. Neural Netw. Learn. Syst. **31**(8), 2857–2867 (2019)
5. Buolamwini, J., Gebru, T.: Gender shades: intersectional accuracy disparities in commercial gender classification. In: Conference on Fairness, Accountability and Transparency, pp. 77–91. PMLR (2018)
6. Burlina, P., Joshi, N., Paul, W., Pacheco, K.D., Bressler, N.M.: Addressing artificial intelligence bias in retinal diagnostics. Transl. Vis. Sci. Technol. **10**(2), 13–13 (2021)
7. Chen, I.Y., Pierson, E., Rose, S., Joshi, S., Ferryman, K., Ghassemi, M.: Ethical machine learning in healthcare. Ann. Rev. Biomed. Data Sci. **4**, 123–144 (2021)
8. Dawid, A.P.: The well-calibrated Bayesian. J. Am. Stat. Assoc. **77**(379), 605–610 (1982)
9. Esteva, A., et al.: Deep learning-enabled medical computer vision. NPJ Digit. Med. **4**(1), 1–9 (2021)
10. Ferrer, L.: Analysis and comparison of classification metrics. arXiv preprint arXiv:2209.05355 (2022)
11. Fitzpatrick, T.B.: The validity and practicality of sun-reactive skin types I through VI. Arch. Dermatol. **124**(6), 869–871 (1988)
12. Groh, M., et al.: Evaluating deep neural networks trained on clinical images in dermatology with the fitzpatrick 17k dataset. In: Proceedings of the IEEE/CVF Conference on Computer Vision and Pattern Recognition, pp. 1820–1828 (2021)
13. Gruber, S., Buettner, F.: Better uncertainty calibration via proper scores for classification and beyond. Adv. Neural. Inf. Process. Syst. **35**, 8618–8632 (2022)
14. Gupta, A.K., Bharadwaj, M., Mehrotra, R.: Skin cancer concerns in people of color: risk factors and prevention. Asian Pac. J. Cancer Prevent.: APJCP **17**(12), 5257 (2016)

15. Kinyanjui, N.M., et al.: Fairness of classifiers across skin tones in dermatology. In: Martel, A.L., et al. (eds.) MICCAI 2020. LNCS, vol. 12266, pp. 320–329. Springer, Cham (2020). https://doi.org/10.1007/978-3-030-59725-2_31
16. Kleinberg, G., Diaz, M.J., Batchu, S., Lucke-Wold, B.: Racial underrepresentation in dermatological datasets leads to biased machine learning models and inequitable healthcare. J. Biomed. Res. **3**(1), 42–47 (2022)
17. Kompa, B., Snoek, J., Beam, A.L.: Second opinion needed: communicating uncertainty in medical machine learning. NPJ Digit. Med. **4**(1), 1–6 (2021)
18. Larrazabal, A.J., Nieto, N., Peterson, V., Milone, D.H., Ferrante, E.: Gender imbalance in medical imaging datasets produces biased classifiers for computer-aided diagnosis. Proc. Natl. Acad. Sci. **117**(23), 12592–12594 (2020)
19. Li, X., Cui, Z., Wu, Y., Gu, L., Harada, T.: Estimating and improving fairness with adversarial learning. arXiv preprint arXiv:2103.04243 (2021)
20. Mehrabi, N., Morstatter, F., Saxena, N., Lerman, K., Galstyan, A.: A survey on bias and fairness in machine learning. ACM Comput. Surv. (CSUR) **54**(6), 1–35 (2021)
21. Mukhoti, J., Kulharia, V., Sanyal, A., Golodetz, S., Torr, P., Dokania, P.: Calibrating deep neural networks using focal loss. Adv. Neural. Inf. Process. Syst. **33**, 15288–15299 (2020)
22. Naeini, M.P., Cooper, G., Hauskrecht, M.: Obtaining well calibrated probabilities using Bayesian binning. In: Twenty-Ninth AAAI Conference on Artificial Intelligence (2015)
23. Ovadia, Y., et al.: Can you trust your model's uncertainty? evaluating predictive uncertainty under dataset shift. arXiv preprint arXiv:1906.02530 (2019)
24. Pacheco, A.G., et al.: PAD-UFES-20: a skin lesion dataset composed of patient data and clinical images collected from smartphones. Data Brief **32**, 106221 (2020)
25. Petersen, E., Ganz, M., Holm, S., Feragen, A.: On (assessing) the fairness of risk score models. In: Proceedings of the 2023 ACM Conference on Fairness, Accountability, and Transparency, pp. 817–829 (2023)
26. Platt, J., et al.: Probabilistic outputs for support vector machines and comparisons to regularized likelihood methods. Adv. Large Margin Classifiers **10**(3), 61–74 (1999)
27. Ricci Lara, M.A., Echeveste, R., Ferrante, E.: Addressing fairness in artificial intelligence for medical imaging. Nat. Commun. **13**(1), 1–6 (2022)
28. Seyyed-Kalantari, L., Zhang, H., McDermott, M., Chen, I.Y., Ghassemi, M.: Underdiagnosis bias of artificial intelligence algorithms applied to chest radiographs in under-served patient populations. Nat. Med. **27**(12), 2176–2182 (2021)
29. Simonyan, K., Zisserman, A.: Very deep convolutional networks for large-scale image recognition. arXiv preprint arXiv:1409.1556 (2014)
30. Zou, J., Schiebinger, L.: AI can be sexist and racist-it's time to make it fair (2018)

Are Sex-Based Physiological Differences the Cause of Gender Bias for Chest X-Ray Diagnosis?

Nina Weng$^{(\boxtimes)}$, Siavash Bigdeli, Eike Petersen, and Aasa Feragen

Technical University of Denmark, Kongens Lyngby, Denmark
{ninwe,sarbi,ewipe,afhar}@dtu.dk

Abstract. While many studies have assessed the fairness of AI algorithms in the medical field, the causes of differences in prediction performance are often unknown. This lack of knowledge about the causes of bias hampers the efficacy of bias mitigation, as evidenced by the fact that simple dataset balancing still often performs best in reducing performance gaps but is unable to resolve all performance differences. In this work, we investigate the causes of gender bias in machine learning-based chest X-ray diagnosis. In particular, we explore the hypothesis that breast tissue leads to underexposure of the lungs and causes lower model performance. Methodologically, we propose a new sampling method which addresses the highly skewed distribution of recordings per patient in two widely used public datasets, while at the same time reducing the impact of label errors. Our comprehensive analysis of gender differences across diseases, datasets, and gender representations in the training set shows that dataset imbalance is not the sole cause of performance differences. Moreover, relative group performance differs strongly between datasets, indicating important dataset-specific factors influencing male/female group performance. Finally, we investigate the effect of breast tissue more specifically, by cropping out the breasts from recordings, finding that this does not resolve the observed performance gaps. In conclusion, our results indicate that dataset-specific factors, not fundamental physiological differences, are the main drivers of male–female performance gaps in chest X-ray analyses on widely used NIH and CheXpert Dataset. The code is available under https://github.com/nina-weng/detecting_causes_of_gender_bias_chest_xrays.

Keywords: Algorithmic fairness · Gender biases · Chest X-ray

1 Introduction

AI fairness receives increased attention with the escalating demand for examining the validity and responsibility of AI methods. This is particularly crucial in the medical field, where automatic and intelligent decision-making algorithms could easily lead to unfair treatment without the awareness of fairness.

Supplementary Information The online version contains supplementary material available at https://doi.org/10.1007/978-3-031-45249-9_14.

A series of studies have assessed fairness in various medical imaging settings, including chest x-rays [14,20], retinal imaging [5], brain MRI [16,23], and cardiac MRI [18]. While different types of bias mitigation techniques have been applied [6,15,25,26], there is currently very limited work that seeks to diagnose the *causes* of bias [17], enabling the targeted selection of bias mitigation methods. Supporting the urgency of this type of *bias reasoning*, [26] shows, by comparing the performance of different bias mitigation strategies, that simply balancing datasets, which targets the representation bias, is still one of the best strategies for mitigating gender bias for chest X-ray diagnosis. At the same time, studies have shown that while the level of group representation affects group-wise performance, balancing datasets alone does not guarantee equal performance across groups [14], see Fig. 1 for an example. This implies that without further investigation of the causes of bias, our

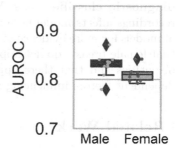

Pneumothorax - NIH

Fig. 1. An excerpt of our results, inspired by [14], showing Pneumothorax diagnosis performance evaluated on men and women for an algorithm trained *solely* on women. In this example, even female over-representation does not yield equal performance. This example led to a hypothesis that female breasts might lead to degraded image quality [7].

attempts to mitigate bias might be very limited in success, as also evidenced by the well-known 'leveling down' phenomenon [27].

In this paper, we investigate causes of gender bias in machine learning-based chest X-ray diagnosis, where significant performance disparities between genders are observed. Previous works [1,3,7,12] have suggested that breast tissue might lead to impaired image quality, and hence lower performance, of chest X-ray-based diagnostic classifiers. Here, we perform a series of experiments to analyze whether the observed gender bias is indeed a result of physiological sex differences. In short, our contributions include:

1. **A new way of sampling** training and test sets from publicly available chest X-ray datasets that reduces the influence of potential confounders, such as a highly skewed distribution of the number of recordings per patient and missing disease labels, on training and analysis. In particular, we propose to sample just a single recording per patient, preferring samples with a disease-positive label. A comparison of different sampling strategies provides further strong evidence of label errors in publicly available datasets.

2. A combination of our proposed sampling method with training sets of varying gender ratios to provide **a comprehensive re-analysis of gender differences** in model performance across multiple diseases in two well-known datasets (CheXpert [11] and NIH ChestX-ray8 [24]). Our results indicate that imbalanced datasets are not the only cause of performance differences, and, crucially, that gender-based performance differences differ between datasets even for the same disease.

3. Further **experiments designed to study whether female breasts cause diagnostic classifier bias.** We assess how cropping out the breasts from recordings affects model performance, and we consider inter-dataset transfer of models for more evidence of inherent gender bias from the dataset. We find that female breasts do not appear to be a strong cause of model performance disparities and that there appear to be other, presently unknown dataset-dependent factors influencing model performance.

2   Related Work

Following claims of radiologist-level performance in machine learning-based chest X-ray disease classification [19], performance disparities of such disease classifiers between patient groups have come under increased scrutiny [14,20,21]. Larraz-abal et al. [14] analyzed the effect of gender imbalance in training sets on performance disparities in the trained classifier, finding a strong link between the two. Moreover, their results show how performance in some groups remains poor even if models are trained *only* on subjects from that group; refer to Fig. 1 for an example. Seyyed-Kalantari et al. [20,21] showed that state-of-the-art classifiers consistently and selective underdiagnosed historically underserved patient populations, such as non-white and female patients. Later, Zhang et al. [26] investigated a range of possible bias mitigation techniques, finding that simple group balancing still appears to be the most successful mitigation technique; this result has also been confirmed in other contexts [10]. Taken together, these results emphasize the importance of well-representative datasets. However, as Larrazabal et al. [14] had shown, group balancing alone cannot alleviate all performance differences, thus emphasizing the urgent need for more nuanced investigations into the *sources* of bias to enable successful bias mitigation.

In the wake of these important studies, there have been several investigations into potential sources of bias in chest X-ray datasets. In response to the study of Seyyed-Kalantari et al. [21], Bernhardt et al. [4] and Glocker et al. [8] pointed out the importance of properly accounting for confounding factors in bias analyses, such as age and disease distributions between patient groups. Bernhardt et al. [4] moreover underlined the challenge of properly evaluating performance differences if *label biases* affect both the training and test sets. There is reason for concern in this regard, since multiple studies have reported high error rates in (NLP-derived [11]) chest X-ray disease labels [22,26]. Our results in this study provide further independent evidence of widespread label errors in these databases.

Separately from these methodological issues, the performance differences between male and female patients led Ganz et al. [7] to speculate that an important cause might lie in female breasts occluding the recordings of important lung regions. Indeed, the confounding effect of female breasts on clinical chest x-ray interpretation is well-known [1,3,12] and physiologically plausible: additional breast tissue results in the underexposure of lung tissue. In this work, we take a step to systematically assess the effect of female breast tissue on machine learning-based chest X-ray diagnosis.

3   Methods

3.1   Datasets

We consider two datasets: ChestX-ray8 (NIH) [24] and CheXpert [11]. As the NIH dataset only contains frontal images, we also only use those views from the CheXpert to enable a fair comparison, resulting in 112,120 recordings from 30,850 patients in the NIH dataset and 190,299 recordings from 64,224 patients in the CheXpert dataset. Both datasets slightly over-represent male subjects (54% males vs. 46% females in NIH, 56% males vs. 45% females in CheXpert); refer to table 1 in the supplementary material for further details.

3.2   Sampling Strategy

Motivation. We observe that in both datasets, *the number of recordings per patient is very uneven*, ranging from 1 to 89 (CheXpert) and 1 to 184 (NIH). In particular, less than 25% of controls and over 50% of patients have more than 5 scans. This results in few patients with many recordings strongly influencing the training process and the final model, as well as strong distribution shifts between different data splits.

In addition, like outlined above, it has been shown that *disease labels automatically derived from patient records are unreliable* [26]. Particularly, it has been observed [2] that the commonly used text mining method worked poorly with the hospital record of "no change from previous", which would be wrongly marked as "no finding". Labels might be especially unreliable in patients with many recordings.

For these reasons outlined above, we select only one sample from each patient to avoid over-representation with a preference for positive labelled samples.

Principles Underlying the Proposed Sampling Strategy. We designed our sampling strategy to conform with the following principles:

- Utilize one sample per patient: reducing risk of distribution shift between splits and over-reliance on individual subjects.
- Prioritize diseased samples when selecting the single sample per patient: reducing risk of label bias towards "no finding" like outlined above.
- Keep disease prevalence constant across splits: reducing risk of distribution shift between splits.
- Allow disease prevalence to vary between protected groups: ensuring that our assessment is realistic, by utilizing the group-specific prevalence.
- For training and validation, draw a fixed-size sample with a predefined percentage of female subjects (0%, 50% or 100%): enabling an assessment of the influence of training set composition on model performance. (For testing, we always draw a fixed-size sample with an equal number of males/females.)
- Use an identical test set when evaluating the same training split at different gender ratios: enabling more reliable performance comparisons.

For further details on our sampling scheme, refer to Algorithm 1 in the supplementary material.

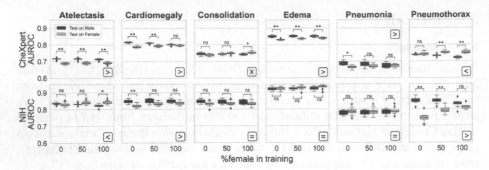

Fig. 2. Model performance (AUROC) in male (blue) and female (yellow) test subjects for training datasets with differing gender representation. Results are only shown for the six disease labels that are present in both datasets; for results on the remaining labels refer to the supplementary material. Icons in the bottom right of each plot indicate the observed performance trends across gender ratios; refer to Sect. 4.1. Statistical significance in this work is all based on Mann-Whitney U test, denoted by ** ($p \leq 0.001$), * ($0.001 < p \leq 0.01$) and ns (not significant, $p > 0.01$), etc. (Color figure online)

3.3   Experimental Settings

All experiments are carried out using the PyTorch framework with a pretrained ResNet50 [9] and the Adam optimizer [13] (learning rate 10^{-6}, 20 epochs, batch size 64). We split the datasets into 60%/10%/30% train, validation, and test sets per gender ten times based on disease prevalence, and then train separate single-label classifiers for all disease labels in both datasets. During training, data augmentation is applied by random horizontal flipping, rotation (degree $\leq 15°$) and scaling (from 0.9 to 1.1) with a probability of 0.5 each. Following [14], experiments are run on three different gender ratios, i.e. 0%, 50% and 100% females. Performances are evaluated by the area under the receiver-operating characteristic curve (AUROC).

4   Results

4.1   Model Performance Across Diseases, Gender Ratios, and Datasets

Figure 2 displays male and female test subject performance under varied in-training gender ratios for the common 6 disease labels in both datasets. More-over, a stylized summary of the observed trends across gender ratios ('×', '>', '<', or '=') is marked in the figure. We highlight three key observations.

Dataset Imbalance is Not the Sole Cause of Performance Differences. In some dataset–disease combinations, such as Pneumothorax–NIH, males keep outperforming females regardless of the proportion of women in the training set. Similar

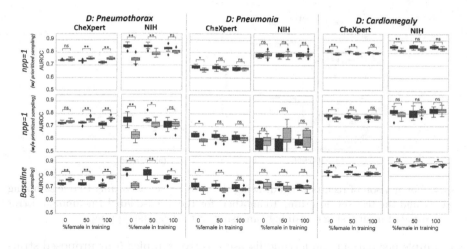

Fig. 3. Comparison of results from 10 varied splits when training and testing using three different sampling strategies for three disease labels. Three rows present one sample per patient (**npp=1**) with/without prioritizing diseased samples, and the original setting from [14] which uses all the samples with multi-label learning. Blue box represents male performance and yellow refers to female. (Color figure online)

and opposite trends (males having worse performance than females across gender ratios) could be observed in other disease labels, marked by '>' and '<' in Fig. 2.

If performance differences were caused solely by training set imbalance, then the majority group should consistently outperform the minority group (resulting in an '×' trend shape), which is not observed in most of the cases.

Performance Trends Sometimes Differ Strongly Between Datasets for the Same Diseases. Consider again the case of Pneumothorax classification, and compare the results between the two datasets: On NIH, males demonstrate significantly better performance regardless of training gender distribution, while on CheXpert, the trend has reversed completely. This suggests that there are dataset-specific factors unrelated to fundamental biological sex differences that strongly influence performance disparities.

Higher Performance and Higher Variance on NIH Compared to CheXpert. Across all six shared disease labels, test-set prediction performance is considerably better on NIH compared to CheXpert, despite the smaller NIH training set size. At the same time, the NIH performance also shows a larger variance across diseases, which appears likely to be related to the smaller test set size.

4.2   Comparison of Different Sampling Strategies

To analyze the effect of our proposed sampling strategy, we first compare the results when training and evaluating using three different sampling strategies: 1)

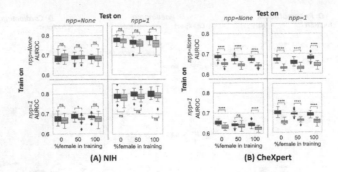

Fig. 4. Comparison of results when drawing the training set according to one sampling strategy and the test set according to another, for the disease label Pneumonia. Error bars computed from a single training run using test set bootstrapping.

one sample per patient, preferring disease-positive samples (our proposed strategy), 2) one sample per patient, drawing the sample randomly and not preferring disease-positive samples, and 3) using all samples per patient with the same split and multi-label training as [14]. Note that in the latter case, we use larger training set sizes. As shown in Fig. 3, in both datasets, test set performance is consistently and considerably improved when using the disease-preferring sampling strategy over a random sampling of one recording per patient; we interpret this as evidence of strong label noise in the "no finding" label like discussed in Sect. 3.2 and like reported previously in other datasets automatically labeled using the CheXpert labeller [26]. The difference between the two setups is especially strong in the NIH dataset, leading to the hypothesis that NIH dataset might suffer from more widespread label errors compared to CheXpert.

Since label noise confounds not only training but also evaluation, we also present results for one disease when the training on one strategy and the test set according to another (Fig. 4). Due to the potential data leakage through train and test set, instead of using the splits from [14], we implemented the experiments with all samples (npp=None) using modified proposed strategy without sampling. The results indicate label errors are more likely occur in NIH than CheXpert, as the results in CheXpert are more stable across sampling strategies while the performance drops when testing on all samples in NIH.

4.3   Breast Cropping Does Not Mitigate Gender Biases

To assess specifically whether differences in male and female breast physiology account for the observer performance differences, we perform an additional experiment. We simply crop the lower two fifths of each recording to ensure that the images contain only the parts above the breast for both genders. An illustration and the results of this experiment are provided in Fig. 5. Compared to

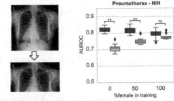

Fig. 5. The illustration of cropped chest x-rays and the results.

Fig. 2, while overall performance drops slightly for both genders, this intervention does not close the performance gap between both genders.

4.4   Dataset Bias v.s. Model Bias

To further investigate the effect of dataset-specific factors, we also evaluated classifiers trained for one disease on CheXpert using NIH test sets, and vice versa (Fig. 6). We observe that large male–female performance gaps only arise when models are tested on NIH, regardless of which dataset the model is trained on.

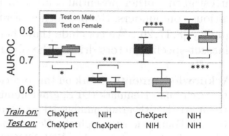

Fig. 6. Inter-dataset inference (50% female in training, *Pneumothorax*) with boost-strapping.

5   Discussion and Conclusions

We interpret the results of this work and address the key findings as follows.

Supporting Evidence of Frequent Label Errors in the "No Finding" Label. In agreement with previous reports on other datasets [26], our analyses provide supporting evidence of the existing hypothesis on frequent mislabeling of records as "no finding" using the proposed simple and easily applicable sampling method.

Male–Female Performance Gaps are Influenced, but Not Fully Explained, by Training Set Representation. In line with the previous results of Larrazabal et al. [14], we find that training set representation does (in some diseases strongly) influence male and female model performance. It does not, however, fully explain the performance gaps: in some cases, even when trained on a fully female dataset, models still perform worse on women (and vice versa).

Male–Female Performance Gaps are Not Primarily Caused by Breast Shadows. It had been previously hypothesized [7] that breast shadows might play an essential role in gender bias in ML-based chest X-ray diagnosis, which is not supported by our findings: breast cropping does not mitigate the performance gaps. Additionally, the varied bias trends between datasets also contradict this hypothesis.

Within Our Studies, We Do Not Find Evidence that Biological Differences Drive the Male–Female Performance Gaps. The performance gaps should be expected to be consistent across datasets if biological sex differences were the main driver of performance gaps, which has not been observed in this work. However, we do observe high noise levels in the datasets, which might mask more subtle differences.

Dataset-Specific Factors Influence Male–Female Performance Gaps. As the previous hypotheses on the origins of male–female performance gaps have been tentatively rejected considering the results, further research should continue investigating other potential sources of bias. Those could be the distribution of various confounders, the prevalence of further label errors, or differing recording quality [8,17,26]. We conclude that there must be further, at present unknown dataset-specific factors driving the observed performance gaps.

Acknowledgements. Work on this project was partially funded by the Independent Research Fund Denmark (DFF, grant number 9131-00097B), Denmark's Pioneer Centre for AI (DNRF grant number P1), a Google Award for inclusion research, the Novo Nordisk Foundation through the Center for Basic Machine Learning Research in Life Science (MLLS, grant number NNF20OC0062606).

References

1. Chest. https://radiologykey.com/chest-11/. Accessed 20 June 2023
2. Half a million x-rays! First impressions of the Stanford and MIT chest x-ray datasets. https://laurenoakdenrayner.com/2019/02/25/half-a-million-x-rays-first-impressions-of-the-stanford-and-mit-chest-x-ray-datasets/. Accessed 20 June 2023
3. Alexander, C.: The elimination of confusing breast shadows in chest radiography. Australas. Radiol. **2**(2), 107–108 (1958)
4. Bernhardt, M., Jones, C., Glocker, B.: Potential sources of dataset bias complicate investigation of under diagnosis by machine learning algorithms. Nat. Med. **28**(6), 1157–1158 (2022). https://doi.org/10.1038/s41591-022-01846-8
5. Burlina, P., Joshi, N., Paul, W., Pacheco, K.D., Bressler, N.M.: Addressing artificial intelligence bias in retinal diagnostics. Transl. Vis. Sci. Technol. **10**(2), 13–13 (2021)
6. Daneshjou, R., et al.: Disparities in dermatology AI performance on a diverse, curated clinical image set. Sci. Adv. **8**(32) (2022). https://doi.org/10.1126/sciadv.abq6147
7. Ganz, M., Holm, S.H., Feragen, A.: Assessing bias in medical AI. In: Workshop on Interpretable ML in Healthcare at International Conference on Machine Learning (ICML) (2021)
8. Glocker, B., Jones, C., Bernhardt, M., Winzeck, S.: Algorithmic encoding of protected characteristics in chest x-ray disease detection models. eBioMedicine **89**, 104467 (2023). https://doi.org/10.1016/j.ebiom.2023.104467
9. He, K., Zhang, X., Ren, S., Sun, J.: Deep residual learning for image recognition. In: Proceedings of the IEEE Conference on Computer Vision and Pattern Recognition, pp. 770–778 (2016)
10. Idrissi, B.Y., Arjovsky, M., Pezeshki, M., Lopez-Paz, D.: Simple data balancing achieves competitive worst-group-accuracy. In: Schölkopf, B., Uhler, C., Zhang, K. (eds.) Proceedings of the First Conference on Causal Learning and Reasoning. Proceedings of Machine Learning Research, vol. 177, pp. 336–351. PMLR. https://proceedings.mlr.press/v177/idrissi22a.html
11. Irvin, J., et al.: Chexpert: a large chest radiograph dataset with uncertainty labels and expert comparison. In: Proceedings of the AAAI Conference on Artificial Intelligence, vol. 33, pp. 590–597 (2019)

12. Jenkins, P.: Making Sense of the Chest X-Ray: A hands-on guide. CRC Press, Boca Raton (2013)
13. Kingma, D.P., Ba, J.: Adam: A method for stochastic optimization. arXiv preprint arXiv:1412.6980 (2014)
14. Larrazabal, A.J., Nieto, N., Peterson, V., Milone, D.H., Ferrante, E.: Gender imbalance in medical imaging datasets produces biased classifiers for computer-aided diagnosis. Proc. Natl. Acad. Sci. **117**(23), 12592–12594 (2020)
15. Pakzad, A., Abhishek, K., Hamarneh, G.: CIRCLe: color invariant representation learning for unbiased classification of skin lesions. In: Karlinsky, L., Michaeli, T., Nishino, K. (eds.) ECCV 2022. LNCS, vol. 13804, pp. 203–219. Springer, Cham (2023). https://doi.org/10.1007/978-3-031-25069-9_14
16. Petersen, E., et al.: Feature robustness and sex differences in medical imaging: a case study in MRI-based Alzheimer's disease detection. In: Wang, L., Dou, Q., Fletcher, P.T., Speidel, S., Li, S. (eds.) MICCAI 2022. LNCS, vol. 13431, pp. 88–98. Springer, Cham (2022). https://doi.org/10.1007/978-3-031-16431-6_9
17. Petersen, E., Holm, S., Ganz, M., Feragen, A.: The path toward equal performance in medical machine learning. Patterns **4**(7) (2023). https://doi.org/10.1016/j.patter.2023.100790
18. Puyol-Antón, E., et al.: Fairness in cardiac magnetic resonance imaging: assessing sex and racial bias in deep learning-based segmentation. Front. Cardiovasc. Med. **9**, 859310 (2022)
19. Rajpurkar, P., et al.: Chexnet: radiologist-level pneumonia detection on chest x-rays with deep learning (2017). https://doi.org/10.48550/ARXIV.1711.05225
20. Seyyed-Kalantari, L., Liu, G., McDermott, M., Chen, I.Y., Ghassemi, M.: CheXclusion: fairness gaps in deep chest x-ray classifiers. In: BIOCOMPUTING 2021: Proceedings of the Pacific Symposium, pp. 232–243. World Scientific (2020)
21. Seyyed-Kalantari, L., Zhang, H., McDermott, M.B.A., Chen, I.Y., Ghassemi, M.: Underdiagnosis bias of artificial intelligence algorithms applied to chest radiographs in under-served patient populations. Nat. Med. **27**(12), 2176–2182 (2021). https://doi.org/10.1038/s41591-021-01595-0
22. Smit, A., Jain, S., Rajpurkar, P., Pareek, A., Ng, A., Lungren, M.: Combining automatic labelers and expert annotations for accurate radiology report labeling using BERT. In: Proceedings of the 2020 Conference on Empirical Methods in Natural Language Processing (EMNLP). Association for Computational Linguistics (2020). https://doi.org/10.18653/v1/2020.emnlp-main.117
23. Stanley, E.A., Wilms, M., Mouches, P., Forkert, N.D.: Fairness-related performance and explainability effects in deep learning models for brain image analysis. J. Med. Imaging **9**(6), 061102–061102 (2022)
24. Wang, X., Peng, Y., Lu, L., Lu, Z., Bagheri, M., Summers, R.M.: Chestx-ray8: hospital-scale chest x-ray database and benchmarks on weakly-supervised classification and localization of common thorax diseases. In: Proceedings of the IEEE Conference on Computer Vision and Pattern Recognition, pp. 2097–2106 (2017)
25. Wu, Y., Zeng, D., Xu, X., Shi, Y., Hu, J.: FairPrune: achieving fairness through pruning for dermatological disease diagnosis. In: Wang, L., Dou, Q., Fletcher, P.T., Speidel, S., Li, S. (eds.) MICCAI 2022. LNCS, vol. 13431, pp. 743–753. Springer, Cham (2022). https://doi.org/10.1007/978-3-031-16431-6_70
26. Zhang, H., Dullerud, N., Roth, K., Oakden-Rayner, L., Pfohl, S., Ghassemi, M.: Improving the fairness of chest x-ray classifiers. In: Conference on Health, Inference, and Learning, pp. 204–233. PMLR (2022)

27. Zietlow, D., et al.: Leveling down in computer vision: Pareto inefficiencies in fair deep classifiers. In: Proceedings of the IEEE/CVF Conference on Computer Vision and Pattern Recognition (CVPR), pp. 10410–10421

Bayesian Uncertainty-Weighted Loss for Improved Generalisability on Polyp Segmentation Task

Rebecca S. Stone[✉], Pedro E. Chavarrias-Solano, Andrew J. Bulpitt, David C. Hogg, and Sharib Ali[✉] [ID]

School of Computing, University of Leeds, Leeds LS2 9JT, UK
{sc16rsmy,s.s.ali}@leeds.ac.uk

Abstract. While several previous studies have devised methods for segmentation of polyps, most of these methods are not rigorously assessed on multi-center datasets. Variability due to appearance of polyps from one center to another, difference in endoscopic instrument grades, and acquisition quality result in methods with good performance on in-distribution test data, and poor performance on out-of-distribution or underrepresented samples. Unfair models have serious implications and pose a critical challenge to clinical applications. We adapt an implicit bias mitigation method which leverages Bayesian epistemic uncertainties during training to encourage the model to focus on underrepresented sample regions. We demonstrate the potential of this approach to improve generalisability without sacrificing state-of-the-art performance on a challenging multi-center polyp segmentation dataset (PolypGen) with different centers and image modalities.

1 Introduction

Colorectal cancer (CRC) is the third most common cancer worldwide [27] with early screening and removal of precancerous lesions (colorectal adenomas such as "polyps") suggesting longer survival rates. While surgical removal of polyps (polypectomy) is a standard procedure during colonoscopy, detecting these and their precise delineation, especially for sessile serrated adenomas/polyps, is extremely challenging. Over a decade, advanced computer-aided methods have been developed and most recently machine learning (ML) methods have been widely developed by several groups. However, the translation of these technologies in clinical settings has still not been fully achieved. One of the main reasons is the generalisability issues with the ML methods [2]. Most techniques are built and adapted over carefully curated datasets which may not match the natural occurrences of the scene during colonoscopy.

Recent literature demonstrates how intelligent models can be systematically unfair and biased against certain subgroups of populations. In medical imaging, the problem is prevalent across various image modalities and target tasks; for example, models trained for lung disease prediction [25], retinal diagnosis [6],

S. Wesarg et al. (Eds.): CLIP/FAIMI/EPIMI 2023, LNCS 14242, pp. 153–162, 2023.
https://doi.org/10.1007/978-3-031-45249-9_15

cardiac MR segmentation [23], and skin lesion detection [1,17] are all subject to biased performance against one or a combination of underrepresented gender, age, socio-economic, and ethnic subgroups. Even under the assumption of an ideal sampling environment, a perfectly balanced dataset does not ensure unbiased performance as relative quantities are not solely responsible for bias [19,31]. This, and the scarcity of literature exploring bias mitigation for polyp segmentation in particular, strongly motivate the need for development and evaluation of mitigation methods which work on naturally occurring diverse colonoscopy datasets such as PolypGen [3].

2   Related Work

Convolutional neural networks have recently worked favourably towards the advancement of building data-driven approaches to polyp segmentation using deep learning. These methods [18,34] are widely adapted from the encoder-decoder U-Net [24] architecture. Moreover, addressing the problem of different polyp sizes using multi-scale feature pruning methods, such as atrous-spatial pyramid pooling in DeepLabV3 [8] or high-resolution feature fusion networks like HRNet [28] have been used by several groups for improved polyp segmentation. For example, MSRFNet [29] uses feature fusion networks between different resolution stages. Recent work on generalisability assessment found that methods trained on specific centers do not tend to generalise well on unseen center data or different naturally occurring modalities such as sequence colonoscopy data [2]. These performance gaps were reported to be large (drops of nearly 20%).

Out-of-distribution (OOD) generalisation and bias mitigation are challenging, open problems in the computer vision research community. While in the bias problem formulation, models wrongly correlate one or more spurious (non-core) features with the target task, the out-of-distribution problem states that test data is drawn from a separate distribution than the training data. Some degree of overlap between the two distributions in the latter formulation exists, which likely includes the core features. Regardless of the perspective, the two problems have clear similarities, and both result in unfair models which struggle to generalise for certain sub-populations. In the literature, many works focus on OOD detection, through normal or modified softmax outputs [13], sample uncertainty thresholds from Bayesian, ensemble, or other models [7,14,20], and distance measures in feature latent space [12]. Other approaches tackle the more difficult problem of algorithmic mitigation through disentangled representation learning, architectural and learning methods, and methods which optimise for OOD generalisability directly [26].

Similarly, several categories of bias mitigation methods exist. Some methods rely on two or more models, one encouraged to learn the biased correlations of the majority, and the other penalised for learning the correlations of the first [16,21]. Other approaches modify the objective loss functions to reward learning core rather than spurious features [22,33], or by neutralising representations to remove learned spurious correlations [10]. Others use data augmentation [6], or explore implicit versions of up-weighting or re-sampling underrepresented samples by discovering sparse areas of the feature space [4] or dynamically identifying samples more likely to be underrepresented [30]. De-biasing methods leveraging Bayesian model uncertainties [5,15,30] provide the added benefits of uncertainty estimations which are useful in clinical application for model interpretability and building user confidence.

To tackle the generalisability problem for polyp segmentation, we consider the diversity of features in a multi-centre polyp dataset [3]. Our contributions can be listed as: 1) adapting an implicit bias mitigation strategy in [30] from a classification to a segmentation task; 2) evaluating the suitability of this approach on three separate test sets which have been shown to be challenging generalisation problems. Our experiments demonstrate that our method is comparable and in many cases even improves the performance compared to the baseline state-of-the-art segmentation method while decreasing performance discrepancies between different test splits.

3 Method

The encoder-decoder architecture for semantic segmentation has been widely explored in medical image analysis. In our approach we have used DeepLabV3 [9] as baseline model that has SOTA performance on the PolypGen dataset [3]. We then apply a probabilistic model assuming a Gaussian prior on all trainable weights (both encoder and decoder) that are updated to the posterior using the training dataset. For the Bayesian network with parameters $\boldsymbol{\theta}$, posterior $p(\boldsymbol{\theta} \mid D)$, training data with ground truth segmentation masks $D = (X, Y)$, and sample x_i, the predictive posterior distribution for a given ground truth segmentation mask y_i can be written as:

$$p(y_i \mid D, x_i) = \int p(y_i \mid \boldsymbol{\theta}, x_i) p(\boldsymbol{\theta} \mid D) d\boldsymbol{\theta} \tag{1}$$

While Monte-Carlo dropout [11] at test-time is a popular approach to approximating this intractable integral, we choose stochastic gradient Monte-Carlo sampling MCMC (SG-MCMC [32]) for a better posterior. Stochastic gradient over mini-batches includes a noise term approximating the gradient over the whole training distribution. Furthermore, the cyclical learning rate schedule introduced in [35] known as cyclical SG-MCMC, or cSG-MCMC, allows for faster convergence and better exploration of the multimodal distributions prevalent in deep neural networks. Larger learning step phases provide a warm restart to the subsequent smaller steps in the sampling phases.

The final estimated posterior of the Bayesian network, $\boldsymbol{\Theta} = \{\boldsymbol{\theta}_1, ...\boldsymbol{\theta}_M\}$, consists of M moments sampled from the posterior taken during the sampling phases of each learning cycle. With functional model $\boldsymbol{\Phi}$ representing the neural network, the approximate predictive mean μ_i for one sample x_i is:

$$\mu_i \approx \frac{1}{M} \sum_{m=1}^{M} \boldsymbol{\Phi}_{\theta_m}(x_i) \tag{2}$$

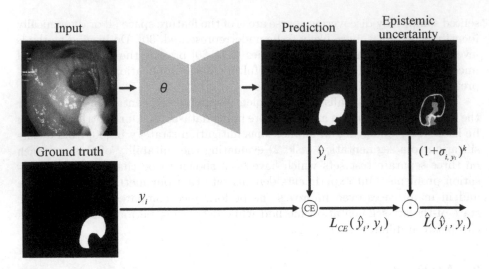

Fig. 1. Pixel-wise weighting of cross entropy (CE) loss contribution based on epistemic uncertainty maps for each training sample; the model is encouraged to focus on regions for which it is more uncertain.

We can derive a segmentation prediction mask $\hat{y}_i$ from μ_i by taking the maximum output between the foreground and background channels. The epistemic uncertainty mask corresponding to this prediction (Eq. 3) represents the *model uncertainty* for the predicted segmentation mask, the variance in the predictive distribution for that sample.

$$\sigma_i \approx \frac{1}{M} \sqrt{\sum_{m=1}^{M} \left(\boldsymbol{\Phi}_{\theta_m}(x_i) - \mu_i \right)^2} \tag{3}$$

We add epistemic uncertainty-weighted sample loss [30] that identifies high-uncertainty sample regions during training. It also scales the pixel-wise contribution of these regions to the loss computation via a simple weighting function (Eq. 4). This unreduced cross-entropy loss is then averaged over each image and batch (see Fig. 1).

$$\hat{L}(\hat{y}_i, y_i) = L_{CE}(\hat{y}_i, y_i) * (1.0 + \sigma_{i,y_i})^{\kappa} \tag{4}$$

The shift by a constant (1.0) normalises the values, ensuring that the lowest uncertainty samples are never irrelevant to the loss term. κ is a tunable debiasing parameter; $\kappa = 1$ being a normal weighting, whereas $\kappa \rightarrow \infty$ increases the importance of high-uncertainty regions. As too large a κ results in degraded performance due to overfitting, the optimal value is determined by validation metrics.

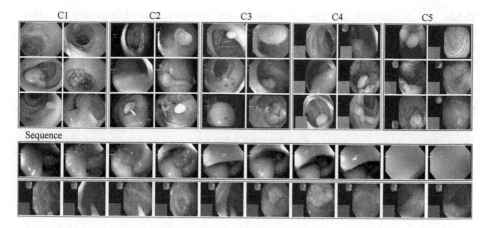

Fig. 2. Samples from the EndoCV2021 dataset; from (*top*) C1-5 single frames and (*bottom*) C1-5-SEQ; (*top*) highlights the data distribution of each center (C1-C5), which consists of curated frames with well-defined polyps; (*bottom*) demonstrates the variability of sequential data due to the presence of artifacts, occlusions, and polyps with different morphology.

4 Experiments and Results

4.1 Dataset and Experimental Setup

PolypGen [3] is an expert-curated polyp segmentation dataset comprising of both single frames and sequence frames (frames sampled at every 10 frames from video) from over 300 unique patients across six different medical centers. The natural data collection format is video from which single frames and sequence data are hand-selected. The single frames are clearer, better quality, and with polyps in each frame including polyps of various sizes (10k to 40k pixels), and also potentially containing additional artifacts such as light reflections, blue dye, partial view of instruments, and anatomies such as colon linings and mucosa covered with stool, and air bubbles (Fig. 2). The sequence frames are more challenging and contain more negative samples without a polyp and more severe artifacts, which are a natural occurrence in colonoscopy. Our training set includes 1449 single frames from five centers (C1 to C5) and we evaluate on the three tests sets used for generalisability assessment in literature [2, 3].

The first test dataset has 88 single frames from an unseen center C6 (C6-SIN), and the second has 432 frames from sequence data also from unseen center C6 (C6-SEQ). Here, the first test data (C6-SIN) comprises of hand selected images from the colonoscopy videos while the second test data (C6-SEQ) includes short sequences (every 10th frame of video) mimicking the natural occurrence of the procedure. The third test dataset includes 124 frames but from seen centers

Table 1. Evaluation of the state-of-the-art deterministic DeepLabV3+, Bay-DeepLabV3+, and our proposed BayDeepLabV3+Unc, showing mean and standard deviations across the respective test dataset samples. **First** and second best results for each metric per dataset formatted.

Dataset	Method	JAC	Dice	F2	PPV	Recall	Accuracy
C6-SIN	SOTA	0.738 ± 0.3	0.806 ± 0.3	0.795 ± 0.3	$\mathbf{0.912 \pm 0.2}$	0.793 ± 0.3	0.979 ± 0.1
	BayDeepLabV3+	0.721 ± 0.3	0.790 ± 0.3	$\mathbf{0.809 \pm 0.3}$	0.836 ± 0.2	$\mathbf{0.843 \pm 0.3}$	0.977 ± 0.1
	Ours	$\mathbf{0.740 \pm 0.3}$	$\mathbf{0.810 \pm 0.3}$	0.804 ± 0.3	0.903 ± 0.1	0.806 ± 0.3	$\mathbf{0.977 \pm 0.1}$
C1-5-SEQ	SOTA	$\mathbf{0.747 \pm 0.3}$	$\mathbf{0.819 \pm 0.3}$	$\mathbf{0.828 \pm 0.3}$	0.877 ± 0.2	0.852 ± 0.3	0.960 ± 0.0
	BayDeepLabV3+	0.708 ± 0.3	0.778 ± 0.3	0.805 ± 0.3	0.784 ± 0.3	$\mathbf{0.885 \pm 0.2}$	$\mathbf{0.963 \pm 0.0}$
	Ours	0.741 ± 0.3	0.810 ± 0.3	0.815 ± 0.3	$\mathbf{0.888 \pm 0.2}$	0.836 ± 0.3	0.961 ± 0.0
C6-SEQ	SOTA	0.608 ± 0.4	0.676 ± 0.4	0.653 ± 0.4	0.845 ± 0.3	0.719 ± 0.3	0.964 ± 0.1
	BayDeepLabV3+	0.622 ± 0.4	0.682 ± 0.4	0.669 ± 0.4	0.802 ± 0.3	$\mathbf{0.764 \pm 0.3}$	0.965 ± 0.1
	Ours	$\mathbf{0.637 \pm 0.4}$	$\mathbf{0.697 \pm 0.4}$	$\mathbf{0.682 \pm 0.4}$	$\mathbf{0.858 \pm 0.3}$	0.741 ± 0.3	$\mathbf{0.967 \pm 0.1}$

C1 - C5; however, these are more challenging as they contain both positive and negative samples with different levels of corruption that are not as present in the curated single frame training set. As no C6 samples nor sequence data are present in the training data, these test sets present a challenging generalisability problem.[1]

Training was carried out on several IBM Power 9 dual-CPU nodes with 4 NVIDIA V100 GPUs. Validation metrics were used to determine optimal models for all experiments with hyper-parameters chosen via grid search. Perhaps due to some frames containing very large polyps with high uncertainties, we found that the gradients of Bayesian models with uncertainty-weighted loss (BayDeepLabV3+Unc) occasionally exploded during the second learning cycle, and clipping the absolute gradients at 1.0 for all weights prevented this issue. All Bayesian DeepLabV3+ (BayDeepLabV3+) models had 2 cycles, a cycle length of 550 epochs, noise control parameter $\alpha = 0.9$, and an initial learning rate of 0.1. For BayDeepLabV3+Unc, we found optimal results with de-biasing tuning parameter $\kappa = 3$. Posterior estimates for BayDeepLabV3+ and BayDeepLabV3+Unc included 6 and 4 samples per cycle, respectively.

4.2 Results

We use the state-of-the-art deterministic model[2] and checkpoints to evaluate on the three test sets, and compare against the baseline Bayesian model BayDeepLabV3+ and BayDeepLabV3+Unc with uncertainty-weighted loss.

---

[1] C1-5-SEQ and C6-SEQ data are referred to as DATA3 and DATA4, respectively, in [2].

[2] https://github.com/sharib-vision/PolypGen-Benchmark.

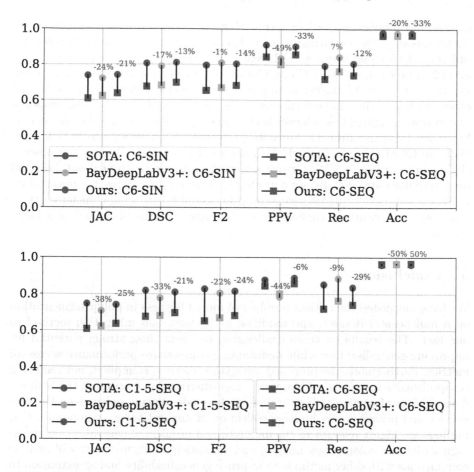

Fig. 3. Performance gaps of the three models (state-of-the-art deterministic DeepLabV3+, BayDeepLabV3+, and BayDeepLabV3+Unc) between the three different test sets; *(top)* comparing performance on single vs. sequence frames from out-of-distribution test set C6 (C6-SIN vs. C6-SEQ), and *(bottom)* sequence frames from C1 - C5 vs. unseen C6 (C1-5-SEQ vs. C6-SEQ). The subtext above bars indicates the percent decrease in performance gap compared to SOTA; a larger percent decrease and shorter vertical bar length indicate better generalisability.

We report results for Jaccard index (JAC), Dice coefficient (Dice), F_β-measure with $\beta = 2$ (F2), positive predictive value (PPV), recall (Rec), and mean pixel-wise accuracy (Acc). PPV in particular has high clinical value as it indicates a more accurate delineation for the detected polyps. Recall and mean accuracy are less indicative since the majority of frames are background in the segmentation task and these metrics do not account for false positives. A larger number of false positive predictions can cause inconvenience to endoscopists during colonoscopic procedure and hence can hinder clinical adoption of methods. Figure 3 illustrates that our approach maintains SOTA performance across most metrics and various test settings, even outperforming in some cases; simultaneously, the performance gaps between different test sets representing different challenging features (1) image modalities (single vs. sequence frames) and (2) source centers (C1 - C5 vs. C6) are significantly decreased. Simply turning the SOTA model Bayesian improves the model's ability to generalise, yet comes with a sacrifice in performance across metrics and datasets. Our proposed uncertainty-weighted loss achieves better generalisability without sacrificing performance (also see Table 1). We note performance superiority to SOTA especially on C6-SEQ, approximately 3% improvement on Dice. We can also observe slight improvement on PPV for test sets with sequence (both held-out data and unseen centre data). Finally, we note that in clinical applications, the uncertainty maps for samples during inference could be useful for drawing clinicians' attention towards potentially challenging cases, increasing the likelihood of a fairer outcome.

5   Conclusion

We have motivated the critical problem of model fairness in polyp segmentation on a multi-center dataset, and modified a Bayesian bias mitigation method to our task. The results on three challenging test sets show strong potential for improving generalisability while maintaining competitive performance across all metrics. Furthermore, the proposed mitigation method is implicit, not requiring comprehensive knowledge of biases or out-of-distribution features in the training data. This is of particular importance in the medical community given the sensitivity and privacy issues limiting collection of annotations and metadata. Our findings are highly relevant to the understudied problem of generalisation across high variability colonoscopy images, and we anticipate future work will include comparisons with other methods to improve generalisability and an extension to the approach. We also anticipate having access to additional test data for more in-depth analysis of the results.

Acknowledgements. R. S. Stone is supported by Ezra Rabin scholarship.

References

1. Abbasi-Sureshjani, S., Raumanns, R., Michels, B.E.J., Schouten, G., Cheplygina, V.: Risk of training diagnostic algorithms on data with demographic bias. In: Cardoso, J., et al. (eds.) IMIMIC/MIL3ID/LABELS -2020. LNCS, vol. 12446, pp. 183–192. Springer, Cham (2020). https://doi.org/10.1007/978-3-030-61166-8_20
2. Ali, S., Ghatwary, N., Jha, D., Isik-Polat, E., Polat, G., Yang, o.: Assessing generalisability of deep learning-based polyp detection and segmentation methods through a computer vision challenge. arXiv preprint: arXiv:2202.12031 (2022)
3. Ali, S., et al.: A multi-centre polyp detection and segmentation dataset for generalisability assessment. Sci. Data **10**(1), 75 (2023)

4. Amini, A., Soleimany, A.P., Schwarting, W., Bhatia, S.N., Rus, D.: Uncovering and mitigating algorithmic bias through learned latent structure. In: Proceedings of the 2019 AAAI/ACM Conference on AI, Ethics, and Society, pp. 289–295 (2019)
5. Branchaud-Charron, F., Atighehchian, P., Rodríguez, P., Abuhamad, G., Lacoste, A.: Can active learning preemptively mitigate fairness issues? arXiv preprint: arXiv:2104.06879 (2021)
6. Burlina, P., Joshi, N., Paul, W., Pacheco, K.D., Bressler, N.M.: Addressing artificial intelligence bias in retinal diagnostics. Transl. Vis. Sci. Technol. **10**(2), 13–13 (2021)
7. Cao, S., Zhang, Z.: Deep hybrid models for out-of-distribution detection. In: Proceedings of the IEEE/CVF Conference on Computer Vision and Pattern Recognition, pp. 4733–4743 (2022)
8. Chen, L.C., Papandreou, G., Kokkinos, I., Murphy, K., Yuille, A.L.: DeepLab: semantic image segmentation with deep convolutional nets, Atrous convolution, and fully connected CRFs. IEEE Trans. Pattern Anal. Mach. Intell. **40**(4), 834–848 (2018)
9. Chen, L.-C., Zhu, Y., Papandreou, G., Schroff, F., Adam, H.: Encoder-decoder with Atrous separable convolution for semantic image segmentation. In: Ferrari, V., Hebert, M., Sminchisescu, C., Weiss, Y. (eds.) ECCV 2018. LNCS, vol. 11211, pp. 833–851. Springer, Cham (2018). https://doi.org/10.1007/978-3-030-01234-2_49
10. Du, M., Mukherjee, S., Wang, G., Tang, R., Awadallah, A., Hu, X.: Fairness via representation neutralization. In: Advances in Neural Information Processing Systems, vol. 34, pp. 12091–12103 (2021)
11. Gal, Y., Ghahramani, Z.: Dropout as a Bayesian approximation: representing model uncertainty in deep learning. In: International Conference on Machine Learning, pp. 1050–1059. PMLR (2016)
12. Gonzalez, C., Gotkowski, K., Bucher, A., Fischbach, R., Kaltenborn, I., Mukhopadhyay, A.: Detecting when pre-trained nnU-Net models fail silently for Covid-19 lung lesion segmentation. In: de Bruijne, M., et al. (eds.) MICCAI 2021. LNCS, vol. 12907, pp. 304–314. Springer, Cham (2021). https://doi.org/10.1007/978-3-030-87234-2_29
13. Hendrycks, D., Mazeika, M., Kadavath, S., Song, D.: Using self-supervised learning can improve model robustness and uncertainty. In: Advances in Neural Information Processing Systems, vol. 32 (2019)
14. Jungo, A., Balsiger, F., Reyes, M.: Analyzing the quality and challenges of uncertainty estimations for brain tumor segmentation. Front. Neurosci. **14**, 282 (2020)
15. Khan, S., Hayat, M., Zamir, S.W., Shen, J., Shao, L.: Striking the right balance with uncertainty. In: Proceedings of the IEEE/CVF Conference on Computer Vision and Pattern Recognition, pp. 103–112 (2019)
16. Kim, N., Hwang, S., Ahn, S., Park, J., Kwak, S.: Learning debiased classifier with biased committee. arXiv preprint: arXiv:2206.10843 (2022)
17. Li, X., Cui, Z., Wu, Y., Gu, L., Harada, T.: Estimating and improving fairness with adversarial learning. arXiv preprint: arXiv:2103.04243 (2021)
18. Mahmud, T., Paul, B., Fattah, S.A.: PolypSegNet: a modified encoder-decoder architecture for automated polyp segmentation from colonoscopy images. Comput. Biol. Med. **128**, 104119 (2021)
19. Mehrabi, N., Morstatter, F., Saxena, N., Lerman, K., Galstyan, A.: A survey on bias and fairness in machine learning. ACM Comput. Surv. (CSUR) **54**(6), 1–35 (2021)

20. Mehrtash, A., Wells, W.M., Tempany, C.M., Abolmaesumi, P., Kapur, T.: Confidence calibration and predictive uncertainty estimation for deep medical image segmentation. IEEE Trans. Med. Imaging **39**(12), 3868–3878 (2020)
21. Nam, J., Cha, H., Ahn, S., Lee, J., Shin, J.: Learning from failure: De-biasing classifier from biased classifier. In: Advances in Neural Information Processing Systems, vol. 33, pp. 20673–20684 (2020)
22. Pezeshki, M., Kaba, O., Bengio, Y., Courville, A.C., Precup, D., Lajoie, G.: Gradient starvation: a learning proclivity in neural networks. In: Advances in Neural Information Processing Systems, vol. 34, pp. 1256–1272 (2021)
23. Puyol-Antón, E., et al.: Fairness in cardiac MR image analysis: an investigation of bias due to data imbalance in deep learning based segmentation. In: de Bruijne, M., et al. (eds.) MICCAI 2021. LNCS, vol. 12903, pp. 413–423. Springer, Cham (2021). https://doi.org/10.1007/978-3-030-87199-4_39
24. Ronneberger, O., Fischer, P., Brox, T.: U-Net: convolutional networks for biomedical image segmentation. In: Navab, N., Hornegger, J., Wells, W.M., Frangi, A.F. (eds.) MICCAI 2015. LNCS, vol. 9351, pp. 234–241. Springer, Cham (2015). https://doi.org/10.1007/978-3-319-24574-4_28
25. Seyyed-Kalantari, L., Liu, G., McDermott, M., Chen, I.Y., Ghassemi, M.: CheXclusion: fairness gaps in deep chest X-ray classifiers. In: BIOCOMPUTING 2021: Proceedings of the Pacific Symposium, pp. 232–243. World Scientific (2020)
26. Shen, Z., et al.: Towards out-of-distribution generalization: a survey. arXiv preprint: arXiv:2108.13624 (2021)
27. Silva, J., Histace, A., Romain, O., Dray, X., Granado, B.: Toward embedded detection of polyps in WCE images for early diagnosis of colorectal cancer. Int. J. Comput. Assist. Radiol. Surg. **9**, 283–293 (2014)
28. Simonyan, K., Zisserman, A.: Very deep convolutional networks for large-scale image recognition. In: 3rd International Conference on Learning Representations, ICLR 2015, San Diego, CA, USA, 7–9 May 2015, Conference Track Proceedings (2015)
29. Srivastava, A., et al.: MSRF-Net: a multi-scale residual fusion network for biomedical image segmentation. IEEE J. Biomed. Health Inform. **26**(5), 2252–2263 (2022)
30. Stone, R.S., Ravikumar, N., Bulpitt, A.J., Hogg, D.C.: Epistemic uncertainty-weighted loss for visual bias mitigation. In: Proceedings of the IEEE/CVF Conference on Computer Vision and Pattern Recognition, pp. 2898–2905 (2022)
31. Wang, M., Deng, W.: Mitigating bias in face recognition using skewness-aware reinforcement learning. In: Proceedings of the IEEE/CVF Conference on Computer Vision and Pattern Recognition, pp. 9322–9331 (2020)
32. Welling, M., Teh, Y.W.: Bayesian learning via stochastic gradient Langevin dynamics. In: Proceedings of the 28th International Conference on Machine Learning (ICML-11), pp. 681–688. Citeseer (2011)
33. Xu, X., et al.: Consistent instance false positive improves fairness in face recognition. In: Proceedings of the IEEE/CVF Conference on Computer Vision and Pattern Recognition, pp. 578–586 (2021)
34. Yeung, M., Sala, E., Schönlieb, C.B., Rundo, L.: Focus U-Net: a novel dual attention-gated CNN for polyp segmentation during colonoscopy. Comput. Biol. Med. **137**, 104815 (2021)
35. Zhang, R., Li, C., Zhang, J., Chen, C., Wilson, A.G.: Cyclical stochastic gradient MCMC for Bayesian deep learning. In: International Conference on Learning Representations (2020)

Mitigating Bias in MRI-Based Alzheimer's Disease Classifiers Through Pruning of Deep Neural Networks

Yun-Yang Huang[(✉)], Venesia Chiuwanara, Chao-Hsuan Lin, and Po-Chih Kuo

National Tsing Hua University, Hsinchu, Taiwan
dan89092989@gapp.nthu.edu.tw, kuopc@cs.nthu.edu.tw

Abstract. With the increasing prevalence of Alzheimer's disease among the elderly population, the development of machine learning models for early identification is crucial. However, it has been observed that these models may exhibit inherent biases, leading to performance disparities across protected attributes such as age and gender. In this study, we propose a model pruning method aimed at reducing bias and achieving model fairness in a pre-trained neural network designed for Alzheimer's disease identification. Our pruning method takes into account both bias reduction and model performance. The results demonstrate that our proposed pruning method significantly reduces disparities between age and gender attribute compared to the baseline model. For disparity in age attribute, our method successfully reduces disparity across multiple metrics, such as true positive rate (-7.3%). Similarly, for gender attribute, our approach mitigated bias in evaluation metrics, such as true negative rate (-5.0%). The results also show that our method maintains or even improves the model's performance and outperforms other methods across most evaluation metrics. These compelling findings underscore the immense potential of our approach in effectively mitigating gender and age disparities in MRI-based Alzheimer's disease identification.

Keywords: Alzheimer's Disease · fairness · bias · model pruning

1 Introduction

Alzheimer's disease (AD) is a neurodegenerative disorder characterized by the degeneration of nerve cells and loss of brain tissue. Research has explored the potential of using machine learning (ML) models on MRI data for the diagnosis of AD. Dufumier et al. [1] combined y-aware metadata and contrastive learning for various binary classification tasks, including AD classification. Zhuang et al. [2] proposed a method inspired by solving a Rubik's Cube, specifically designed for 3-dimensional (3D) medical images. Liu et al. [3] trained a stacked autoencoder to learn hidden representations followed by a softmax output layer for classification.

While ML models have shown promise in Alzheimer's disease (AD) diagnosis using MRI data, it is essential to address potential biases in the training data. ML

© The Author(s), under exclusive license to Springer Nature Switzerland AG 2023
S. Wesarg et al. (Eds.): CLIP/FAIMI/EPIMI 2023, LNCS 14242, pp. 163–171, 2023.
https://doi.org/10.1007/978-3-031-45249-9_16

models heavily rely on the training data to minimize prediction errors, but biases related to race, gender, and age can impact predictions, leading to performance disparities. In some cases, the model may even worsen these biases, making it untrustworthy. To mitigate bias in ML models, various approaches have been proposed, including adversarial learning [4,5], model pruning [6,11], perturbation of loss functions [7], data preprocessing techniques [8,9], usage of multi-exit framework [12,13], and the use of multi-source datasets [10]. However, simply reducing bias is not sufficient if it comes at the cost of a significant decrease in the overall performance of the model.

Our work extends the pruning method proposed in [6] to address its limitations. The existing method employs a stopping criterion based on bias reaching 0, but we found that insufficiently large pruning may have minimal impact on testing data. Additionally, the method incorporates another stopping criterion involving the model's performance dropping below a predetermined threshold, but it does not ensure bias mitigation before the model's performance deteriorates beyond the threshold. In our extension, we propose modifications to the pruning algorithm and fairness evaluation metric to enhance the method's effectiveness in mitigating bias while maintaining model performance.

Our proposed method aims to maintain the model's performance while allowing for the pruning of more neurons, which directly enhances the model's debiasing capability. We measure fairness by evaluating the disparities across the protected attributes using metrics such as accuracy, area under the receiver operating characteristics (AUC), precision, true positive rate (TPR), true negative rate (TNR), and equalized odds. Additionally, we conduct a comparative analysis of our results with other model debiasing methods.

2   Materials and Methods

2.1   Data and Preprocess

Our study used the T1-weighted structural MRI dataset in the NIfTI format, which was obtained from the ADNI database (https://adni.loni.usc.edu/). The dataset is summarized in Table 1. All scans underwent preprocessing steps including multiplanar reconstruction (MPR), Gradwarp, and B1 non-uniformity correction. We labeled subjects in the dataset as either AD or Cognitively Normal (CN). We focused on gender and age as protected attributes in this study. To prevent any potential data leakage and ensure unbiased evaluation, the data split was performed at the subject level. The data were split into training (70%), validation (15%), and testing (15%) sets. We performed skull stripping on the MRI scans using the CAT12 segmentation toolkit, and all images were cropped to (96, 96, 96) from the center. To prevent overfitting, random augmentations, including rotation and shearing, were applied during the training process.

2.2   Debiasing by Pruning

In this study, we extended an existing pruning technique to mitigate unwanted bias while maintaining optimal performance, which increased the number of

Table 1. Demographic information of subjects from ADNI dataset.

Diagnose	Male	Female	Age > 75	Age ≤ 75	Total
AD	48.0%	52.0%	44.3%	55.7%	558
CN	48.2%	51.8%	38.9%	61.0%	936
Total	48.1%	51.9%	41.0%	59.0%	1494

prunable neurons, and we applied this method to the 3D convolutional neural network (CNN) pre-trained model which is trained by training dataset. Figure 1 illustrates an overview of the proposed method. Our approach involved evaluating individual neurons through validation dataset based on their contributions to the disparities and their impact on the model's performance in predicting AD/CN, and identifying which neurons should be pruned.

Bias Influence. One widely adopted metric for assessing classification parity is equal opportunity difference (EOD) [14, 15], which measures TPR disparities across protected attributes. However, focusing solely on TPR disparities during pruning may not achieve full debiasing. To overcome this, we introduced balanced accuracy (BA) disparities across protected attributes, which represents the average of TPR and TNR, as additional evaluation metrics:

$$BA(a) = \frac{1}{2}Pr(\hat{Y} = 1|Y = 1, A = a) + \frac{1}{2}Pr(\hat{Y} = 0|Y = 0, A = a), \quad (1)$$

$$BA\ Disparity = BA(0) - BA(1), \quad (2)$$

where Y is the targeted variable and $\hat{Y}$ is the model prediction given by input X, and A represents the protected attribute given by X (e.g.: age and gender).

Performance Influence. To avoid risking the removal of important nodes and prevent a decline in the model's performance, we considered both bias influence and each node's impact on the model's performance before pruning. We utilized cross-entropy, which was also used during pre-training, to accurately evaluate the classification performance after pruning.

Neuron Importance. We extended the differentiable proxy function proposed in [6], which originally calculated the neuron importance solely based on the bias influence. In our extension, we further enhanced the function to compute neuron importance by combining both the bias influence and the performance influence. Given sets of N data points, $X = \{x_i\}_{i=1}^{N}$, $Y = \{y_i\}_{i=1}^{N}$, and $A = \{a_i\}_{i=1}^{N}$, with a differentiable bias measure function μ (e.g. Eq. 2), and a model performance measure function ν (e.g. cross-entropy), we could get the neuron importance of the k-th unit in the j-th layer using the following equation:

$$I_{j,k} = \frac{1}{N} \sum_{i=1}^{N} \frac{\partial \mu(f_\theta, X, Y, A)}{\partial z_k^j(x_i)} - \frac{1}{N} \sum_{i=1}^{N} \frac{\partial \nu(f_\theta, X, Y)}{\partial z_k^j(x_i)}, \qquad (3)$$

where z_k^j stands for the unit's output at input x_i and f_θ stands for the target pruning model. The result obtained from Eq. 3 can effectively characterize neurons that exhibit high bias.

Pruning Process. The pruning process involved four steps (see Fig. 1 for illustration). (*i*) Selection of Targeted Layers: To enhance efficiency and effectiveness, we strategically identified specific layers for pruning, enabling us to focus on the most impactful areas of the model. (*ii*) Neuron Importance Evaluation: For the chosen layers, we assessed the neuron importance $I_{j,k}$ of each neuron (see Eq. 3) using validation dataset. (*iii*) Neuron Pruning: Neuron pruning was performed based on the calculated neuron importance, with the pruning order determined by the bias sign. For the bias sign, if it was the first time entering step (*ii*), the initial model's bias sign was used; otherwise, it was based on the pruned model's bias sign from step (*iii*). The objective was to drive bias towards 0, accomplished by increasing or decreasing it during pruning. To strike a balance between reducing bias and preserving model performance, the number of neurons pruned was determined empirically. Once the neurons were identified for pruning, their outputs were set to zero. (*iv*) Evaluation and Iteration: After the pruning step, we reevaluated the pruned model's bias using Eq. 2 and conducted pruned model's performance evaluation. We iteratively repeated steps (*ii*) − (*iv*), pruning additional neurons progressively. The number of iterations was either determined empirically or continued until performance dropped below the predetermined threshold, even with biases reaching 0. Considering performance during pruning prevented rapid performance decline, enabling strategic pruning of more neurons.

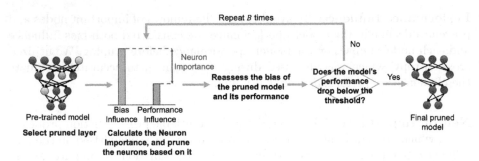

Fig. 1. Illustration of our proposed pruning process.

3   Experiment

All experiments were conducted on an NVIDIA GeForce RTX 3090 graphics card with 24 GB of memory.

3.1   Implementation and Evaluation

Initially, we trained a model without protected attributes using the 3D ResNet-18 architecture. The training utilized adaptive moment estimation (Adam) with a learning rate of 10^{-5} for 50 epochs, a batch size of 16, and employed cross-entropy as the loss function. The training process was completed in approximately 1 h. After obtaining the pre-trained model, we performed the pruning process on the validation dataset. In this study, we selected six layers as the target layers and pruned 0.001% of the neurons in each pruning step, and repeat the pruning step 25 times.

In our study, we employed a five-fold cross-validation approach to evaluate the performance of the model. To assess the model's performance, we utilized various metrics including accuracy, AUC, precision, TPR, TNR, and equalized odds. To assess gender disparity, we calculated the absolute difference in performance between males and females. For age disparity, we computed the absolute difference in performance between individuals younger than 75 years old and those older than 75 years old, considering the roughly even distribution within these two categories.

3.2   Comparison

We compared our method with the baseline model, which represented the biased model without undergoing any debiasing methods. Additionally, to further evaluate the effectiveness of our proposed method, our method was compared with other debiasing techniques. These techniques included training the model using two-step adversarial debiasing [5] preventing the model from learning the features related to protected attributes, penalizing the Maximum Mean Discrepancy distance or mean distributions across protected attributes [7], and training the model on a balanced dataset by upsampling the unprivileged groups with and without data augmentation.

4   Result

We measured the models' output results by five-fold cross-validation through the testing dataset. As shown in Fig. 2, our proposed method effectively maintained or even improved the model's performance across various metrics compared to the baseline model. Specifically, for age attribute, our method successfully improved the model's performance in accuracy (+0.1%), AUC (+0.1%), precision (+1.1%), TNR (+0.1%), and equalized odds (+1.8%). As for gender attribute, our method effectively maintained the model's performance in

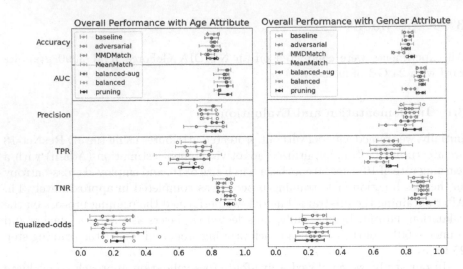

Fig. 2. Comparison of the overall performance across different debiasing methods. The debiasing models were trained using age (left) and gender (right) information. The red dot corresponds to the male or younger subgroup, while the blue dot corresponds to the female or older subgroup. (Color figure online)

accuracy (+1.9%), AUC (+0.9%), precision (+1.3%), TPR (+0.9%), and TNR (+2.4%).

As shown in Fig. 3, the baseline model exhibited significant disparities across various metrics. Compared with the disparities in the baseline model, our proposed method effectively mitigated these disparities across protected groups for multiple metrics. Specifically, for age attribute, our method successfully reduced bias in accuracy (−3.9%), AUC (−0.7%), precision (−3.2%), TPR (−7.3%), and equalized odds (−3.0%). As for gender attribute, our approach mitigated bias in accuracy (−2.3%), AUC (−1.9%), precision (−4.8%), TPR (−1.0%), and TNR (−5.0%). Notably, our pruning method demonstrated competitive performance compared to other debiasing techniques. We also performed a paired t-test and calculated p-value to assess the disparity before and after the pruning process. For age attribute, there were significant differences (p-value < 0.05) in ACC (0.013) and TPR (0.045). As for gender attribute, there was a significant difference in AUC (0.031).

We conducted comparisons between our approach and the original pruning techniques, both with and without the three adjustments (see Fig. 4). For age attribute, compared with the original pruning method, our proposed method further reduced bias in accuracy (−5.2%), AUC (−0.9%), precision (−7.3%), TPR (−9.7%), TNR (−1.7%), and equalized odds (−7.2%). Similarly, for gender attribute, our proposed method further mitigated bias in accuracy (−6.2%), AUC (−2.0%), precision (−3.7%), and TNR (−7.3%).

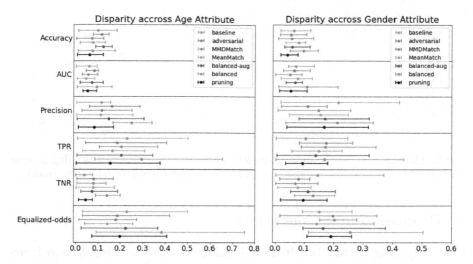

Fig. 3. Comparison of performance disparities across different debiasing methods, across age attribute (left) and across gender attribute (right). Lower disparities indicated similar performance across protected groups.

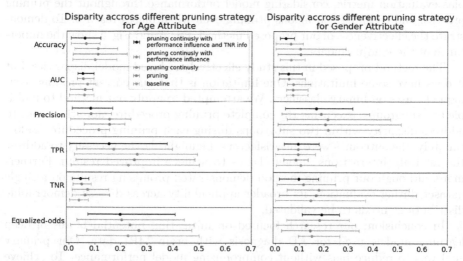

Fig. 4. Comparison of performance disparities across different pruning methods, across age attribute (left) and across gender attribute (right).

As shown in Fig. 5, we presented the changes in the bias and the model's performance during the pruning process. In contrast, the pruning procedure without considering performance influence, and stopping when bias reached 0 had a limited impact on the testing dataset. By maintaining the model's performance while pruning a greater number of neurons, our approach effectively increased the debiasing capability.

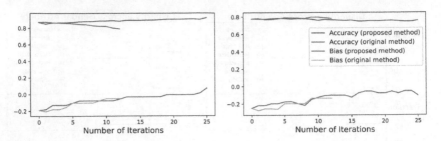

Fig. 5. The changes in bias and the model's performance during the pruning process were observed for both the validation dataset (left) and the testing dataset (right).

5   Discussion and Conclusion

This study mitigates bias in MRI-Based Alzheimer's Disease Classifiers by Pruning of DNN. In comparison to the original pruning method proposed by [6], we have made three major adjustments that have proven to enhance pruning effectiveness. These adjustments included incorporating TNR information into the bias evaluation metric, considering model performance throughout the pruning process, and enabling continuous pruning even when biases reach 0. To demonstrate the effectiveness of our proposed method, Fig. 4 and Fig. 5 show the importance of these adjustments.

The method proposed in this study effectively reduced model disparities, but it does have some limitations. One limitation is that our current pruning technique focuses on binary classifiers. We attempted to apply our method to multiple age subgroups, executing the complete pruning procedure more than once to address disparity within two subgroups during each pruning procedure. Unfortunately, the outcomes were unsatisfactory, highlighting the necessity to address the intricate interactions among classes to achieve improved results. Furthermore, although our pruning method demonstrated promising results on a single dataset during evaluation, its broader applicability across diverse demographic distributions needs to be validated.

In conclusion, our research focused on mitigating potential biases in deep learning models used for AD diagnosis with brain MRI data. Our primary goal was to reduce bias without compromising model performance. To achieve this, we introduced an extended pruning method that incorporated additional bias evaluation metrics and considered model performance during pruning. The results demonstrated the feasibility of achieving fairness without sacrificing overall model performance. By striking a delicate balance between performance and fairness, our study significantly contributed to the field of deep learning and highlighted the potential for ethically sound and reliable AD diagnosis models.

The source code of the proposed method is available at code.

References

1. Dufumier, B., et al.: Contrastive learning with continuous proxy meta-data for 3D MRI classification. http://arxiv.org/abs/2106.08808 (2021). https://doi.org/10.48550/arXiv.2106.08808

2. Zhuang, X., Li, Y., Hu, Y., Ma, K., Yang, Y., Zheng, Y.: Self-supervised feature learning for 3D medical images by playing a Rubik's cube http://arxiv.org/abs/1910.02241 (2019). https://doi.org/10.48550/arXiv.1910.02241

3. Liu, S., Liu, S., Cai, W., Pujol, S., Kikinis, R., Feng, D.: Early diagnosis of Alzheimer's disease with deep learning. In: 2014 IEEE 11th International Symposium on Biomedical Imaging (ISBI), pp. 1015–1018 (2014). https://doi.org/10.1109/ISBI.2014.6868045

4. Zhang, B.H., Lemoine, B., Mitchell, M.: Mitigating unwanted biases with adversarial learning. In: Proceedings of the 2018 AAAI/ACM Conference on AI, Ethics, and Society, New Orleans, LA, USA, pp. 335–340. ACM (2018). https://doi.org/10.1145/3278721.3278779

5. Correa, R., Jeong, J.J., Patel, B., Trivedi, H., Gichoya, J.W., Banerjee, I.: Two-step adversarial debiasing with partial learning - medical image case-studies. http://arxiv.org/abs/2111.08711 (2021)

6. Marcinkevics, R., Ozkan, E., Vogt, J.E.: Debiasing deep chest X-ray classifiers using intra- and post-processing methods. In: Proceedings of the 7th Machine Learning for Healthcare Conference, pp. 504–536. PMLR (2022)

7. Pfohl, S.R., Foryciarz, A., Shah, N.H.: An empirical characterization of fair machine learning for clinical risk prediction. J. Biomed. Inform. **113**, 103621 (2021). https://doi.org/10.1016/j.jbi.2020.103621

8. Larrazabal, A.J., Nieto, N., Peterson, V., Milone, D.H., Ferrante, E.: Gender imbalance in medical imaging datasets produces biased classifiers for computer-aided diagnosis. Proc. Natl. Acad. Sci. **117**, 12592–12594 (2020). https://doi.org/10.1073/pnas.1919012117

9. Petersen, E., et al.: Feature robustness and sex differences in medical imaging: a case study in MRI-based Alzheimer's disease detection http://arxiv.org/abs/2204.01737 (2022)

10. Seyyed-Kalantari, L., Liu, G., McDermott, M., Chen, I.Y., Ghassemi, M.: CheXclusion: fairness gaps in deep chest X-ray classifiers. In: Biocomputing 2021, Kohala Coast, Hawaii, USA, pp. 232–243. World Scientific (2020). https://doi.org/10.1142/9789811232701_0022

11. Lin, X., Kim, S., Joo, J.: FairGRAPE: fairness-aware GRAdient pruning mEthod for face attribute classification. In: Avidan, S., Brostow, G., Cissé, M., Farinella, G.M., Hassner, T. (eds.) Computer Vision - ECCV 2022. LNCS, vol. 13673, pp. 414–432. Springer, Cham (2022). https://doi.org/10.1007/978-3-031-19778-9_24

12. Chiu, C.-H., Chung, H.-W., Chen, Y.-J., Shi, Y., Ho, T.-Y.: Toward fairness through fair multi-exit framework for dermatological disease diagnosis http://arxiv.org/abs/2306.14518 (2023). https://doi.org/10.48550/arXiv.2306.14518

13. Chiu, C.-H., Chung, H.-W., Chen, Y.-J., Shi, Y., Ho, T.-Y.: Fair multi-exit framework for facial attribute classification http://arxiv.org/abs/2301.02989 (2023)

14. Hardt, M., Price, E., Srebro, N.: Equality of opportunity in supervised learning http://arxiv.org/abs/1610.02413 (2016). https://doi.org/10.48550/arXiv.1610.02413

15. Savani, Y., White, C., Govindarajulu, N.S.: Intra-processing methods for debiasing neural networks. In: Advances in Neural Information Processing Systems, pp. 2798–2810. Curran Associates, Inc. (2020)

Auditing Unfair Biases in CNN-Based Diagnosis of Alzheimer's Disease

Vien Ngoc Dang[1]($\boxtimes$)(iD), Adrià Casamitjana[2](iD), Martijn P. A. Starmans[1,3](iD),
Carlos Martín-Isla[1](iD), Jerónimo Hernández-González[1](iD), Karim Lekadir[1](iD),
and for the Alzheimer's Disease Neuroimaging Initiative

[1] Departament de Matemàtiques i Informàtica, Facultat de Matemàtiques i
Informàtica, Universitat de Barcelona, Barcelona, Spain
dangn@ub.edu
[2] Departament de Teoria del Senyal i Comunicacions, Universitat Politècnica de
Catalunya, Barcelona, Spain
[3] Department of Radiology and Nuclear Medicine, Erasmus University Medical
Center, Rotterdam, The Netherlands

Abstract. Convolutional neural networks (CNN), while effective in
medical diagnostics, have shown concerning biases against underrepresented patient groups. In this study, we provide an in-depth exploration of these biases in the realm of image-based Alzheimer's disease
(AD) diagnosis using state-of-the-art CNNs, building upon and extending prior investigations. We dissect performance-based and calibration-
based biases across patient subgroups differentiated by sex, ethnicity,
age, educational qualifications, and APOE4 status. Our findings reveal
substantial disparities in model performance and calibration, underscor-
ing the challenges intersectional identities impose. Such biases highlight
the importance of fairness analysis in fostering equitable AI applications
in the AD domain. Appropriate mitigation actions should be carried out
to ensure that, those who need it, receive healthcare attention independently from the subgroup they belong to.

Keywords: fairness · bias · medical imaging · convolutional neural
networks · Alzheimer's disease

1 Introduction

Alzheimer's disease (AD) is a critical global health challenge which requires
robust diagnostic approaches [1]. Recently, the application CNNs in medical
imaging presents a potential solution in enhancing early detection and classification of AD [2]. Particularly, 3D CNN-based diagnosis models excel in handling

Data used in preparation of this article was obtained from the Alzheimer's Disease Neuroimaging Initiative (ADNI) database (http://www.adni-info.org/). The investigators
within the ADNI contributed to the design and implementation of ADNI and/or provided data, but did not participate in analysis or writing of this report.

S. Wesarg et al. (Eds.): CLIP/FAIMI/EPIMI 2023, LNCS 14242, pp. 172–182, 2023.
https://doi.org/10.1007/978-3-031-45249-9_17

high-dimensional neuroimaging datasets and enhancing our understanding of complex AD pathologies and various stages of AD by preserving spatial information from input images and presenting results as a volume [3]. Despite the advantages, these models may encounter biases against certain population subgroups. These biases can arise from sources such as data imbalance, attribute-class imbalance, spurious correlations, anatomical differences, and inherent pathological differences among subpopulations [4,5]. Such factors may increase the likelihood of misclassification and miscalibration, particularly among underrepresented subgroups. This study focuses on how they affect model performance and calibration across different subpopulations.

Classification Bias. Occurs when there are disparities in predictive performance across subgroups. It offers a notion of how accurately the model operates across these different groups. Such disparities can lead to unequal provision of care and may increase existing health inequalities. *Calibration bias*, on the other hand, is characterized by a discrepancy between observed and predicted probabilities across different subgroups. It offers a notion of how reliable the model's predictions are for these various groups. Notably, features that modernize CNNs such as batch normalization, weight decay, and deep layers can actually contribute to their miscalibration [6]. This issue is critical in healthcare applications, where biases can skew the predicted probabilities for specific subpopulations, potentially leading to unequal treatment. Auditing and understanding the distinct and overlapping impacts of these biases is crucial for enhancing the fairness and reliability of AI diagnostic tools.

Fairness analysis studies in the recent literature cover various medical disciplines, such as radiology, primarily focusing on chest X-ray and mammography imaging, dermatology, ophthalmology, and cardiology [5]. Nonetheless, this type of analysis is scarce in neuroimaging studies, including AD [5]. Most research in the AD domain has overlooked the need for fairness analysis, potentially leading to biased results [7,8]. Performance disparities in CNNs based on sex and age, specifically within the context of AD, were studied by [9] when varying the proportion of female subjects within the training data. Furthermore, Mendelson et al. [10] shed light on the considerable impact of selection bias, particularly in small AD datasets, revealing how these biases can skew performance enhancements during pipeline optimizations. We present the first work that, to our knowledge, examines the potential biases that 3D CNNs might exhibit when diagnosing AD across several key attributes (sex, ethnicity, age, educational qualifications, and APOE4 status). Ideally, equitable diagnosis outcomes should be provided across the subpopulations defined according to these protected attributes. We analyze performance and calibration disparities across subgroups and intersectional identities, identify unfair biases, and hypothesize about their possible sources based on the dataset's characteristics, known flaws of CNN-based methods and known aspects of AD described in the related scientific literature.

2   Materials and Methods

2.1   Data Description and Preprocessing

In our study, we utilize magnetic resonance imaging (MRI) volumes from a combined set of ADNI-1, ADNI-2, and ADNI-GO cohorts in the ADNI dataset [11], comprising 336, 330, and 487 subjects diagnosed with AD, CN, and MCI, respectively. While recognizing the critical nature of early AD detection encompassing MCI stages, we opted to distinguish between CN and AD subjects because it is the most commonly researched task when using end-to-end CNN frameworks [2]. The dataset presents a balance in terms of sex, a high predominance of white subjects (>90%), and a primary age concentration within the 65–84 years range. Education level is notably higher in the CN group, and the APOE4 allele status distribution is notoriously different between CN and AD subjects. See Table 1 for a full characterization.

The T1w MRI images were preprocessed. First, N4ITK bias field correction method was applied to adjust for signal variations due to potentially inconsistent MRI scanners [12]. Then, a linear (affine) registration was performed using the SyN algorithm to align images to the MNI space, a standard coordinate system in neuroimaging [13]. Images were centrally cropped to remove the background, resulting in a final size of $169 \times 208 \times 179$ with $1\,mm^3$ isotropic voxels. We then performed min/max intensity rescaling and applied the QC framework [14] to automatically verify the quality of the resulting images.

Table 1. Distribution of subjects per protected attribute and CN/MCI/AD diagnostic categories. Here, 'qualifications' refers to 'years of education'.

Subgroup	Attribute	CN (N%)	AD (N%)	MCI (N%)
Sex	Male	160 (48.5)	185 (55.1)	287 (58.9)
	Female	170 (51.5)	151 (44.9)	200 (41.1)
Ethnicity	White	299 (90.6)	311 (92.6)	452 (92.8)
	Other	31 (9.4)	25 (7.4)	35 (7.2)
Age at baseline	<65	11 (3.3)	44 (13.1)	69 (14.2)
	65–74	163 (49.4)	99 (29.5)	182 (37.4)
	75–84	117 (35.5)	136 (40.5)	179 (36.8)
	85+	39 (11.8)	57 (17.0)	57 (11.8)
Qualifications*	8–12	31 (9.4)	80 (23.8)	83 (17.0)
	13–15	69 (20.9)	68 (20.2)	100 (20.5)
	16+	227 (68.8)	184 (54.8)	303 (62.2)
APOE4*	0	245 (74.2)	111 (33.0)	224 (46.0)
	1	76 (23.0)	157 (46.7)	204 (41.9)
	2	7 (2.0)	64 (19.0)	55 (11.3)

* Partial counts associated with protected attributes may differ from the total due to missing APOE4 data and the exclusion of scarce samples with qualifications < 8.

2.2 Models

We opted to use the already-build model shared by [2], ensuring we replicated its specific pre-processing requirements, as previously detailed, to guarantee its optimal performance. This model consists of five convolutional blocks (each comprising four successively connected layers: convolutional, batch normalization, ReLU activation, and max pooling) and three fully connected layers with 1300, 50, and 2 hidden units, respectively. A probabilistic output is obtained by applying a softmax function. Transfer learning (TL) was implemented using an autoencoder (AE), comprised of an encoder (designed similarly to the convolutional blocks of the CNN) and a decoder that inverts the encoder's sequence. Although TL could potentially introduce bias, particularly if training data is unbalanced, analyzing this potential effect is outside the scope of our work. This model adheres to state-of-the-art performance standards in CNN-based AD classification and shows no suspected data leakage [2]. Thus, we are exploring potential biases in a cutting-edge CNN-based AD diagnosis model.

Model Training and Evaluation. The CNN models were trained with Adam optimizer ($\beta_1 = 0.9$, $\beta_2 = 0.999$, $\epsilon = 1e - 8$), a learning rate of 1e-4 and a batch size of 12. The training objective was to minimize binary cross-entropy loss without regularization over 50 epochs, with a dropout rate of 0.5. Early stopping was applied based on validation loss. For AE pre-training, the same setting was used while the objective shifted to minimizing the mean squared error (MSE). A set of 200 randomly selected, age- and sex-matched subjects (100 CN and 100 AD patients) are reserved for testing. With the remaining data, five distinct CNN models were generated through 5-fold cross-validation using AD, CN, and MCI (for pre-training the AE) subjects.

2.3 Bias Evaluation Metrics

The AD diagnosis model is audited for potential bias by analyzing both performance and calibration disparities across subgroups and intersectional identities.

Performance-Based Bias Analysis. We apply the *equalized odds* criterion, a widely accepted fairness metric, to evaluate potential biases in model's performance across various groups [15]. This criterion aims for equal false negative rates (FNR) and false positive rates (FPR) across subgroups to ensure equitable outcomes [16]. FNR measures the proportion of AD patients who are incorrectly classified as CN and, in our study, reflects underdiagnosis. Similarly, FPR measures the proportion of CN individuals who are misclassified as having AD and reflects overdiagnosis. These metrics are examined at subgroup (e.g., 'female') and intersectional identity (e.g., 'white female') levels. Let (X, Y, C) be the variables describing our problem, with the protected variables X, descriptive variables Y, and the class variable C ($C = 1$ represents a patient with AD, whereas $C = 0$ means CN). Let us denote the predictions of classifier M as $\hat{C}$, the predicted distribution for (x, y) as $p_M(C|x, y)$, and the predicted class as $\hat{c} = \arg\max_c p_M(C = c|x, y)$ Let Ω_X be the set of all possible values of the protected variable X. The FPR and FNR are defined as:

$$FPR_x = P[\hat{C} = 1|X = x, C = 0], FNR_x = P[\hat{C} = 0|X = x, C = 1] \qquad (1)$$

where x is a vector of elements for a set of protected attributes. When it comprises just a single element $x \in (\Omega_X)$, such as ('female'), or two attributes $x \in (\Omega_{X_1}, \Omega_{X_2})$, such as ('female', 'white'), this equation provides the subgroup or intersectional identity measurements, respectively.

Calibration-Based Bias Analysis. We use calibration curves to visually assess model calibration across subgroups. These diagrams group into bins the model's predicted probabilities of AD (we use 10 bins, as a balance between granularity and interpretability). The mean predicted probability of each bin is then plotted against the observed proportion of positive samples in the bin. A perfect model would show a diagonal line, indicating that its predictive probabilities match the observed frequencies. We compute these lines for each group to compare calibration per group. Thus, the closer a subgroup's curve is to the diagonal line, the better the calibration for that subgroup. To quantify calibration, we calculate the Expected Calibration Error (ECE) for each subgroup $x \in \Omega_X$:

$$ECE = \sum_{m=1}^{M} \frac{|D_m^t|}{|D^t|} \left| \text{prop}(D_m^t) - \text{prprob}(D_m^t) \right|, \tag{2}$$

where D_m^t is the m-th bin in the test set D^t, $\text{prop}(D_m^t)$ and $\text{prprob}(D_m^t)$ denote the proportion of positive samples ($\frac{1}{|D_m^t|} \sum_{(x,y,c) \in D_m^t} I[c = 1]$), and the mean predicted probability for the samples in the bin ($\frac{1}{|D_m^t|} \sum_{(x,y,c) \in D_m^t} p_M(C = 1|x,y)$), respectively. It is crucial to note that ECE can be particularly sensitive to sample size. In subgroups with limited samples, its reliability might be compromised [17,18]. As a result, interpreting calibration biases as a difference of ECE across subgroups might be misleading in these situations. In our efforts to perform a comprehensive evaluation, we incorporate the Brier score per subgroup $x \in \Omega_X$:

$$\text{Brier} = \frac{1}{|D^t|} \sum_{(x,y,c) \in D^t} (c - p_M(C = 1|x,y))^2, \tag{3}$$

This metric computes the mean squared differences between the predicted probability of the positive class and actual labels across test points, under the simplifying assumption that the 'real' probability of a sample labeled as positive is 1 and that of negative samples is 0. It provides insights into calibration and discrimination ability offering a richer view of predictive accuracy [19].

3   Results and Discussion

Here we analyze fairness regarding predictive performance and model calibration.

3.1   Auditing Fairness with Respect to Model Performance

In Table 2, we use several metrics to depict the performance of the CNN model for AD diagnosis as the average value on the test set over the five runs, along

Table 2. Mean performance and 95% confidence interval of the model on the test set

ROC-AUC	Balanced Accuracy	Specificity	Sensitivity
0.90 [0.88,0.92]	0.81 [0.79,0.84]	0.84 [0.81,0.87]	0.79 [0.72,0.85]

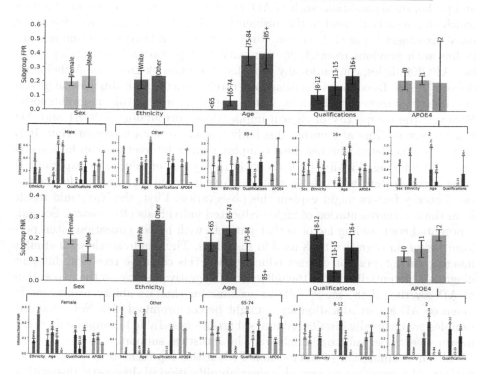

Fig. 1. Results in terms of FPR (overdiagnosis, upper panel) and FNR (underdiagnosis, lower panel) per subgroup of protected attributes sex, ethnicity, age, qualifications, and APOE4 status. Each panel shows, in the first row, the values per subgroup and, for the subgroup with the highest value, in the second row, intersectional measurements.

with the corresponding 95% confidence intervals. Figure 1 illustrates disparities in misdiagnosis rates across subgroups. Specifically, the upper panel of Fig. 1 illustrates FPR (overdiagnosis) across subgroups of the same attribute, along with intersectional measurements for the subgroup with the highest FPR. The lower panel, on the other hand, depicts FNR (underdiagnosis) and intersectional measurements for the subgroup with the highest FNR. We find that females and patients labeled as 'other' for ethnicity suffered from higher underdiagnosis rates compared to males and white patients, respectively. A more uneven distribution is observed in overdiagnosis rates. It is noteworthy that males and white individuals are overrepresented in the training dataset, which might be compromising the 3D CNN's performance. We also observe uneven diagnosis rates as the chronological age of the patients increases. We detect a trend of overdiagnosis in cognitively healthy older patients and underdiagnosis in younger ones, indicating a possible systematic bias in the model that mirrors real-world diagnostic challenges clinicians face at the extremes of age [20,21]. This bias could be potentially due to the overlap between imaging signatures of healthy aging and pathological conditions such as AD, and/or the coexistence of comorbidities which can potentially lead to the misidentification of AD in patients with other more prominent types of dementia. Our observations related to sex and age are in line with previous research [9]. Contrary to expectations, patients carrying two APOE4 alleles, conventionally considered a high-risk group for Alzheimer's disease [22,23], faced high underdiagnosis rates. Patients with this genetic profile are underrepresented in the data, which might be causing this underdiagnosis. However, the complexity of clinical implications surrounding APOE4 and AD risk prevent us from attributing this finding solely to data representation. It is still to be elucidated whether these complex genetic interactions can be learned from 3D imaging. Finally, patients with 16+ years of education experienced more overdiagnosis, while those with 13–15 years faced increased underdiagnosis. Two key factors might explain this observation. First, the bias could result from the overrepresentation of highly educated individuals (16+ years). Second, a potential confounding factor is that patients with higher education often perform better on cognitive tests used in diagnosis. Thus, individuals with similar imaging phenotypes but distinct educational levels could be receiving different diagnoses, potentially due to the influence of cognitive reserve on the perception of AD severity. If the model fails to account for this, its ability to accurately diagnose AD across education levels might be compromised [24]. Note that the samples are unequally distributed into subgroups (see Table 1), emphasizing that measurements per subgroup might vary in statistical support.

In our examination of intersectional identities represented by two distinct patient subgroups, we uncovered a clear amplification of diagnostic disparities. This escalation can be observed in the second and fourth rows of Fig. 1. Notably, the 'other-female' group, an under-served patient subpopulation in the training dataset, faced a greater risk of underdiagnosis compared to their 'white-female' counterparts. Misdiagnosis is exacerbated when we consider intersectional identities combining two underrepresented subgroups such as 'female' and 'other'. This trend, suggesting an ethnic bias in the diagnostic algorithm, echoes findings in previous works on algorithmic bias [25,26], marking the first time such bias has been identified in the context of Alzheimer's disease.

3.2 Auditing Fairness with Respect to Model Calibration

Figure 2 shows the results of model calibration, which uncover distinct calibration biases across patient subgroups. Despite similar overall calibration trends for both male and female subgroups, the calibration curve for females showed a slight over-estimation in the upper quadrant and under-estimation in some lower quadrant bins, with high standard deviation indicating variability among folds. Note that samples are not uniformly distributed through all the bins.

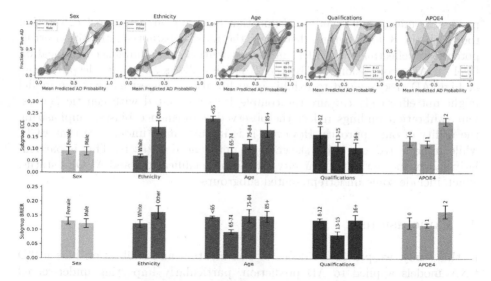

Fig. 2. Results of model calibration per subgroup based on protected attributes sex, ethnicity, age, qualifications, and APOE4 status. We report results in terms of calibration curves (plots in the first row), ECE (second row) and Brier score (third row). Points in the calibration curves are displayed with different sizes, proportional to the number of samples used to calculate them.

Non-uniform sample distribution complicates the interpretation of the calibration curves and the relationship of results in Fig. 1 and Fig. 2. In particular, the high FNR for the female group observed in Fig. 1 aligns with the calibration curve of Fig. 2 when we note that, in general, the number of samples in the upper quadrant bins is more sparse than that of the lower quadrant bins. In addition, substantial variations emerged when examining ethnicity subgroups. The white group demonstrated well-calibrated predictions, whereas the 'other' group seemed to display over-forecasting in prediction probabilities around 0.5 and a notable peak around 0.25. Note the unbalanced distribution of points in the curve for the 'other' group: over-forecasted points in prediction probabilities around 0.5 are scarce and points over the diagonal in the first quadrant (including the referred peak at 0.25) are, in comparison, larger. The unbalanced distribution and the presence of this peak are instrumental in the high FNR observed for this group in Fig. 1. Age-based subgroups also exhibited distinct miscalibration patterns. Particularly, the <65 age group shows an underestimation in their prediction probabilities, indicating that the model might be underconfident when predicting positive cases for younger patients. Conversely, the 85+ group, linked with a high overdiagnosis rate, exhibits an over-forecasting trend, suggesting the model might be overconfident when predicting positive cases in this age group. Consequently, these variations in calibration across subgroups, consistent with the results observed in Fig. 1, may result in different optimal decision thresholds for each group, which suggests a potentially successful post-processing mitigation technique. Education level appeared to affect calibration too, with the 8–12 subgroup displaying the most substantial calibration error, potentially reflecting the model's misinterpretation of the influence of education level on AD progression. Lastly, the 2 APOE4 alleles subgroup, despite being at high AD risk, seemed to exhibit miscalibration, suggesting that the model might not effectively capture the complexities associated with genetic factors. Our calibration findings reflect the observed performance biases, emphasizing the utility of our approach. However, it is important to underscore that ECE, while informative, can show skewness in unbalanced datasets. This is particularly salient for attributes ethnicity, education qualification, and APOE4 status, which include some underrepresented subgroups.

4   Conclusions

In this study, we provide evidence of classification and calibration bias in 3D CNN models applied to AD prediction, particularly impacting under-served patient subgroups. We depict the influence of demographic characteristics such as ethnicity, sex, and age on diagnostic and calibration disparities. We show that biases are increased at the intersection of some of the unfavored subgroups according to these attributes. In each case, we identified possible sources of the bias in relation to the training data, the complexity of the task, and the scientific knowledge about AD. It is crucial for future model developments to carry out a detailed and thorough evaluation and to address these limitations to ensure equitable healthcare, emphasizing the need for models that perform fairly across all patient subgroups.

Acknowledgements. VND, JHG, and KL received funding from the EU's Horizon 2020 research and innovation programme under Grant Agreement No 848158, EarlyCause, and MPAS under Grant Agreement No 952103, EuCanImage. AC received funding from Ministry of Universities and Recovery, Transformation and Resilience Plan, through UPC (Grant No 2021UPC-MS-67573). JHG is a Serra Húnter fellow.

References

1. Alzheimer's disease facts and figures. Alzheimers Dement **18**(4), 700–789 (2022)
2. Wen, J., et al.: Convolutional neural networks for classification of Alzheimer's disease: overview and reproducible evaluation. Med. Image Anal. **63**, 101694 (2020)
3. Illakiya, T., Karthik, R.: Automatic detection of Alzheimer's disease using deep learning models and neuro-imaging: current trends and future perspectives. Neuroinformatics **21**, 339–364 (2023)
4. Seyyed-Kalantari, L., Liu, G., McDermott, M.B., Ghassemi, M.: CheXclusion: fairness gaps in deep chest X-ray classifiers. In: Pacific Symposium on Biocomputing. Pacific Symposium on Biocomputing, vol. 26, pp. 232–243 (2020)
5. Ricci, L.M.A., Echeveste, R., Ferrante, E.: Addressing fairness in artificial intelligence for medical imaging. Nat. Commun. **13**, 4581 (2022)

6. Guo, C., Pleiss, G., Sun, Y., Weinberger, K.Q.: On calibration of modern neural networks. In: Proceedings of the 34th International Conference on Machine Learning, pp. 1321–1330 (2017)
7. Bae, B., et al.: Identification of Alzheimer's disease using a convolutional neural network model based on T1-weighted magnetic resonance imaging. Sci. Rep. **10**, 1–10 (2020)
8. El-Sappagh, S., et al.: Trustworthy artificial intelligence in Alzheimer's disease: state of the art, opportunities, and challenges. Artif. Intell. Rev. (2023)
9. Petersen, E., et al.: Feature robustness and sex differences in medical imaging: a case study in MRI-based Alzheimer's disease detection. In: MICCAI 2022: 25th International Conference, pp. 88–98 (2022)
10. Mendelson, A.F., Zuluaga, M.A., Lorenzi, M., Hutton, B.F.: Selection bias in the reported performances of AD classification pipelines. NeuroImage Clin. **14**, 400–416 (2017)
11. Jack, C.R., et al.: The Alzheimer's disease neuroimaging initiative (ADNI): MRI methods. J. Magn. Resonan. Imaging **27**(4), 685–691 (2008)
12. Tustison, N.J., et al.: N4ITK: improved N3 bias correction. IEEE Trans. Med. Imaging **29**, 1310–1320 (2010)
13. Avants, B.B., Epstein, C.L., Grossman, M., Gee, J.C.: Symmetric diffeomorphic image registration with cross-correlation: evaluating automated labeling of elderly and neurodegenerative brain. Med. Image Anal. **12**(1), 26–41 (2008)
14. Fonov, V., Dadar, M.: The PREVENT-AD research group. In: Louis Collins, D. (ed.) Deep Learning of Quality Control for Stereotaxic Registration of Human Brain MRI. bioRxiv (2018)
15. Chen, I.Y., Johansson, F.D., Sontag, D.: Why is my classifier discriminatory? In: Proceedings of the 32nd International Conference on Neural Information Processing Systems (NIPS'18), pp. 3543–3554. Curran Associates Inc., Red Hook (2018)
16. Hardt, M., Price, E., Srebro, N.: Equality of opportunity in supervised learning. In: Proceedings of the 30th International Conference on Neural Information Processing Systems, pp. 3315–3323
17. Gruber, S.G., Buettner, F.: Better uncertainty calibration via proper scores for classification and beyond. In: Oh, A.H., Agarwal, A., Belgrave, D., Cho, K. (eds.) NeurIPS 2022: Advances in Neural Information Processing Systems (2022)
18. Roelofs, R., Cain, N., Shlens, J., Mozer, M.C.: Mitigating bias in calibration error estimation. In: Camps-Valls, G., Ruiz, F.J.R., and Valera, I. (eds.) AISTATS 2022: The 25th International Conference on Artificial Intelligence and Statistics (Proceedings of Machine Learning Research), vol. 151, pp. 4036–4054. PMLR (2022)
19. Murphy, A.H.: A new vector partition of the probability score. J. Appl. Meteorol. **12**(4), 595–600 (1973)
20. Knopman, D.S., et al.: Age and neurodegeneration imaging biomarkers in persons with Alzheimer disease dementia. Neurology **87**(7), 691–698 (2016)
21. Dukart, J., Schroeter, M.L., Mueller, K.: Alzheimer's disease neuroimaging initiative: age correction in dementia-matching to a healthy brain. PLoS ONE **6**(7), e22193 (2011)
22. ten Kate, M., et al.: Impact of APOE-ε4 and family history of dementia on gray matter atrophy in cognitively healthy middle-aged adults. Neurobiol. Aging **38**, 14–20 (2016)
23. Cacciaglia, R., et al.: Effects of APOE-ε4 allele load on brain morphology in a cohort of middle-aged healthy individuals with enriched genetic risk for Alzheimer's disease. Alzheimers Dement. **14**(7), 902–912 (2018)

24. Stern, Y.: Cognitive reserve. Neuropsychologia **47**(10), 2015–2028 (2009)
25. Buolamwini, J., Gebru, T.: Gender shades: intersectional accuracy disparities in commercial gender classification. In: Proceedings of Machine Learning Research, vol. 81, pp. 77–91 (2018)
26. Seyyed-Kalantari, L., et al.: Underdiagnosis bias of artificial intelligence algorithms applied to chest radiographs in under-served patient populations. Nat. Med. **27**, 2176–2182 (2021)

Distributionally Robust Optimization and Invariant Representation Learning for Addressing Subgroup Underrepresentation: Mechanisms and Limitations

Nilesh Kumar, Ruby Shrestha$^{(\boxtimes)}$, Zhiyuan Li, and Linwei Wang

Rochester Institute of Technology, Rochester, NY, USA
rs9466@rit.edu

Abstract. Spurious correlation caused by subgroup underrepresentation has received increasing attention as a source of bias that can be perpetuated by deep neural networks (DNNs). Distributionally robust optimization has shown success in addressing this bias, although the underlying working mechanism mostly relies on upweighting under-performing samples as surrogates for those underrepresented in data. At the same time, while invariant representation learning has been a powerful choice for removing nuisance sensitive features, it has been little considered in settings where spurious correlations are caused by significant underrepresentation of subgroups. In this paper, we take the first step to better *understand* and *improve* the mechanisms for debiasing spurious correlation due to subgroup underrepresentation in medical image classification. Through a comprehensive evaluation study, we first show that 1) generalized reweighting of under-performing samples can be problematic when bias is not the only cause for poor performance, while 2) naive invariant representation learning suffers from spurious correlations itself. We then present a novel approach that leverages robust optimization to facilitate the learning of invariant representations at the presence of spurious correlations. Finetuned classifiers utilizing such representation demonstrated improved abilities to reduce subgroup performance disparity while maintaining high average and worst-group performance.

Keywords: Spurious correlations · DRO · Invariant representations

1 Introduction

As deep neural networks (DNNs) continue to demonstrate successes in various tasks [7,18], their potential to generate and perpetuate bias also received growing attention [14,16]. One source of biases being increasingly discussed is

---

N. Kumar and R. Shrestha—These authors contributed equally to this work

S. Wesarg et al. (Eds.): CLIP/FAIMI/EPIMI 2023, LNCS 14242, pp. 183–193, 2023.
https://doi.org/10.1007/978-3-031-45249-9_18

the presence of *spurious correlation* due to *subgroup underrepresentation*, where features irrelevant to a task happen to co-exist with a decision label in the majority of the samples (*bias-aligning* samples) whereas samples that do not exhibit such correlation are underrepresented (*bias-conflicting* samples). Examples may include the presence of treatment features in positive disease samples (*e.g.*, the presence of drain as a treatment was observed in the majority of pneumothorax samples in CXR-14 chest X-rays [16]), or a higher prevalence of disease in certain demographic subgroups (*e.g.*, malignant skin cancer was more commonly reported in lighter skin tones [9]). With the standard *empirical risk minimization* (ERM), [23], the DNN is trained to exploit such spurious correlation for reducing the *average* loss of all training samples [14]. Such DNN will struggle with bias-conflicting subgroups.

Most works have approached this bias from the lens of distributionally robust optimization (DRO), a classic optimization concept that focuses on minimizing worst-case losses over possible test distributions [1]. The recent integration of DRO into DNN was done by describing possible test distributions with importance sampling of the observed subgroups [20]: the group DRO (GDRO) is then solved by iteratively optimizing the importance weights of all subgroups to maximize the DNN loss while optimizing the DNN to minimize such worst-case loss. This min-max formulation enables GDRO to pay equal or even greater attention to underrepresented subgroups, although it requires the subgroup labels to be known *a priori*. To remove this need, various approaches have followed to identify bias-conflicting samples by, for instance, considering high-loss samples in a biased model [12,15] or clustering the biased representations [22]. At the core of all of these approaches is *a generalized re-weighting scheme to achieve DRO, i.e.*, minimizing worst-case losses by upweighting the contribution from bias-conflicting subgroups. A fundamental working mechanism behind these approaches, however, assumes that high training loss of the ERM model can be used to determine the upweighting of bias-conflicting samples. While this was shown to be empirically effective on carefully designed benchmark datasets where spurious correlation is the dominating cause for poor performance, it is not clear how this working mechanism may generalize in real-world datasets where the high loss of a training sample may be due to other challenges such as the difficulty to extract discriminative semantics for certain disease categories.

In parallel, *invariant representation learning* has been a mainstream approach for removing confounding semantics not responsible for a task, especially for training DNNs that can adapt or generalize across domains [4]. It has also been widely used in *fair representation learning* to obfuscate sensitive group attributes from latent representations [3,13,17,21,27]. These works however either do not deal with spurious correlations due to underrepresented subgroups, deal with limited spurious correlation resulting from insubstantial bias ratio, or make the assumption that there exists a curated subset of data (for example, a training class) that does not exhibit subgroup underrepresentation. Within limited attempts of using invariant representations to handle spurious correlation, there are often specific assumptions of what may be the spurious factors (*e.g.* texture of the image) and thus lacks generality [24]. To the best of our knowledge, there is very limited work on general purpose methods to learn representations invariant to spurious correlations [26]. This leaves another intriguing open question: what may be the role and challenges of invariant representation learning in addressing spurious correlation caused by underrepresentation?

In a nutshell, despite increasing interest in DNN fairness and bias [25], approaches to specifically address spurious correlation caused by subgroup representation have been limited in medical image classification [8,22] and their working debiasing mechanisms and limitations are not well understood. In this paper, we take the first step to better *understand* and *improve* the debiasing mechanism for addressing spurious correlation in medical image classification. To understand the potential mechanisms for debiasing and corresponding limitations (Sect. 2), we conduct a comprehensive evaluation study in which we examine the aforementioned families of approaches to address spurious correlation on two skin lesion datasets. Results suggested that 1) generalized reweighting based on under-performing samples can be problematic in real-world datasets where bias is not the only cause for low performance, and 2) invariant representation learning suffers from spurious correlation and subgroup underrepresentation itself. As a first step to address these observed limitations, we further show that – while GDRO in itself has little influence on representation learning – its combination with domain adversarial loss can enable invariant representation learning in the presence of spurious correlation (Sect. 3). Fine-tuning the classifier using such representation on balanced validation dataset, as inspired by deep feature reweighting (DFR) [10], further improved the reduction of subgroup disparity while achieving high average performance.

2 Assessing Debiasing Mechanisms

2.1 Methodology

Data: We consider two skin lesion datasets that exhibit different levels of complexity in the underlying tasks, biases, and the number of subgroups.

ISIC [2]: The spurious correlation in ISIC was originally caused by the presence of bandages in benign examples and the absence of bandages in malignant examples [5]. Here we introduce a small subgroup of such malignant samples in the training to assess bias due to such underrepresentation. This is achieved by artificially adding bandages to malignant examples using the segmentation masks for bandages from [19]. Table 1 shows the distribution of training data.

Fitzpatrick [6]: As summarized in Table 1, in Fitzpatrick, there is a higher prevalence of malignant skin cancers in lighter skin tones (Fitzpatrick skin type 1–2) versus a higher prevalence of non-neoplastic lesions in darker skin tones (Fitzpatrick skin type 5–6). This shift of label distribution among skin-type subgroups may create a bias of under-diagnosis of malignant skin cancers in individuals with darker skin tones. To focus on this potential bias, we reduce overall skin-type imbalance by downsampling samples from skin types 1–4, while maintaining the same label distribution within each skin-type subgroup.

Table 1. Subgroup distribution in training. Bias-conflicting subgroups shaded gray.

Dataset	Subgroup Distribution in Training Data			
ISIC	Benign-no bandage	Benign-bandage	Malignant-no bandage	Malignant-bandage
	4843	4890	5205	100

Fitzpatrick Skin type Class	1	2	3	4	5	6
Benign	15.07%	13.96%	14.36%	13.20%	10.37%	6.93%
Malignant	15.37%	15.43%	13.78%	10.82%	9.59%	9.61%
Non-neoplastic	69.56%	70.61%	71.86%	75.98%	80.04%	83.46%

Models: We consider three families of models to provide complete coverage of existing strategies focused on DRO and/or representation learning.

Generalized reweighting approaches: Representing approaches using known subgroup labels, we consider simple importance weighting (with fixed subgroup weights) and GDRO [20] (with dynamic weights). Representing approaches with unknown subgroups, we consider JTT [12] where bias-conflicting samples are identified as under-performing samples in ERM, followed by upweighting.

Invariant representation learning: We consider a classic invariant representation learning strategy that removes domain information from the latent representation by a reversal gradient layer (known as domain adversarial training of neural networks/DANN) [4]. We use the bias-inducing factor as the domain, *i.e.* the presence and absence of bandage in ISIC and the skin types in Fitzpatrick. We also consider the only existing general-purpose invariant representation learning approach reported for spurious correlation, where contrastive learning is used to push representations of samples of the same class closer (CnC) [26].

Separated representation & decision-boundary learning: We include deep feature reweighting (DFR) [10] that first obtains biased ERM representation, and then re-trains a classifier using such representations on a balanced validation set.

Evaluation Metrics: We consider three types of evaluation metrics.

To *evaluate bias in representations,* we leverage self-organizing maps (SOM) [11] to cluster and visualize the latent representations. We also adopt a quantitative metric of *cluster purity* as a surrogate measure of biases in the latent representations. This is motivated by the concept that, in an ideal scenario, an unbiased representation should discard and thus not be separable by spuriously-correlated features.

To *evaluate bias in the decision boundary,* we consider two general types of quantitative metrics: 1) test performance for the worst-performing subgroup, the anticipated bias-conflicting subgroups, and averages across all subgroups; 2) The performance disparity (Δ) among all subgroups: in ISIC, this is measured as performance difference between worst-performing subgroups to best-performing subgroups ($\Delta_{best-worst}$) and to group-average performance ($\Delta_{avg-worst}$), respectively. In Fitzpatrick, because the number of subgroups is large (3×6), this is measured by Δ between best- and worst-performing subgroups in each class:

Table 2. Accuracy of benign *vs.* malignant classification on ISIC. Malignant samples with a bandage is the underrepresented bias-conflicting group as expected from Table 1. Columns 2–4 list accuracy of averaged and individual subgroups. Columns 5–6 measure the performance disparity among subgroups. Best performance is shaded gray, with bolded performance closely behind.

	Avg	Worst	Malignant Bandage	$\Delta_{best-worst}$	$\Delta_{avg-worst}$
ERM	0.803 ± 0.005	0.556 ± 0.072	0.556 ± 0.072	0.443 ± 0.072	0.246 ± 0.066
Important weighting	0.793 ± 0.011	0.593 ± 0.040	0.593 ± 0.040	0.406 ± 0.040	0.200 ± 0.030
JTT	0.790 ± 0.011	0.550 ± 0.051	0.550 ± 0.051	0.450 ± 0.051	0.240 ± 0.041
DANN	0.843 ± 0.005	0.730 ± 0.017	0.730 ± 0.017	0.266 ± 0.023	0.113 ± 0.011
DFR	0.871 ± 0.004	0.758 ± 0.008	0.865 ± 0.010	0.232 ± 0.010	0.112 ± 0.010
GDRO	0.846 ± 0.011	0.760 ± 0.017	0.763 ± 0.015	0.240 ± 0.017	0.086 ± 0.005
GDRO with group adjustment	$\mathbf{0.866 \pm 0.005}$	0.786 ± 0.011	0.803 ± 0.020	0.213 ± 0.011	0.080 ± 0.005
Proposed method	$\mathbf{0.863 \pm 0.005}$	$\mathbf{0.800 \pm 0.000}$	$\mathbf{0.813 \pm 0.023}$	0.190 ± 0.000	0.063 ± 0.005

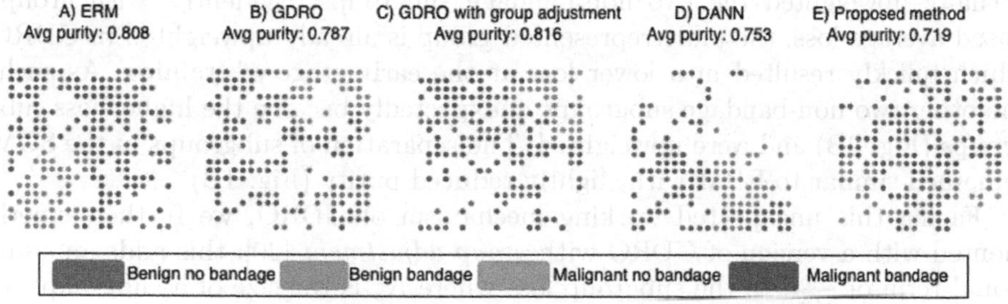

Fig. 1. SOM plots and their averaged subgroup purities.

malignant, non-neoplastic, and benign ($\Delta_{mn}, \Delta_{nn}, \Delta_{bg}$) as well as their average (Δ_{avg}).

To *assess the debiasing mechanisms,* for generalized reweighting with and without known subgroups, we track their importance weights, or the subgroups identified for up-weighting, respectively. For learning invariant representations, we track the accuracy of the domain classifier in removing spurious features.

2.2 Experiments and Results

Following convention [20, 26], we used ResNet-50 from torch-vision with pre-trained weights from Imagenet. We used held-out validation set to choose the best model unless the method is using the validation set for fine-tuning (*e.g.,* DFR).

ISIC: As summarized in Table 2, with ERM, the DNN suffered from a significant performance drop for the bias-conflicting subgroup (malignant samples with bandage). This bias is also evident in the SOM (Fig. 1A), which shows clearly separated subgroups and a relatively high purity score.

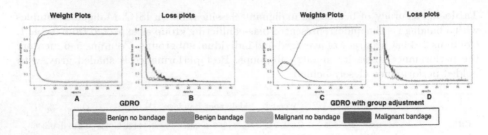

Fig. 2. Weight and loss plots for GDRO with and without group adjustment.

Generalized reweighting approaches: Simple importance weighting (to equally weight all subgroups) improved the test accuracy on the bias-conflicting group but to a limited extent. GDRO more significantly improved the test performance of the bias-conflicting subgroup. Interestingly, a closer inspection on the subgroup weights obtained by the GDRO (Fig. 2A) showed that GDRO actually upweighted the two non-bandage subgroups. Evidently, with group-based average loss, the underrepresented group is already upweighted in GDRO which quickly resulted in a lower loss at the early stage of training. As such, the other two non-bandage subgroups unexpectedly became the higher-loss subgroups (Fig. 2B) and were upweighted. The separation of subgroups in the SOM remained similar to ERM with slightly reduced purity (Fig. 1B).

Facing this unexpected working mechanism of GDRO, we further experimented with a version of GDRO with *group adjustment* [20]: this adds an additional term of $\frac{1}{\sqrt{N_s}}$ to the subgroup loss where N_s is the size of a subgroup. As expected, this term dominates the loss used to optimize the weights (Fig. 2D) and resulted in a heavy upweighting of the bias-conflicting group (Fig. 2C). This resulted in further improvement on this subgroup. Clustering of the latent representation and its purity however remained similar to the ERM (Fig. 1C).

These results revealed that the use of group adjustment played a significant role in the weighting mechanism of GDRO. Without this, the upweighting was influenced by other factors contributing to low performance and the overall results appeared to be adversely affected.

Invariant representation approaches: DANN was trained with upsampling of the underrepresented subgroup. Its performance was better than simple importance weighting, but inferior to GDRO. A closer look at the obtained representation clusters (Fig. 1D) and the behaviour of the domain classifier (Fig. 3A) suggested that DANN was not successful in removing bandage information from the bias-aligning subgroup (benign with bandage). This failure suggested that domain-invariant representation learning also suffers from spurious correlations.

Approaches without subgroup labels: For the remaining models that do not require the use of subgroup labels, the performance was suboptimal. JTT showed minor improvement over ERM, while CnC under-performed than ERM (results thus not included). Common to both approaches was the reliance on a biased ERM model to identify bias-conflicting examples. A closer inspection showed that, among the under-performing samples selected from the biased ERM, only 1.05–2.75% belonged to the bias-conflicting subgroup. This explained why the subsequent learning approaches may not be successful. Finally, DFR achieved an impressive average performance as well as on the bias-conflicting subgroup, albeit at the expense of creating a new worst-performing subgroup.

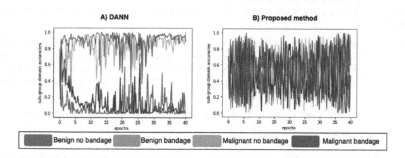

Fig. 3. Subgroup accuracy of domain classifier in A) DANN and B) as proposed.

Fitzpatrick: Table 3 shows the average AUC, worst-AUC among all 18 subgroups, and AUCs for four selected subgroups that are most indicative of the bias present: with ERM, the subgroups with darker skin tones are disadvantaged in malignant cancer detection (MN-56, AUC of 0.81 *vs.*, 0.88 for MN-12), whereas the subgroups with lighter skin tones are disadvantaged in non-neoplastic lesion detection (NN-12, AUC of 0.78 *vs.*, 0.86 for NN-56). This disparity in performance is perfectly associated with the subgroup label distributional shifts shown in Table 1, an intriguing non-artificial bias that may be challenging to address.

The trend of performance of all models was similar to that observed in ISIC with some distinction. JTT and CnC continued to under-perform and their results are not included. Unlike on ISIC, here both DANN and DFR obtained negligible or even negative gains over ERM. In comparison, various reweighting strategies were all effective in addressing the bias against MN-56, albeit 1) at the expense of decreased average AUC, and 2) ineffective in addressing the bias against NN-12. The measures of performance disparity within and averaged across classes (Table 4) suggested the same trend.

Note that the worst-performance subgroups from all models were different from the bias-conflicting subgroups, which may be why this particular bias was challenging to remove.

3   Improving the Debiasing of Spurious Correlations

Methodology: The investigations in Sect. 2 provided two important findings. First, the reliance on low training performance for identifying or upweighting bias-conflicting samples may face challenges where bias is not the only cause

Table 3. AUC of benign, malignant, and non-neoplastic skin lesion classification on Fitzpatrick. MN-12/-56: malignant with skin type 1–2/5–6. NN-12/56: non-neoplastic with skin type 1–2/5–6. MN-56 and NN-12 are the primary bias-conflicting subgroups.

	Average	Worst	MN-12	MN-56	NN-12	NN-56
ERM	0.810 ± 0.000	0.617 ± 0.040	0.878 ± 0.008	0.810 ± 0.018	0.777 ± 0.010	0.855 ± 0.005
Importance weighting	0.803 ± 0.015	0.673 ± 0.035	0.878 ± 0.010	0.853 ± 0.008	0.780 ± 0.001	0.853 ± 0.010
DANN	0.787 ± 0.006	0.607 ± 0.015	0.857 ± 0.010	0.758 ± 0.014	0.785 ± 0.005	0.800 ± 0.015
DFR	0.790 ± 0.010	0.660 ± 0.036	0.863 ± 0.003	0.813 ± 0.020	0.778 ± 0.008	0.828 ± 0.010
GDRO	0.800 ± 0.010	0.640 ± 0.030	0.868 ± 0.008	0.848 ± 0.055	0.788 ± 0.012	0.825 ± 0.020
GDRO with group adjustment	0.780 ± 0.010	0.647 ± 0.006	0.852 ± 0.008	0.877 ± 0.016	0.782 ± 0.015	0.788 ± 0.022
Proposed method	0.810 ± 0.000	0.680 ± 0.026	0.883 ± 0.003	0.837 ± 0.003	0.803 ± 0.003	0.830 ± 0.009

Table 4. Difference in AUCs between best- and worst-performing subgroups in each class, from left to right: malignant, non-neoplastic, benign, and averaged.

	Δ_{mn}	Δ_{nn}	Δ_{bg}	Δ_{avg}
ERM	0.140 ± 0.026	0.083 ± 0.015	0.223 ± 0.032	0.149 ± 0.020
Importance Weighting	0.073 ± 0.021	0.080 ± 0.020	0.153 ± 0.006	0.102 ± 0.010
DANN	0.250 ± 0.036	0.040 ± 0.010	0.233 ± 0.015	0.174 ± 0.018
DFR	0.157 ± 0.040	0.053 ± 0.006	0.150 ± 0.030	0.120 ± 0.021
GDRO	0.090 ± 0.087	0.053 ± 0.025	0.197 ± 0.055	0.113 ± 0.055
GDRO with group adjustment	0.053 ± 0.015	0.030 ± 0.010	0.163 ± 0.021	**0.082 ± 0.004**
Proposed method	0.090 ± 0.010	**0.037 ± 0.006**	0.117 ± 0.031	0.081 ± 0.008

for under-performing samples. Second, while invariant representation in concept appears a natural candidate for removing confounders, its learning ironically also suffers from spurious correlation caused by underrepresentation. This raises a critical question: is our best hope upweighting known bias-conflicting subgroups?

We re-examine DANN's failure to confuse the domain classifier on the bias-aligned subgroups (Fig. 3A). A possible explanation is that the main task classifier was exploiting spurious correlations and overpowers the domain classifier. Based on this, we hypothesize that a robust main classifier with a reduced tendency to exploit spurious correlation may better support DANN to learn representations invariant to spurious correlations. To this end, we propose to optimize the main classifier in DANN with GDRO and, with the learned invariant representation, fine-tune only the classifier on a small validation set.

Experiments and Results: On ISIC, the proposed model was successful in extracting invariant representations, evidenced both by the confused DANN domain classifier (Fig. 3B) and the substantially reduced SOM purity (Fig. 1E). After fine-tuning this invariant representation on a small balanced validation set as used in DFR, the proposed approach achieved the highest worst-performance among all models considered (Table 2). Its performance on the bias-conflicting subgroup was better than GDRO with group adjustment, and its average performance on par. Note that while DFR exhibited a stronger average performance on the bias-conflicting subgroup, its worst-group performance was much lower. In comparison, the proposed method was the most successful in reducing subgroup disparity (last two columns in Table 2) with strong average performance.

Evidence of improved invariant representations was similar on Fitzpartrick (results shown in supplemental materials).

As summarized in Tables 3–4, compared to reweighting methods, the fine-tuned classifier as proposed was the only one that was able to reduce the bias against NN-12, and delivered the best worst-group as well as average performance. Overall, its performance in removing subgroup disparity was on par with GDRO with group adjustment (Table 4) while delivering significantly higher average and worst-group AUCs (Table 3).

Conclusions and Discussion: We presented an evaluation study that derived important new insights into the working mechanisms and limitations of DRO and invariant representation learning to address spurious correlation caused by underrepresentation. The findings motivated us to present a novel approach that leverages robust optimization to facilitate the learning of invariant representations at the presence of spurious correlations. Finetuned classifiers utilizing such representation demonstrated an improved ability to reduce subgroup performance disparity while maintaining high average and worst-group performance. Future investigations will include a broader spectrum of approaches including those utilizing data augmentation, as well as extending to a wider range of medical image datasets exploring potential hidden biases.

Acknowledgments. This work is supported by the National Institute of Nursing Research (NINR) of the National Institutes of Health (NIH) under Award Number R01NR018301.

References

1. Ben-Tal, A., Den Hertog, D., De Waegenaere, A., Melenberg, B., Rennen, G.: Robust solutions of optimization problems affected by uncertain probabilities. Manage. Sci. **59**(2), 341–357 (2013)
2. Codella, N., et al.: Skin lesion analysis toward melanoma detection 2018: a challenge hosted by the international skin imaging collaboration (ISIC). arXiv preprint: arXiv:1902.03368 (2019)
3. Deng, W., Zhong, Y., Dou, Q., Li, X.: On Fairness of medical image classification with multiple sensitive attributes via learning orthogonal representations. In: Frangi, A., de Bruijne, M., Wassermann, D., Navab, N. (eds.) IPMI 2023. Lecture Notes in Computer Science, vol. 13939. Springer, Cham (2023). https://doi.org/10.1007/978-3-031-34048-2_13
4. Ganin, Y., et al.: Domain-adversarial training of neural networks (2015). https://doi.org/10.48550/arXiv.1505.07818

5. Goel, K., Gu, A., Li, Y., Ré, C.: Model patching: closing the subgroup performance gap with data augmentation. arXiv preprint: arXiv:2008.06775 (2020)
6. Groh, M., et al.: Evaluating deep neural networks trained on clinical images in dermatology with the fitzpatrick 17k dataset. In: Proceedings of the IEEE/CVF Conference on Computer Vision and Pattern Recognition, pp. 1820–1828 (2021)
7. He, K., Zhang, X., Ren, S., Sun, J.: Deep residual learning for image recognition. In: Proceedings of the IEEE Conference on Computer Vision and Pattern Recognition, pp. 770–778 (2016)
8. Jiménez-Sánchez, A., Juodelye, D., Chamberlain, B., Cheplygina, V.: Detecting shortcuts in medical images-a case study in chest X-rays. arXiv preprint: arXiv:2211.04279 (2022)
9. Kinyanjui, N.M., et al.: Fairness of classifiers across skin tones in dermatology. In: Martel, A.L., et al. (eds.) MICCAI 2020. LNCS, vol. 12266, pp. 320–329. Springer, Cham (2020). https://doi.org/10.1007/978-3-030-59725-2_31
10. Kirichenko, P., Izmailov, P., Wilson, A.G.: Last layer re-training is sufficient for robustness to spurious correlations. arXiv preprint: arXiv:2204.02937 (2022)
11. Kohonen, T.: The self-organizing map. Proc. IEEE **78**(9), 1464–1480 (1990)
12. Liu, E.Z., et al.: Just train twice: Improving group robustness without training group information. In: Meila, M., Zhang, T. (eds.) Proceedings of the 38th International Conference on Machine Learning. Proceedings of Machine Learning Research, vol. 139, pp. 6781–6792. PMLR (2021). https://proceedings.mlr.press/v139/liu21f.html
13. Louizos, C., Swersky, K., Li, Y., Welling, M., Zemel, R.: The variational fair autoencoder (2015). https://doi.org/10.48550/arXiv.1511.00830
14. McCoy, T., Pavlick, E., Linzen, T.: Right for the wrong reasons: diagnosing syntactic heuristics in natural language inference. In: Proceedings of the 57th Annual Meeting of the Association for Computational Linguistics, pp. 3428–3448. Association for Computational Linguistics, Florence (2019). https://doi.org/10.18653/v1/P19-1334, https://aclanthology.org/P19-1334
15. Nam, J., Cha, H., Ahn, S., Lee, J., Shin, J.: Learning from failure: De-biasing classifier from biased classifier. In: Advances in Neural Information Processing Systems, vol. 33, pp. 20673–20684 (2020)
16. Oakden-Rayner, L., Dunnmon, J., Carneiro, G., Ré, C.: Hidden stratification causes clinically meaningful failures in machine learning for medical imaging. In: Proceedings of the ACM Conference on Health, Inference, and Learning, pp. 151–159 (2020)
17. Park, S., Hwang, S., Kim, D., Byun, H.: Learning disentangled representation for fair facial attribute classification via fairness-aware information alignment. In: Proceedings of the AAAI Conference on Artificial Intelligence, vol. 35, pp. 2403–2411 (2021)
18. Rajpurkar, P., et al.: CheXNet: radiologist-level pneumonia detection on chest X-rays with deep learning. arXiv preprint: arXiv:1711.05225 (2017)
19. Rieger, L., Singh, C., Murdoch, W., Yu, B.: Interpretations are useful: penalizing explanations to align neural networks with prior knowledge. In: International Conference on Machine Learning, pp. 8116–8126. PMLR (2020)
20. Sagawa, S., Koh, P.W., Hashimoto, T.B., Liang, P.: Distributionally robust neural networks for group shifts: on the importance of regularization for worst-case generalization. arXiv preprint: arXiv:1911.08731 (2019)
21. Sarhan, M.H., Navab, N., Eslami, A., Albarqouni, S.: Fairness by Learning orthogonal disentangled representations. In: Vedaldi, A., Bischof, H., Brox, T., Frahm,

J.-M. (eds.) ECCV 2020. LNCS, vol. 12374, pp. 746–761. Springer, Cham (2020). https://doi.org/10.1007/978-3-030-58526-6_44

22. Sohoni, N., Dunnmon, J., Angus, G., Gu, A., Ré, C.: No subclass left behind: Fine-grained robustness in coarse-grained classification problems. In: Advances in Neural Information Processing Systems, vol. 33, pp. 19339–19352 (2020)

23. Vapnik, V.: Principles of risk minimization for learning theory. in: Advances in Neural Information Processing Systems, vol. 4 (1991)

24. Wang, H., He, Z., Lipton, Z.C., Xing, E.P.: Learning robust representations by projecting superficial statistics out. arXiv preprint: arXiv:1903.06256 (2019)

25. Wu, Y., Zeng, D., Xu, X., Shi, Y., Hu, J.: FairPrune: achieving fairness through pruning for dermatological disease diagnosis. In: Wang, L., Dou, Q., Fletcher, P.T., Speidel, S., Li, S. (eds.) Medical Image Computing and Computer Assisted Intervention - MICCAI 2022. Lecture Notes in Computer Science, pp. 743–753. Springer, Cham (2022). https://doi.org/10.1007/978-3-031-16431-6_70

26. Zhang, M., Sohoni, N.S., Zhang, H.R., Finn, C., Re, C.: Correct-N-contrast: a contrastive approach for improving robustness to spurious correlations. In: Chaudhuri, K., Jegelka, S., Song, L., Szepesvari, C., Niu, G., Sabato, S. (eds.) Proceedings of the 39th International Conference on Machine Learning. Proceedings of Machine Learning Research, vol. 162, pp. 26484–26516. PMLR (2022). http://proceedings.mlr.press/v162/zhang22z.html

27. Zhao, Q., Adeli, E., Pohl, K.M.: Training confounder-free deep learning models for medical applications. Nat. Commun. 11(1), 6010 (2020)

Analysing Race and Sex Bias in Brain Age Prediction

Carolina Piçarra[(✉)] and Ben Glocker

Department of Computing, Imperial College London, London, UK
c.picarra@imperial.ac.uk

Abstract. Brain age prediction from MRI has become a popular imaging biomarker associated with a wide range of neuropathologies. The datasets used for training, however, are often skewed and imbalanced regarding demographics, potentially making brain age prediction models susceptible to bias. We analyse the commonly used ResNet-34 model by conducting a comprehensive subgroup performance analysis and feature inspection. The model is trained on 1,215 T1-weighted MRI scans from Cam-CAN and IXI, and tested on UK Biobank (n=42,786), split into six racial and biological sex subgroups. With the objective of comparing the performance between subgroups, measured by the absolute prediction error, we use a Kruskal-Wallis test followed by two post-hoc Conover-Iman tests to inspect bias across race and biological sex. To examine biases in the generated features, we use PCA for dimensionality reduction and employ two-sample Kolmogorov-Smirnov tests to identify distribution shifts among subgroups. Our results reveal statistically significant differences in predictive performance between Black and White, Black and Asian, and male and female subjects. Seven out of twelve pairwise comparisons show statistically significant differences in the feature distributions. Our findings call for further analysis of brain age prediction models.

1 Introduction

The global population growth and longer life expectancy are linked to the rising prevalence of age-related neurodegenerative and neuropsychiatric diseases [1–3]. As a result, there is an increasing need to establish connections between brain ageing and disease processes, to better understand their mechanisms and enable early detection and diagnosis. Significant research efforts have focused on investigating the potential of brain-predicted age as an indicator of how an individual's brain health may deviate from the norm [4,5]. As a neuroimaging-driven biomarker, it has the potential of containing a broad spectrum of brain characteristics in a single measurement [6]. Several studies have proposed brain age prediction for the characterisation of neuropathology [7,8], epilepsy [9], as well as an indicator of clinical risk factors [10,11]. Most studies used structural MRI, due to its common use in clinical settings and high resolution, capturing even

© The Author(s), under exclusive license to Springer Nature Switzerland AG 2023
S. Wesarg et al. (Eds.): CLIP/FAIMI/EPIMI 2023, LNCS 14242, pp. 194–204, 2023.
https://doi.org/10.1007/978-3-031-45249-9_19

small structural variations in brain anatomy. Deep learning (DL), and in particular convolutional neural networks (CNNs), are widely used models for brain age prediction from MRI [12,13]. Studies rely on well-established datasets, including the UK Biobank [14], the Cambridge Centre for Ageing Neuroscience (Cam-CAN) dataset [15], IXI [16], the Alzheimer's Neuroimaging Initiative (ADNI) dataset [17], the The Open Access Series of Imaging Studies (OASIS) [18], among others. These datasets tend to be skewed and biased regarding ethnic and racial diversity, with a majority of White subjects. When models are trained on data with unbalanced demographics, the performance may degrade in relevant subgroups [19]. Thus, it is important to test such models for potentially disparate performance across subgroups. In this study, we analyse a ResNet-34 brain age prediction model by conducting a comprehensive statistical subgroup performance analysis and feature inspection.

2   Materials and Methods

Datasets. For training and validation of the brain age prediction model, we used the Cam-CAN [15] and the IXI dataset with healthy volunteers. For testing, the UK Biobank dataset was selected due to its size and availability of race and biological sex information. The demographics for each dataset are available in Table 1. Patient racial information is not provided for the Cam-CAN dataset. However, considering that the data collection took place in Cambridge (United Kingdom), we assume the majority of volunteers were White. All scans from Cam-CAN and IXI were pre-processed by us using the following steps: 1) Lossless image reorientation using the direction information from the image header; 2) Skull stripping with ROBEX v1.2[1] [20]; 3) Intensity-based rigid registration to MNI atlas ICBM 152 2009a Nonlinear Symmetric[2]; 4) Bias field correction with N4ITK[3] [21]. The UK Biobank images were already skull-stripped and bias field corrected, and only the registration to MNI space was performed by us.

Model. We adapted the conventional ResNet-34 model [22] for age regression from 3D images. ResNet stands for Residual Network and is a type of CNN model with residual connections, a distinctive architecture designed to address the vanishing gradient problem during deep network training. We trained this model with whole preprocessed T1-weighted MRI images. The data was augmented through a composition of transformations, including random horizontal flip, contrast change, addition of Gaussian noise with random parameters and motion artifacts.

2.1   Bias Analysis

We divided our statistical bias analysis into two parts, each focusing on a specific aspect. The first part aimed to assess bias in predictive performance, while the

---

[1] https://www.nitrc.org/projects/robex.

[2] http://nist.mni.mcgill.ca/?p=904.

[3] https://itk.org.

Table 1. Demographic information of all datasets used.

	Cam-CAN	IXI	UK Biobank
N	652	563	42,786
Age (years)			
Mean ± SD	54.3 ± 18.6	48.6 ± 16.5	64.0 ± 7.7
Range	18–88	20–86	44–82
Sex			
Female/Male	330/332	313/250	20,206/22,580
Race			
White	—	451	41,417
Black	—	14	286
Asian	—	50	454
Chinese	—	14	122
Other	—	34	507

latter delved deeper into the model to examine biases in the generated features. To ensure a sufficient sample size for each subgroup, we considered the Chinese subjects to be part of the Asian group, and excluded all subjects with race classified as "Other" (which includes "Mixed"). We then further divided each racial subgroup ("White", "Asian" and "Black") into "Female" and "Male", resulting in six test set subgroups.

Absolute Performance Assessment. We calculated the absolute error of prediction, using it as the main performance metric. With the goal of comparing the performance between all subgroups, we then progressed by verifying the assumptions necessary to perform an Analysis of Variance (ANOVA), i.e. assumption of normality - through visual inspection of the absolute error distribution and Shapiro-Wilk tests - and the assumption of homogeneity of variances, through the Levene's test. The assumption of sample independence is met from the experimental design, as all subgroups are constituted by different subjects. Given that not all assumptions were met, we progressed by using the non-parametric Kruskal-Wallis test to compare the absolute error medians of all subgroups. Further pairwise comparisons were completed using the post-hoc Conover-Iman test. Since the Kruskal-Wallis test is the non-parametric equivalent of the one-way ANOVA, we conducted two Conover-Iman tests in order to take into consideration our two factors, race and biological sex. Although the Kruskal-Wallis test can handle unbalanced data, when the differences are large its power is reduced, which may lead to inconsistent/intransitive results [23]. In order to ensure the validity and consistency of our results, we balanced the data by randomly selecting a sample from each subgroup with equal sample size. The sample size chosen for each subgroup was 126, i.e. the size of the smallest group (Black female subjects). After calculating the mean absolute error for each subgroup sample, we repeated the random sampling ten times to estimate the standard deviation. We then repeated the statistical procedure, verifying ANOVA's assumptions, and upon rejection of normality following with the Kruskal-Wallis test and corresponding post-hoc Conover-Iman tests.

Model Features Assessment. Additionally, we assessed if the features generated by the model were biased using the framework for feature exploration proposed by Glocker et al. [24]. This strategy consists of passing each test set scan through our model up to the penultimate layer, extracting its output features and subsequently inputting them to a principal component analysis (PCA) model in order to reduce their dimensionality. The PCA projections consist of a new set of dimensions (also called "modes") which capture the directions of the largest variation in the high-dimensional feature space. Given that our model was trained to predict age, it is expected that the strongest separation for samples in different age groups is seen in the first PCA modes. We plotted the distribution of samples in PCA space (first four modes) through kernel density estimation plots, split by the three demographic attributes of interest (age, biological sex, and race). Age was divided into five brackets to facilitate visual analysis. This was accompanied by two-sample Kolmogorov-Smirnov tests to compare the feature distributions of all possible pairwise combinations across race, sex, and age subgroups, in the first four modes of PCA. To decrease the number of statistical tests to be completed, for this step age was divided into only two brackets, [40–60] and [60–90]. To account for multiple testing and the consequent type 1 error inflation, the p-values were adjusted using the Benjamini-Yekutieli procedure. We considered statistical significance at a 95% confidence level.

3   Results

Figure 1 shows the age distribution for each subgroup, including all samples available in the test set. Within White subjects, we can observe a tendency of younger male and older female subjects, whereas for Black subjects, we find the opposite.

Absolute Performance Assessment. Our first analysis involved conducting a Shapiro-Wilk test to evaluate whether the subgroup's prediction errors followed a normal distribution, and the Levene's test to verify whether all groups to be compared had equal variance. The resulting p-values from the six Shapiro-Wilk tests were below the defined significance level of 0.05. This indicates that we can confidently reject the null hypothesis that the population from which each sample is drawn follows a normal distribution. For visual confirmation, the distribution of the absolute error for each subgroup, along with the corresponding p-values from the Shapiro-Wilk tests are given in the Appendix (Fig. 5). On the other hand, the Levene's test returned a p-value of 5.82×10^{-52}, confirming that we have sufficient evidence to reject the null hypothesis and conclude that not all samples come from populations with equal variances. Upon the rejection of both assumptions for two-way ANOVA, we proceeded by conducting a Kruskal-Wallis test. The resulting p-value was 6.99×10^{-116}, leading us to reject the null

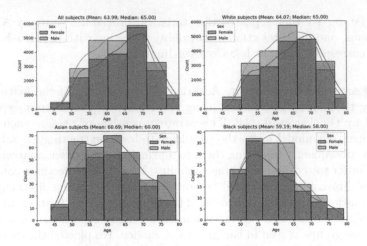

Fig. 1. Age distribution of all subjects in the test set and of each racial subgroup, separated by biological sex. Overlapping lines show the probability density curves.

hypothesis, i.e. that the population medians are all equal. The resulting p-values from the two Conover-Iman tests conducted for further pairwise comparisons were as follows: White vs Asian: 0.022; White vs Black: 0.017; Black vs Asian: 0.0015; female vs male: 4.20×10^{-118}. As suspected, the p-values from both the Kruskal-Wallis and the Conover-Iman test were notably low. The plot in Fig. 2 shows the age distribution of the random samples from each subgroup (n=126), taken to ensure the robustness of our statistical procedure. It reveals a similar pattern to that observed when examining all subjects in the test set, showing higher prevalence of younger White males and older White females, while the opposite is seen for Black subjects.

The first plot in Fig. 3 shows the mean absolute error (MAE) for each subgroup sample, accompanied by its corresponding error bar, calculated from ten random samples of the same dimension. The adjacent plot illustrates the disparity in absolute error concerning the average absolute error across all subjects. Notably, we see a considerable under-performance of the model for Black male subjects. The model achieves its highest performance on White female and Asian subjects.

Similarly to the tests conducted with all test set subjects, the p-values resulting from the Shapiro-Wilk tests for normality testing indicated that we could reject the null hypothesis that the samples come from a normally-distributed population, with a significance level of 0.05. Contrarily, the Levene's test produced a p-value of 0.78, suggesting that we cannot reject the null hypothesis of equal variances across all samples. However, as not all ANOVA assumptions were met, we proceeded with the Kruskal-Wallis test, which yielded a p-value of 0.0015. With a p-value below our defined significance level (0.05), we reject the

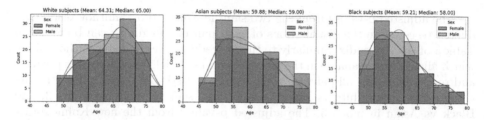

Fig. 2. Age distribution of a random sample (n=126) from each racial subgroup, separated by biological sex. Overlapping lines show the probability density curves.

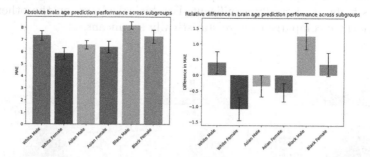

Fig. 3. Left: MAE, considering only a random sample of 126 subjects for each subgroup. Right: Relative difference in brain age prediction performance across patient subgroups. Difference calculated in relation to the average absolute error across all subjects (i.e. all random samples, with a total of 756 subjects). For both plots, the error bars were created by repeating the random sampling ten times and calculating the standard deviation.

null hypothesis and have sufficient evidence to suggest that the differentiating factors among subgroups lead to statistically significant differences in the model's performance. The resulting p-values from the two post-hoc Conover-Iman tests were the following: White vs Asian: 0.76; White vs Black: 0.022; Black vs Asian: 0.013; female vs male: 0.008. These outcomes reveal statistically significant disparities in the model's performance between White and Black subjects and Black and Asian subjects, as well as between female and male subjects.

Model Features Assessment. Proceeding to the examination of bias in the model's features, the kernel density estimation plots presented in Fig. 4 show the density distribution of each age, race, and biological sex subgroup as generated by PCA in its first four modes. These plots include all subjects available in our test set. Here, we can infer that the PCA modes of primary interest are modes 1, 2 and 3, as they show the strongest separation between age groups, aligning with the model's training objective. Therefore, we are particularly interested in examining subgroup differences in these modes. It is nevertheless worth noting that in PCA mode 4 there is a clear separation between racial subgroups. However, these disparities might not be of primary concern as these features may not be informative for age prediction.

The adjusted p-values from the two-sample Kolmogorov-Smirnov tests, conducted to compare the distributions of each pair of subgroups, can be found in Table 3 of the Appendix. Similarly to the procedure described above, the tests in which all subjects available in the test set were used yielded notably low p-values and rendered almost all pairwise comparisons statistically significant, with only three exceptions: White vs Black in mode 2, Asian vs White in mode 3, and Black vs Asian in mode 3. The adjusted p-values from the new Kolmogorov-Smirnov tests, including only a equal-sized sample of each subgroup, are shown in Table 2. When looking at the first three PCA modes and comparing racial subgroups, we find five of the nine pairwise comparisons between marginal distributions to be statistically significant. For biological sex, on the other hand, two of the three comparisons were statistically significant.

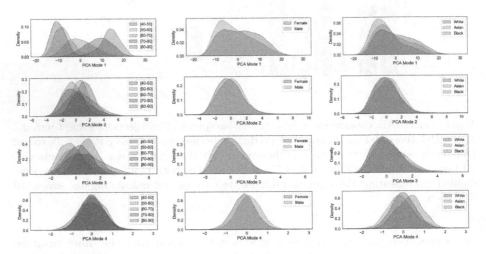

Fig. 4. Kernel density estimation plots depicting the density distribution of each age, race, and biological sex subgroup across the first four PCA modes of the feature space.

4   Discussion and Conclusion

In this study, we aimed to thoroughly investigate the potential race and biological sex bias in a model for brain age prediction from MRI predominantly trained on White subjects. The statistical tests conducted to evaluate the model's absolute performance reveal statistically significant differences for Black subjects, compared to both Asian and White subjects, as well as differences between male and female subjects. When looking back at the model's average performance per

Table 2. P-values resulting from two-sample Kolmogorov-Smirnov tests which compared marginal distributions from the pairs of subgroups indicated, across the first four PCA modes. Results including a random sample of each subgroup, all with equal size. The p-values were adjusted for multiple testing using the Benjamini-Yekutieli procedure. Significance level is set to 0.05. Statistically significant results are coloured with red.

	Age 40-60/60-90	Asian/White	Black/Asian	White/Black	Female/Male
PCA mode 1	<0.0001	0.0065	0.011	<0.0001	0.39
PCA mode 2	<0.0001	0.011	0.0065	1	0.006
PCA mode 3	<0.0001	0.52	0.52	0.13	0.0037
PCA mode 4	0.09	0.25	<0.0001	<0.0001	0.025

subgroup (Fig. 3), we can observe the same evidence, concluding that the negatively affected groups are Black and male subjects. One possible explanation might be the fact that these are the two most underrepresented groups in our training set, which contained 582 male (versus 643 female) subjects, and only 14 Black (versus 451 White and 64 Asian) subjects. The imbalance in racial distribution for the training set only include the IXI dataset as race information was not available for the Cam-CAN dataset. Here, we assumed that Cam-CANis predominantly White. Additionally, the results of our statistical analysis over the model's feature inspection suggests that some of the features that encode information useful for age prediction, also allow for the separation of both racial and biological sex subgroups.

In practicality, recent research primarily focuses on assessing the correlation between brain age gap and neurological disorders/clinical risks. This gap represents the model's prediction error, which can be attributed to noise (model accuracy, data quality) and physiology. When evaluating the latter, it's crucial to distinguish disorder-related changes from inherent biological differences due to sex or ethnicity. This study reveals an average of 1 to 2-year statistically significant disparities among ethnicity subgroups. Depending on biomarker application, these deviations hold significance. For instance, Cole et al. (2018) [10] found a 6.1% rise in mortality risk between ages 72–80 per extra predicted brain year. Accounting for ethnicity could be vital in such cases.

Lange et al. [6] have previously reported that the metrics used to evaluate brain age prediction performance, including MAE, are significantly affected by discrepancies in the age range of the training and testing datasets. One limitation of our study is hence the limited age range of UK Biobank (44–82) - our test set - when compared to the broad range encompassed by the training set (18–88), which is desired for an age prediction model. As a consequence, we might observe a lower overall age prediction performance than the state-of-the-art. Nevertheless, given that our primary goal was to compare the model's performance across subgroups, and that the age range is similar across the random samples of each test subgroup (Fig. 2), we can assume that this evaluation remains meaningful despite age range variations in training and test sets.

Another limitation of our study is the use of a single model type, one combination of datasets and a specific type of input features (T1-weighted MRI scans). However, we believe that our findings are relevant to further motivate a systematic bias assessment, including a diverse range of commonly employed models, such as other CNN models, ensembles, or simpler machine learning models like XGBoost, as these have been shown to have comparable performance to more complex DL models [25]. Another interesting avenue for exploration would be to examine whether the similar biases persist when employing MRI-derived features, e.g. white and grey matter maps or volumes of subcortical structures.

Our results suggest that training brain age prediction on imbalanced data leads to significant differences in subgroup performance. We call for comprehensive bias assessment in other brain age prediction models, as these have emerged as important diagnostic and prognostic clinical tools.

Acknowledgments. B.G. is grateful for the support from the Royal Academy of Engineering as part of his Kheiron Medical Technologies/RAEng Research Chair in Safe Deployment of Medical Imaging AI. C.P gratefully reports financial support provided by UKRI London Medical Imaging & Artificial Intelligence Centre for Value Based Healthcare.

A   Appendix

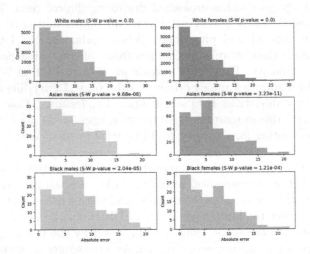

Fig. 5. Histograms showcasing the distribution of the absolute prediction error for each subgroup, including all subjects in test set.

Table 3. P-values resulting from two-sample Kolmogorov-Smirnov tests which compared marginal distributions from the pairs of subgroups indicated, across the first four PCA modes. Results including all samples available in the test set. The p-values were adjusted for multiple testing using the Benjamini-Yekutieli procedure. Significance level is set to 0.05. Statistically significant results are coloured with red.

	Age 40-60/60-90	Asian/White	Black/Asian	White/Black	Female/Male
PCA mode 1	<0.0001	<0.0001	<0.0001	<0.0001	<0.0001
PCA mode 2	<0.0001	<0.0001	0.0071	1	<0.0001
PCA mode 3	<0.0001	0.78	0.44	0.039	<0.0001
PCA mode 4	0.025	<0.0001	<0.0001	<0.0001	<0.0001

References

1. Hou, Y., et al.: Ageing as a risk factor for neurodegenerative disease. Nat. Rev. Neurol. **15**(10), 565–581 (2019)
2. Deuschl, G., et al.: The burden of neurological diseases in Europe: an analysis for the global burden of disease study 2017. Lancet Public Health **5**(10), e551–e567 (2020)
3. Dumurgier, J., Tzourio, C.: Epidemiology of neurological diseases in older adults. Revue Neurologique **176**(9), 642–648 (2020)
4. Baecker, L. et al.: Machine learning for brain age prediction: introduction to methods and clinical applications. EBioMedicine **72** (2021)
5. Cole, J.H., et al.: Predicting brain age with deep learning from raw imaging data results in a reliable and heritable biomarker. Neuroimage **163**, 115–124 (2017)
6. de Lange, A.M.G., et al.: Mind the gap: performance metric evaluation in brain-age prediction. Hum. Brain Mapp. **43**(10), 3113–3129 (2022)
7. Cole, J.H., et al.: Longitudinal assessment of multiple sclerosis with the brain-age paradigm. Ann. Neurol. **88**(1), 93–105 (2020)
8. Rokicki, J., et al.: Multimodal imaging improves brain age prediction and reveals distinct abnormalities in patients with psychiatric and neurological disorders. Hum. Brain Mapp. **42**(6), 1714–1726 (2021)
9. Sone, D., et al.: Neuroimaging-based brain-age prediction in diverse forms of epilepsy: a signature of psychosis and beyond. Mol. Psychiatry **26**(3), 825–834 (2021)
10. Cole, J.H., et al.: Brain age predicts mortality. Mol. Psychiatry **23**(5), 1385–1392 (2018)
11. Beck, D., et al.: Cardiometabolic risk factors associated with brain age and accelerate brain ageing. Hum. Brain Mapp. **43**(2), 700–720 (2022)
12. Tanveer, M., et al.: Deep learning for brain age estimation: a systematic review. Inf. Fusion (2023)
13. Peng, H., et al.: Accurate brain age prediction with lightweight deep neural networks. Med. Image Anal. **68**, 101871 (2021)
14. Sudlow, C., et al.: UK biobank: an open access resource for identifying the causes of a wide range of complex diseases of middle and old age. PLoS Med. **12**(3), e1001779 (2015)
15. Taylor, J.R., et al.: The Cambridge centre for ageing and neuroscience (Cam-CAN) data repository: structural and functional MRI, MEG, and cognitive data from a cross-sectional adult lifespan sample. Neuroimage **144**, 262–269 (2017)

16. IXI dataset. http://brain-development.org/ixi-dataset/. Accessed 29 June 2023
17. Weiner, M.W., et al.: Recent publications from the Alzheimer's disease neuroimaging initiative: reviewing progress toward improved ad clinical trials. Alzheimer's Dementia **13**(4), e1–e85 (2017)
18. Marcus, D.S., et al.: Open access series of imaging studies (OASIS): cross-sectional MRI data in young, middle aged, nondemented, and demented older adults. J. Cogn. Neurosci. **19**(9), 1498–1507 (2007)
19. Castro, D.C., Walker, I., Glocker, B.: Causality matters in medical imaging. Nat. Commun. **11**(1), 3673 (2020)
20. Iglesias, J.E., et al.: Robust brain extraction across datasets and comparison with publicly available methods. IEEE Trans. Med. Imaging **30**(9), 1617–1634 (2011)
21. Tustison, N.J., et al.: N4ITK: improved N3 bias correction. IEEE Trans. Med. Imaging **29**(6), 1310–1320 (2010)
22. He, K., et al.: Deep residual learning for image recognition. In: Proceedings of the IEEE Conference on Computer Vision and Pattern Recognition, pp. 770–778 (2016)
23. Brunner, E., Bathke, A.C., Konietschke, F.: Rank and Pseudo-Rank Procedures for Independent Observations in Factorial Designs. Springer, Cham (2018)
24. Glocker, B., et al.: Algorithmic encoding of protected characteristics in chest X-ray disease detection models. Ebiomedicine **89** (2023)
25. More, S., et al.: Brain-age prediction: a systematic comparison of machine learning workflows. Neuroimage **270**, 119947 (2023)

Studying the Effects of Sex-Related Differences on Brain Age Prediction Using Brain MR Imaging

Mahsa Dibaji[1]([⊠])(iD), Neha Gianchandani[2](iD), Akhil Nair[3](iD), Mansi Singhal[4](iD), Roberto Souza[1](iD), and Mariana Bento[2](iD)

[1] Department of Electrical and Software Engineering, University of Calgary, Calgary, AB, Canada
seyedemahsa.dibaji@ucalgary.ca
[2] Department of Biomedical Engineering, University of Calgary, Calgary, AB, Canada
[3] Department of Aerospace Engineering, Indian Institute of Technology Kharagpur, Kharagpur, India
[4] Department of Electrical Engineering, Dayalbagh Educational Institute (Deemed University), Agra, India

Abstract. While utilizing machine learning models, one of the most crucial aspects is how bias and fairness affect model outcomes for diverse demographics. This becomes especially relevant in the context of machine learning for medical imaging applications as these models are increasingly being used for diagnosis and treatment planning.

In this paper, we study biases related to sex when developing a machine learning model based on brain magnetic resonance images (MRI). We investigate the effects of sex by performing brain age prediction considering different experimental designs: model trained using only female subjects, only male subjects and a balanced dataset. We also perform evaluation on multiple MRI datasets (Calgary-Campinas(CC359) and CamCAN) to assess the generalization capability of the proposed models.

We found disparities in the performance of brain age prediction models when trained on distinct sex subgroups and datasets, in both final predictions and decision making (assessed using interpretability models). Our results demonstrated variations in model generalizability across sex-specific subgroups, suggesting potential biases in models trained on unbalanced datasets. This underlines the critical role of careful experimental design in generating fair and reliable outcomes.

Keywords: Sex · Brain Aging · Magnetic Resonance Imaging · Convolutional Neural Network

1 Introduction

Machine Learning (ML) has shown great promise in healthcare applications, from assisting with diagnoses to informing treatment strategies. However, when

S. Wesarg et al. (Eds.): CLIP/FAIMI/EPIMI 2023, LNCS 14242, pp. 205–214, 2023.
https://doi.org/10.1007/978-3-031-45249-9_20

deploying ML models in such critical areas, we need to ensure that the algorithms are reliable and do not perpetuate existing biases [2]. If an ML model is systematically biased towards a specific demographic group, it can lead to unfair outcomes, and in the worst cases, even harmful consequences [14]. Such biases could arise from the development data, either due to existing historical biases or the under-representation of certain groups. Even using unbiased data, the outcome might be unfair if protected features are used for prediction. Recognizing and minimizing these biases is essential for creating fair and trustworthy ML models, particularly in healthcare where stakes are high [12,19].

Machine learning's (ML) increasing application in predicting brain age through T1-weighted MRI scans has become a key area of focus [13], related to brain development, cognitive decline, and neurodegenerative diseases [5]. ML models often aim for a 'global' brain age index, reflecting brain maturity and serving as a biomarker to assess structural changes and aging [3,21]. Exploring what these models learn and the significant brain regions they identify may offer insights into individual brain variations.

A crucial aspect to consider when developing these models is the sex differences in brain volumes, which could significantly influence their predictions. Existing literature has demonstrated considerable sex-related differences in the total and regional brain volumes, including Gray Matter, White Matter, and Cerebrospinal Fluid [6,22]. These differences, and their interaction with age, can have significant impacts on cognitive impairment, especially in the elderly [9]. Therefore, ensuring our ML models account for these variations and perform equally well across different sex subgroups is crucial for fairness and reliability.

In this paper, we aim to investigate how the performance of brain age prediction models varies across sex subgroups: males only, females only, and balanced datasets. Our goal is to develop accurate and interpretable ML models, while also ensuring their fairness. By understanding how predictions vary across these different subgroups, we seek to offer valuable insights for the development of more reliable and fair ML models, and to promote transparency in ML processes. While focusing on brain age prediction using MRI, our experimental design also can be translatable for other tasks, examining sex differences and dataset variations, enhancing transparency and reliability.

Our experimental design followed a similar approach presented in [10]. We trained and validated different models to perform brain age prediction, considering the following experimental design: a) using only female samples, b) using only male samples, and c) using a balanced mix of both sexes. We also performed experiments using different datasets, with varying data distributions, evaluating inconsistencies for different populations or equity-deserving groups.

The rest of the paper unfolds as follows: Sect. 2 covers materials and methodologies. Section 3 outlines our experimental results. Section 4 discusses these findings and their implications. The paper concludes with a summary and potential future work.

2   Materials and Methods

2.1   Brain MR Datasets

In this study, we utilized the Calgary-Campinas-359 (CC359) [17] and the Cambridge Centre for Ageing and Neuroscience (CamCAN) [16,20] datasets. CC359 is a multi-vendor (General Electric (GE), Philips, and Siemens), multi-field strength (1.5 T and 3 T magnetic field strengths) volumetric brain MRI dataset, comprising 359 T1-weighted three-dimensional (3D) volumes. It has balanced sex distribution, with 183 (50.97%) female and 176 (49.03%) male healthy subjects, aged 29 to 80 years. Brain masks are also available for this dataset.

The CamCAN data set comprises MR images that were collected at a single site (MRC-CBSU) using a Siemens TIM Trio scanner at 3 T magnetic field strength. The dataset is composed of T1-weighted reconstructed brain MR volumes and segmentation masks for certain structures. From the total number of 651 Samples 329 (50.54%) correspond to female subjects, and 322 (49.46%) correspond to male subjects aged 18 to 88 years. Figure 1 depicts the distribution of age and sex subgroups in CC359 and CamCAN datasets.

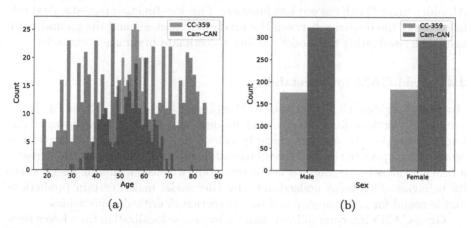

(a)   (b)

Fig. 1. Age and sex distribution across the CC359 and CamCAN datasets: (a) The CC359 dataset exhibits an age distribution close to normal and a smaller age range compared to CamCAN. (b) Sex distribution in both datasets reveals a balanced representation of male and female samples.

2.2   Pre-processing

In this study, the preprocessing pipeline consisting of skull stripping, registration, and intensity scaling. We trained a UNet model using the CC359 images and their corresponding brain masks for the skull stripping step. The UNet architecture was chosen for its effectiveness in handling high-resolution brain MR images. This trained model achieved 97% dice score on the validation set and was then used to perform brain extraction on the CamCAN dataset, thereby effectively stripping non-brain tissues.

We registered the skull-stripped brain MR images to the MNI152 standard atlas [4]. This registration was performed using FSL's FLIRT tool [7,8], which allows for a 6 Degrees-of-Freedom rigid registration. This process involves only rotations and translations without distorting the brain's shape and size. This step ensures that the brain images are comparable and consistent for further analyses. Image intensities were also scaled to fall between 0 and 1.

2.3   Brain Age Prediction Task

We utilized a Convolutional Neural Network (CNN) architecture, the Simple Fully Convolutional Network (SFCN) proposed in [13] for estimating brain age based on 3D T1-weighted images. This model comprises seven convolutional blocks. The initial five blocks down-sample the input after each $3 \times 3 \times 3$ convolutional layer, followed by a $1 \times 1 \times 1$ convolutional block and a classification head. To stabilize the training process, batch normalization is incorporated. The only modification to this model was replacing the classification head with a ReLU-activated linear regressor for brain age prediction.

During the training and validation stages of our study, we employed Mean Absolute Error (MAE) as our loss function. This loss function played a vital role in assessing the disparity between the predicted brain age and the ground truth age labels, evaluating the model's ability to estimate brain age accurately.

2.4   Grad-CAM Interpretability

Gradient-weighted Class Activation Mapping (Grad-CAM) [15] is an interpretability method for deep learning models, especially CNNs. It highlights important regions in an input image by calculating gradients of the target class score with respect to the final convolutional layer's feature maps. It then creates a heatmap showing regions crucial for the model's decision, offering insights into its behavior. This helps understand why the model makes certain predictions and is useful for transparency and bias detection in critical applications.

Grad-CAM's interpretability is limited by coarse localization from lower resolution in deeper layers [11]. To overcome this limitation, we averaged maps from both early-stage and final convolutional layers. By blending low-level features like textures with high-level insights, this approach offers a multifaceted interpretation of the model's reasoning, and enhances the robustness of visualizations by potentially mitigating noise.

2.5   Experimental Setting

We established test sets to evaluate the performance of our models through a stratified sampling approach based on vendor and magnetic field, while taking into consideration both sex and dataset. A total of four test sets were constructed in this manner: 30 females from CamCAN (Cam_F); 30 males from CamCAN (Cam_M); 30 females from CC359 (CC_F) and 30 males from CC359 (CC_M).

The remaining samples from each dataset (591 CamCAN and 299 CC359 samples) were used to create three distinctdevelopment sets: female subjects only, male subjects only, and larger balanced sets combining both male and female subjects. This resulted in a total of 6 development datasets. Each of these sets was then stratified by vendor and magnetic field strength, and divided into 80% training and 20% validation sets.

This process led to the generation of six separate models. Three were developed using the CC359 dataset, named as CC359-F (trained on female data), CC359-M (trained on male data), and CC359-A (trained on data from all subjects). The remaining three models were trained on the CamCAN dataset, similarly named as CamCAN-F, CamCAN-M, and CamCAN-A. An illustration of this design is provided in Fig. 2.

To mitigate overfitting, we employed an augmentation step where 50% of the training and validation samples were randomly subjected to a 15° rotation on-the-fly. Training was carried out using the Adam optimizer with a Mean Absolute Error (MAE) loss function and a batch size of 8. The learning rate was initially set to 0.001, then halved every 10 epochs, across a total of 50 epochs.

For our experiments, we implemented data transforms, and Grad-CAM heat maps using MONAI [1] which is a robust, open-source platform developed by NVIDIA built upon PyTorch. The University of Calgary Advanced Research Computing (ARC) cluster, specifically gpu-a100 and gpu-v100, were also utilized.

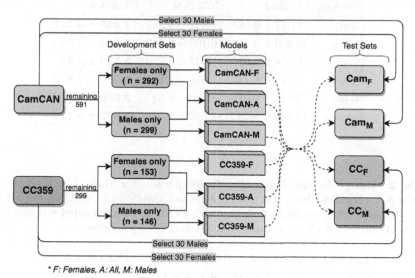

Fig. 2. Overview of the experimental design. We utilized 3 development sets: Females only, Males only, and All Subjects, extracted from the CamCAN and CC359 datasets separately. These sets were split in a stratified manner based on vendor and magnetic field strength, resulting in training (80%) and validation (20%) subsets. Subsequently, each model was evaluated on four test sets: Cam_F, Cam_M, CC_F, and CC_M

3   Results

Table 1 provides a summary of the results (MAE) of models trained on CC359 dataset (CC359-F, CC359-M, and CC359-A), evaluated against the defined test sets: Cam_F, Cam_M, CC_F, and CC_M (detailed in Sect. 2.5). To compare models performances across sex subgroups regardless of dataset source, we created two aggregated test sets, one each for Males (CC_M and Cam_M combined) and Females (CC_F and Cam_F combined). Table 2 presents similar results for models trained using CamCAN dataset (CamCAN-F, CamCAN-M, and CamCAN-A).

In addition, we have produced averaged Grad-CAM maps on each model's predictions for all test sets, using only the aggregated Females test set for visualization and comparisons purposes (Fig. 3). These maps represent the mean saliency maps across all samples in this test set. They highlight the regions that each model, on average, considered significant for prediction [18]. The same approach was applied to all test sets, yielding similar outcomes across the board. However, we focused on the Females test set to discuss the observed differences.

Table 1. Mean Absolute Error (MAE) comparison for three CC359 model: CC359-F (Females Only), CC359-M (Males Only), and CC359-A (All Subjects) across 4 test sets Cam_F, Cam_M, CC_F, CC_M, and 2 aggregated sets combining female (Cam_F and CC_F) and male (Cam_M and CC_M) samples.

Test Sets	CC359-F	CC359-M	CC359-A
Cam_F	17.91 ± 8.864	17.347 ± 9.387	$\mathbf{14.355 \pm 8.117}$
Cam_M	16.269 ± 9.452	16.408 ± 8.983	$\mathbf{13.804 \pm 7.564}$
CC_F	5.588 ± 5.929	6.47 ± 5.238	$\mathbf{5.428 \pm 5.291}$
CC_M	8.311 ± 6.432	7.056 ± 4.883	$\mathbf{5.535 \pm 4.844}$
Males	12.29 ± 9.011	11.732 ± 8.61	$\mathbf{9.669 \pm 7.578}$
Females	11.749 ± 9.737	11.909 ± 9.346	$\mathbf{9.892 \pm 8.177}$

Table 2. Mean Absolute Error (MAE) comparison for three CamCAN models: CamCAN-F (Females Only), CamCAN-M (Males Only), and CamCAN-A (All Subjects), across 4 test sets CamF, CamM, CCF, CCM, and 2 aggregated sets combining females (CamF and CCF) and males (CamM and CCM) samples.

Test Sets	CamCAN-F	CamCAN-M	CamCAN-A
Cam_F	7.801 ± 6.186	$\mathbf{6.105 \pm 4.018}$	6.366 ± 5.386
Cam_M	7.002 ± 4.69	6.504 ± 4.721	$\mathbf{5.763 \pm 4.403}$
CC_F	9.973 ± 6.913	9.571 ± 6.581	$\mathbf{9.018 \pm 5.963}$
CC_M	$\mathbf{10.049 \pm 5.949}$	12.614 ± 10.144	10.549 ± 7.264
Males	8.526 ± 5.569	9.559 ± 8.481	$\mathbf{8.156 \pm 6.466}$
Females	8.77 ± 6.63	7.838 ± 5.721	$\mathbf{7.692 \pm 5.834}$

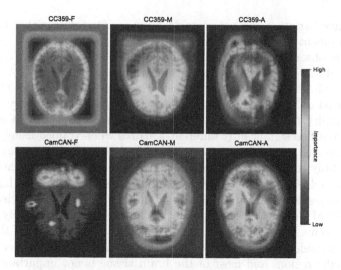

Fig. 3. Averaged Grad-CAM heatmaps represents a comparative snapshot of significant regions for brain age prediction between models trained on distinct datasets and sex-specific groups, predicted on Females test set (CC_F and Cam_F combined). Although our model is 3D based, we are showing a 2D slice for visualization and comparison purposes (same slice is shown for each model variation.). Models are identified above each map: CC359-F, CC359-M, CC359-A, CamCAN-F, CamCAN-M, CamCAN-A.

4   Discussion

The observed results in Table 1, which relate to CC359 models, and Table 2, which relate to CamCAN models, present a considerable degree of variability. Despite being trained for the same task, using the same architecture, and undergoing the same preprocessing within a closely aligned experimental design, divergent outcomes emerged due solely to differences in the development sets.

Our experiments suggest that models trained on a specific dataset fail to generalize effectively to an external dataset. For instance, in Table 1, the MAE of all three models was significantly lower on CC_F and CC_M (sourced from CC359) compared to Cam_F and Cam_M. This pattern is also observed on Table 2, although the difference in performance is less significant.

The improved generalizability of CamCAN models can likely be attributed to development set size, as shown in Fig. 2. While increased diversity in data (e.g., different vendors and magnetic field strengths) typically improves generalization, this could lead to less accurate predictions if the training samples are insufficient to capture these variations. Therefore, a balance is necessary between dataset size and variability. With increased variability, a larger sample size is required for optimal model function.

The evaluation outcomes on the combined Male and Female test sets demonstrates no consistent trend. Certain models exhibited superior performance on female data, while others were more effective with male data. Interestingly, these performance variations do not seem to be associated with either the origin of the development data or the sample size.

A closer observation of the model performances on sex-specific test sets (especially females) reveals minimal variation among the F, M, and A model variants, as depicted in each table. To better comprehend whether these models were focused on similar features, we used the GradCAM saliency maps as an interpretability method. Figure 3 shows that even models trained on a specific dataset identified significantly different features when trained using a single-sex subgroup versus a mixed-sex group. This highlights the importance of model's interpretabiliy to ensure that they are learning the appropriate features.

One model (CC359-F) considered regions outside of the brain highly important (red edges in Fig. 3), which ideally should not have affected decision-making. For CC359-M, almost all the regions in the brain were relatively important. CC359-A seems to be focused on smaller regions, however, one of those regions is almost outside of the brain. Similarly, CamCAN-F model is also focused on more specific regions and most of the brain tissue is not important in predicting brain age. Interestingly, CamCAN-M seems to be focused on more specific regions compared to CamCAN-A, despite the fact that CamCAN-A had twice as many training samples and was thus anticipated to learn more specific features.

Our proposal has limitations related to the size of the development sets, potentially impacting the comparisons. However, the performance of models trained on combined males and females was not very different from models trained using only half of the samples (single-sex subgroups). Another limitation is the disregard of age distribution during the split of training, validation, and test sets. Since the CC359 and CamCAN have different age distribution (Fig. 1a), an improved way of splitting training, validation, and test sets should consider the stratified age distribution. Additionally, the use of Grad-CAM, which generates relatively coarse saliency maps may impact the detailed comparison between models and restrict the interpretation of intricate relationships between specific brain regions and predictions. Lastly, the straightforward averaging used to produce an aggregated map for a population presents some challenges, such as potential loss of specificity and alignment complexities. However, these challenges do not diminish the method's promising potential to improve interpretability.

5   Conclusion

In this study, we examined the influence of sex and dataset variations on brain age prediction. Our findings emphasize that thoughtful experimental design is crucial in shaping the performance and feature learning of models, leading to outcomes that are both reliable and fair. This underscores the broader need for interpretability methods to ensure trustworthy results. We aimed to evaluate how these variations and sex differences impact the model's performance and generalizability, rather than achieving state-of-the-art accuracy. We intend to make our code available for easy reproduction and benchmarking of our findings.

Future work should delve into two key areas: a more rigorous examination of performance disparity through statistical tests (e.g., Wilcoxon signed-rank test), complemented by using more precise saliency maps and a more reliable method for aggregating these maps. This will foster greater confidence in conclusions. Concurrently, efforts must be directed towards designing and optimizing predictive models that specifically address sex-related differences. This dual focus aims to reduce biases and ensure reliable and consistent results across varied populations, strengthening the overall impact of the study.

References

1. Cardoso, M.J., et al.: Monai: an open-source framework for deep learning in healthcare. arXiv preprint arXiv:2211.02701 (2022)
2. Chen, R.J., et al.: Algorithmic fairness in artificial intelligence for medicine and healthcare. Nature Biomed. Eng. **7**(6), 719–742 (2023)
3. Cole, J.H., Franke, K.: Predicting age using neuroimaging: innovative brain ageing biomarkers. Trends Neurosci. **40**(12), 681–690 (2017)
4. Fonov, V.S., Evans, A.C., McKinstry, R.C., Almli, C.R., Collins, D.: Unbiased nonlinear average age-appropriate brain templates from birth to adulthood. NeuroImage (47), S102 (2009)
5. Franke, K., Gaser, C.: Ten years of brainage as a neuroimaging biomarker of brain aging: what insights have we gained? Front. Neurol. 789 (2019)
6. Gur, R.C., et al.: Sex differences in brain gray and white matter in healthy young adults: correlations with cognitive performance. J. Neurosci. **19**(10), 4065–4072 (1999)
7. Jenkinson, M., Bannister, P., Brady, M., Smith, S.: Improved optimization for the robust and accurate linear registration and motion correction of brain images. Neuroimage **17**(2), 825–841 (2002)
8. Jenkinson, M., Smith, S.: A global optimisation method for robust affine registration of brain images. Med. Image Anal. **5**(2), 143–156 (2001)
9. Jäncke, L., Mérillat (-Koeneke), S., Liem, F., Hänggi, J.: Brain size, sex, and the aging brain. Hum. Brain Mapp. **36**, 150–169 (2014)
10. Larrazabal, A.J., Nieto, N., Peterson, V., Milone, D.H., Ferrante, E.: Gender imbalance in medical imaging datasets produces biased classifiers for computer-aided diagnosis. Proc. Natl. Acad. Sci. **117**(23), 12592–12594 (2020)
11. McAllister, D., Mendez, M., Bermúdez, A., Tyrrell, P.: Visualization of layers within a convolutional neural network using gradient activation maps. J. Undergraduate Life Sci. **14**(1), 6 (2020)
12. Mehrabi, N., Morstatter, F., Saxena, N., Lerman, K., Galstyan, A.: A survey on bias and fairness in machine learning. ACM Comput. Surv. (CSUR) **54**(6), 1–35 (2021)
13. Peng, H., Gong, W., Beckmann, C.F., Vedaldi, A., Smith, S.M.: Accurate brain age prediction with lightweight deep neural networks. Med. Image Anal. **68**, 101871 (2021)
14. Rajkomar, A., Hardt, M., Howell, M.D., Corrado, G., Chin, M.H.: Ensuring fairness in machine learning to advance health equity. Ann. Intern. Med. **169**(12), 866–872 (2018)
15. Selvaraju, R.R., Cogswell, M., Das, A., Vedantam, R., Parikh, D., Batra, D.: Gradcam: visual explanations from deep networks via gradient-based localization. In: Proceedings of the IEEE ICCV, pp. 618–626 (2017)

16. Shafto, M.A., et al.: The Cambridge centre for ageing and neuroscience (cam-can) study protocol: a cross-sectional, lifespan, multidisciplinary examination of healthy cognitive ageing. BMC Neurol. **14**, 1–25 (2014)
17. Souza, R., et al.: An open, multi-vendor, multi-field-strength brain MR dataset and analysis of publicly available skull stripping methods agreement. Neuroimage **170**, 482–494 (2018)
18. Stanley, E.A., Wilms, M., Mouches, P., Forkert, N.D.: Fairness-related performance and explainability effects in deep learning models for brain image analysis. J. Med. Imaging **9**(6), 061102 (2022)
19. Suresh, H., Guttag, J.: A framework for understanding sources of harm throughout the machine learning life cycle. In: Equity and Access in Algorithms, Mechanisms, and Optimization, pp. 1–9, New York, NY, USA. ACM (2021)
20. Taylor, J.R., et al.: The Cambridge centre for ageing and neuroscience (cam-can) data repository: structural and functional MRI, meg, and cognitive data from a cross-sectional adult lifespan sample. Neuroimage **144**, 262–269 (2017)
21. Wang, J., et al.: Gray matter age prediction as a biomarker for risk of dementia. Proc. Natl. Acad. Sci. **116**(42), 21213–21218 (2019)
22. Wang, Y., Xu, Q., Luo, J., Hu, M., Zuo, C.: Effects of age and sex on subcortical volumes. Front. Aging Neurosci. **11**, 259 (2019)

An Investigation into the Impact of Deep Learning Model Choice on Sex and Race Bias in Cardiac MR Segmentation

Tiarna Lee[1(✉)], Esther Puyol-Antón[1,4], Bram Ruijsink[1,2], Keana Aitcheson[1], Miaojing Shi[3], and Andrew P. King[1]

[1] School of Biomedical Engineering & Imaging Sciences, King's College London, London, UK
tlarna.lee@kcl.ac.uk
[2] St Thomas' Guy's and St Thomas' Hospital, London, UK
[3] Department of Informatics, King's College London, London, UK
[4] HeartFlow Inc, London, UK

Abstract. In medical imaging, artificial intelligence (AI) is increasingly being used to automate routine tasks. However, these algorithms can exhibit and exacerbate biases which lead to disparate performances between protected groups. We investigate the impact of model choice on how imbalances in subject sex and race in training datasets affect AI-based cine cardiac magnetic resonance image segmentation. We evaluate three convolutional neural network-based models and one vision transformer model. We find significant sex bias in three of the four models and racial bias in all of the models. However, the severity and nature of the bias varies between the models, highlighting the importance of model choice when attempting to train fair AI-based segmentation models for medical imaging tasks.

Keywords: Segmentation · Fairness · CNN · Cardiac MRI

1 Introduction

The popularity and increasing clinical translation of artificial intelligence (AI) techniques in medical image analysis have prompted many to investigate their fairness. Fairness can be defined in a number of ways [11] but in simple terms, it refers to the absence of inexplicable bias, or disparate performance, between different protected groups. Biases, which cause performances to be unfair, can have several causes. One such cause is training models using datasets that are imbalanced by protected group and recent work has shown that this can lead to bias in a range of medical imaging applications [9,10,12,15,17]. For example, [9] found that accuracy for a chest X-ray classification task was significantly lower for a protected group, in this case sex, that was under-represented in the training dataset.

Supplementary Information The online version contains supplementary material available at https://doi.org/10.1007/978-3-031-45249-9_21.

Although under-representation in the training data can be a root cause of bias in AI, there has been relatively little research into the impact of other design choices on bias. One such choice is model architecture. To date, we are aware of no work in medical image analysis that has compared biases between different models trained for the same task with the same data. Furthermore, we are aware of no work that has aimed to assess the potential for bias in the latest vision transformer based architectures in medical image analysis. Therefore, in this work we aim to perform a comparative investigation of the impact of model architecture on bias, including both convolutional neural network (CNN)-based models and vision transformers.

We investigate the potential for race and sex bias in deep learning-based cardiac magnetic resonance (CMR) segmentation, an area where bias has previously been reported [10,14,15] for a CNN-based model (nnU-Net [8]). We perform a systematic analysis of the impact of training set imbalance on segmentation performance, similar to [10]. However, we perform these experiments for four different deep learning-based architectures, including one based on vision transformers, and analyse their differences, both in terms of overall performance and bias.

To summarise, our key contributions are:

1. We perform the first investigation into the impact of model architecture on AI model bias in medical image analysis.
2. We perform the first investigation into potential bias in vision transformer-based models in medical image analysis.

2   Materials

In this work, we used CMR images from the UK Biobank [13]. The dataset consists of end diastolic (ED) and end systolic (ES) cine short-axis images from 661 subjects. The demographic data for these subjects were also gathered from the UK Biobank database and can be seen in Table 1 for the subjects used.

Table 1. Clinical characteristics of subjects used in the experiments. Mean values are presented for each characteristic with standard deviations given in brackets. Statistically significant differences are indicated with an asterisk * ($p < 0.05$) and were determined using a two-tailed Student's t-test.

Health measure	Overall	Male	Female	White	Black
#Subjects	661	321	340	427	234
Age (years)	60.9 (8.1)	61.0 (8.2)	60.9 (8.0)	62.1 (8.4)*	58.8 (6.9) *
Weight (kg)	79.3 (16.1)	86.0 (14.9) *	72.9 (14.4) *	78.0 (16.0)	81.6 (16.0)
Standing height (cm)	169.8 (9.5)	176.7 (6.6) *	163.2 (6.9) *	170.3 (9.6)	168.8 (9.3)
Body Mass Index (kg)	27.4 (4.7)	27.5 (4.2)	27.4 (5.1)	26.8 (4.4)*	28.6 (5.1) *

Manual segmentation of the left ventricular blood pool (LVBP), left ventricular myocardium (LVM), and right ventricular blood pool (RVBP) was performed for the ED and ES images of each subject. This was done by outlining the LV endocardial and epicardial borders and the RV endocardial border using cvi42 (version 5.1.1, Circle Cardiovascular Imaging Inc., Calgary, Alberta, Canada). A panel of ten experts was provided with the same guidelines and one expert annotated each ground truth image. The selection of images for annotation was randomised and included subjects with different sexes and races. The experts were not provided with demographic information about the subjects such as their race or sex.

3   Methods

3.1   Dataset Sampling

We investigated the effect of varying the proportions of training subjects based on two protected attributes: male *vs.* female (Experiment 1) and White *vs.* Black (Experiment 2). In Experiment 1, the race of the subjects was controlled so that 50% of male and female subjects were Black and 50% were White. Similarly, in Experiment 2, the sex of the subjects was controlled so that 50% of subjects from each race were female and 50% were male. For each experiment, a group of 176 subjects were chosen for training from each protected group and the combination of these two groups was sampled to create five training sets which varied the proportion of subjects with the selected protected attribute from 0%/100% to 100%/0% as in [10]. The total number of subjects in each of the five training sets was constant at 176.

One difference in our experimental setup compared to [10] was that, to remove the effect of a potential confounder in our analysis, the age of the subjects was controlled so that each subject in a protected group was matched to a subject in the other protected group whose age was within ±1 year. For both of the experiments, the test sets comprised 84 subjects and contained 50% Black and 50% White subjects, and 50% male and 50% female subjects. These subjects were also controlled for age.

3.2   Model Architecture and Implementation

All experiments were performed using four separate segmentation models: U-Net [16], nnU-Net [8], Swin-Unet [2] and DeepLabv3+ [4]. For nnU-Net, we used the same training parameters as in [10]. Swin-Unet is a U-Net-like transformer model which uses an encoder-decoder structure based on Swin Transformer blocks and skip connections to merge the multi-scale features [2]. Training hyperparameters from [2] were used, which had been tuned using cine short axis CMR images from the ACDC dataset [1]. The model was pre-trained using data from ImageNet [5]. The U-Net model was trained using the U-Net64 model from [3] and used the hyperparameters from their work which were chosen using CMR data from the

UK Biobank. The final model was DeepLabv3+, an encoder-decoder architecture which utilises atrous spatial pyramid pooling and atrous convolutions [4]. The model was trained using a ResNet-50 backbone and hyperparameters were those reported in [4] which were chosen using the PASCAL VOC dataset [7].

A summary of the training setup for each of the four models is provided in Table S1. The same data augmentation methods were used for each of the models and are as described in [10]. For each of the models, the images were cropped to 224×224. All models were optimised using Stochastic Gradient Descent. The models were implemented using Python and PyTorch, and were trained on one NVIDIA A100 GPU.

3.3 Model Evaluation

Model performance was assessed using the *Dice similarity coefficient* (DSC), which measures the spatial overlap between two sets. For a ground truth segmentation A and its corresponding prediction B, the DSC is given by $DSC = \frac{2|A \cap B|}{|A|+|B|}$.

To compare the fairness of the different models, the median DSC values for each protected group in the test set were first calculated. We denote these median DSC values by $\mathbf{D}_{A_1} = [D_{A_1}^0, D_{A_1}^{25}, ... D_{A_1}^{100}]$ and $\mathbf{D}_{A_2} = [D_{A_2}^{100}, D_{A_2}^{75}, ... D_{A_2}^{0}]$ where A_1 and A_2 are the protected groups and the superscripts indicate the percentage of the protected group in the training set. Based upon $\mathbf{D}_{A_1}$ and $\mathbf{D}_{A_2}$, the following metrics were calculated:

- **Fairness gap**: For a given model and level of imbalance, the fairness gap is found by subtracting the median DSCs of the model evaluated on the two protected groups. We report the maximum and minimum fairness gap across all levels of imbalance, i.e. $FG_{max} = \max(\mathbf{D}_{A_1} - \mathbf{D}_{A_2})$ and $FG_{min} = \min(\mathbf{D}_{A_1} - \mathbf{D}_{A_2})$. These metrics quantify the level of performance disparity *between* protected groups in both directions, i.e. in favour of A_1 (the maximum fairness gap) and in favour of A_2 (the minimum fairness gap).
- **Performance range**: The performance range for a given protected group is found by calculating the difference between the minimum and maximum median DSC scores across all levels of imbalance, i.e. $PR_{A_1} = \max(\mathbf{D}_{A_1}) - \min(\mathbf{D}_{A_1})$ and $PR_{A_2} = \max(\mathbf{D}_{A_2}) - \min(\mathbf{D}_{A_2})$. These metrics quantify differences in performance *within* protected groups.
- **Skewed error ratio**: For protected groups A_1 and A_2, the skewed error ratio is defined by $SER_{A_1} = \frac{\max(1-\mathbf{D}_{A_1})}{\min(1-\mathbf{D}_{A_1})}$ and $SER_{A_2} = \frac{\max(1-\mathbf{D}_{A_2})}{\min(1-\mathbf{D}_{A_2})}$. This represents the ratio between the error rate of the best performing model and the worst performing model across all levels of imbalance. A high value of the skewed error ratio indicates strong bias and the lowest value (i.e. equal performance) is 1.
- **Standard deviation of performance**: For a given protected group, the standard deviation of median DSC values across all levels of imbalance was computed, i.e. $SD_{A_1} = std(\mathbf{D}_{A_1})$ and $SD_{A_2} = std(\mathbf{D}_{A_2})$. Again, a high value indicates strong bias, but this time the lowest value is 0.

– **Bias trend**: For a given protected group, the trend in bias across the different levels of imbalance was quantified by calculating a line of best fit across the five levels of imbalance. The gradient G for this line is reported. A positive value of G indicates an increase in performance as the representation of the first protected group increases from 0% to 100%.

4 Results

Experiment 1 - Male *vs*. Female: The results for Experiment 1 investigating the effect of imbalances in subjects' sex can be seen in Fig. 1. The full set of DSC scores for both of the protected groups can be found in Table S2. Table S4 shows the fairness gap for all of the models at each level of imbalance. For nnU-Net, there were no statistical differences in performance as the level of imbalance of the subjects changed. However, for the other three models, male subjects had statistically higher DSC scores than the females when the males were in the majority of the training set and when the training set was evenly balanced. For these models, as the proportion of female subjects in the dataset increased, accuracy parity was achieved.

Table 2 presents the summary of bias statistics. We can see that, for the U-Net and Swin-Unet models, PR_{male} was 1.4 and 1.7 times larger than PR_{female} respectively. The SD of the DSC scores and SER were also higher for U-Net compared to the other models and it had the largest FG_{max}. DeepLabv3+, which also showed an increase in performance as the proportion of females increased, had a larger PR, SER and SD for the female subjects. nnU-Net, which had accuracy parity for all proportions of female subjects in the dataset, further displayed by the smallest FG_{max} and FG_{min} for all of the models.

Table 2. Statistics for Experiment 1 investigating the effect of varying subjects by sex. PR = performance range, SER = skewed error ratio, SD = standard deviation, G = gradient of bias trend, FG = fairness gap. The FG was calculated by finding $\mathbf{D}_{female} - \mathbf{D}_{male}$. G was calculated by finding the gradient of the line of best fit from 0% females/100% males to 100% females/0% males.

Model	PR		SER		SD		G		FG	
	Female	Male	Female	Male	Female	Male	Female	Male	Max	Min
nnU-Net	0.0083	0.0020	1.11	1.03	0.0032	0.00079	0.00089	−0.00015	−0.0041	0.00043
U-Net	0.026	0.036	1.18	1.36	0.011	0.014	−0.0054	−0.0072	−0.042	−0.024
Swin-Unet	0.0039	0.0068	1.00	1.07	0.0016	0.0027	0.000041	−0.0015	−0.015	−0.0067
Deeplabv3+	0.012	0.0082	1.14	1.11	0.0044	0.0032	0.0016	−0.0018	−0.021	−0.0057

Experiment 2 - Black *vs*. White: The results for Experiment 2 investigating the effect of imbalances in subjects' race can be seen in Fig. 2. The full set of DSC scores for both of the protected groups can be found in Table S3. Table S5

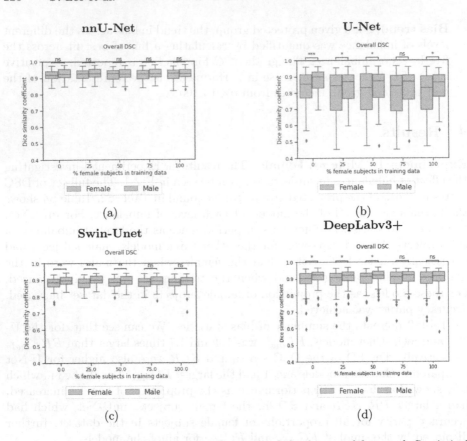

Fig. 1. Overall DSC for Experiment 1 for each of the four models tested. Statistical significance was found using a Mann-Whitney U test and is denoted by **** ($p \leq 0.0001$), *** ($0.001 < p \leq 0.0001$), ** ($0.01 < p \leq 0.001$), * ($0.01 < p \leq 0.05$), ns ($0.05 \leq p$).

shows the fairness gap for all of the models at each level of imbalance. As the proportion of Black subjects increased, both nnU-Net and Swin-Unet models produced significantly different DSC scores between the Black and White subjects, with the DSC scores for the Black subjects being significantly higher when their proportion of these subjects was above 25%. However, for the DeepLabv3+ models, the performances were only significantly different at the extremes of the training dataset imbalance *i.e.* 0%/100%. Lastly, for the U-Net models, only one of the performances, with 100% White subjects and 0% Black subjects, was significantly different.

Table 3 shows the bias statistics for this experiment. For each of the models, the PR, SER and SD are higher for the Black subjects than the White subjects. Swin-Unet had the largest PR_{Black}, SER_{Black}, SD_{Black} and FG_{max} which can also be seen in the significantly different performances in Fig. 2c. Despite finding more significant differences in performance than the U-Net model, DeepLabv3+ had the smallest PR, SER_{Black}, FG_{max} and FG_{min}.

Table 3. Statistics for Experiment 2 investigating the effect of varying subjects by race. PR = performance range, SER = skewed error ratio, SD = standard deviation, G = gradient of bias trend, FG = fairness gap. The fairness gap was calculated by finding $\mathbf{D}_{White} - \mathbf{D}_{Black}$. G was calculated by finding the gradient of the line of best fit from 0% Black/100% White subjects to 100% Black/0% White subjects.

Model	PR		SER		SD		G		FG	
	Black	White	Black	White	Black	White	Black	White	Max	Min
nnU-Net	0.071	0.036	2.61	1.44	0.029	0.015	0.016	−0.0075	−0.074	−0.018
U-Net	0.15	0.073	2.52	1.80	0.059	0.027	0.029	0.0082	0.088	−0.0075
Swin-Unet	0.21	0.023	3.25	1.20	0.090	0.0096	0.042	−0.0057	0.18	−0.014
Deeplabv3+	0.048	0.023	1.63	1.29	0.020	0.0088	0.0099	−0.0053	0.044	−0.00067

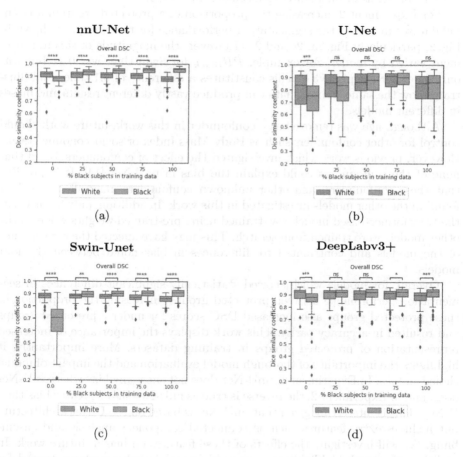

Fig. 2. Overall DSC for Experiment 2 for each of the four models tested. Statistical significance was found using a Mann-Whitney U test and is denoted by **** ($p \leq 0.0001$), *** ($0.001 < p \leq 0.0001$), ** ($0.01 < p \leq 0.001$), * ($0.01 < p \leq 0.05$), ns ($0.05 \leq p$).

5   Discussion

To the best of our knowledge, this work represents the first analysis and comparison of the effect of dataset imbalance on medical imaging segmentation performance in different segmentation models. Our results show that imbalances in dataset composition affect segmentation performance for under-represented protected groups and that these biases vary for different segmentation models.

Unlike previous work [10,15] which only considered nnU-Net, significant differences were found in the other three models for Experiment 1 investigating the effect of varying the proportion of subjects by sex. Despite accuracy parity being achieved for the Swin-Unet and Deeplabv3+ models where the female subjects comprised 75% of the training set, the male subjects had a higher median DSC score at all levels of imbalance, as discussed in Sect. 4.

For Experiment 2, increasing the proportion of a protected group in a training dataset increased the segmentation performance for this group, as shown in Fig. 2, particularly Fig. 2a, 2c and 2d. However, the magnitude of this improvement varied by model. For example, PR_{Black} for Swin-Unet was 0.21 but was only 0.048 for DeepLabv3+. This constitutes a difference of 9.1 times and illustrates how the same training data can produce vastly different results and biases in different models.

Although age was removed as a confounder in this work, future work should control for other confounders such as Body Mass Index or socioeconomic status. However, previous work which investigated the effect of confounders found that none of the confounders could explain the bias in nnU-Net [14]. It is possible that these confounders, and other unknown confounders, may affect the bias found in the other models investigated in this work. In addition, only Swin-Unet, the transformer-based model, was trained using pre-trained weights whereas the other models were trained from scratch. This may have affected the convergence of the models and contributed to differences in bias noted between the four models.

Overall, for models which showed statistically significant differences in segmentation performance between protected groups, increasing the proportion of these protected groups also increased DSC scores for under-represented groups and resulted in accuracy parity. This work displays the importance of increased representation of protected groups in training datasets. More importantly, it highlights the importance of thorough model evaluation and the impact of model choice on bias. In Experiment 1, nnU-Net does not show any bias whereas U-Net does, and in Experiment 2, the reverse is true as nnU-Net shows greater bias than U-Net. This is interesting given that nnU-Net is based on the U-Net architecture but includes extra features such as connected component analysis and ensembling. We will investigate the effects of these features on bias in future work. In addition, this work highlights the effect of dataset imbalance and potential for bias in vision transformers, which will prove to be an important consideration as vision transformers are used increasingly in medical imaging applications.

Acknowledgements. This work was supported by the Engineering & Physical Sciences Research Council Doctoral Training Partnership (EPSRC DTP) grant EP/T517963/1. This research has been conducted using the UK Biobank Resource under Application Number 17806.

References

1. Bernard, O., et al.: Deep learning techniques for automatic MRI cardiac multi-structures segmentation and diagnosis: is the problem solved? IEEE Trans. Med. Imaging **37**(11), 2514–2525 (2018). https://doi.org/10.1109/TMI.2018.2837502. ISSN: 1558254X

2. Cao, H., et al.: Swin-Unet: Unet-like pure transformer for medical image segmentation. In: Karlinsky, L., Michaeli, T., Nishino, K. (eds.) Computer Vision – ECCV 2022 Workshops. ECCV 2022. Lecture Notes in Computer Science, vol. 13803. Springer, Cham (2023). https://doi.org/10.1007/978-3-031-25066-8_9

3. Chen, C., et al.: Improving the generalizability of convolutional neural network-based segmentation on CMR images. Front. Cardiovasc. Med. **7**, 105 (2020). https://doi.org/10.3389/FCVM.2020.00105. ISSN: 2297055X

4. Chen, L.-C., Zhu, Y., Papandreou, G., Schroff, F., Adam, H.: Encoder-decoder with Atrous separable convolution for semantic image segmentation. In: Ferrari, V., Hebert, M., Sminchisescu, C., Weiss, Y. (eds.) ECCV 2018. LNCS, vol. 11211, pp. 833–851. Springer, Cham (2018). https://doi.org/10.1007/978-3-030-01234-2_49

5. Deng, J., et al.: ImageNet: a large-scale hierarchical image database. In: 2009 IEEE Conference on Computer Vision and Pattern Recognition (2010). https://doi.org/10.1109/cvpr.2009.5206848

6. Drozdzal, M., Vorontsov, E., Chartrand, G., Kadoury, S., Pal, C.: The importance of skip connections in biomedical image segmentation. In: Carneiro, G., et al. (eds.) LABELS/DLMIA -2016. LNCS, vol. 10008, pp. 179–187. Springer, Cham (2016). https://doi.org/10.1007/978-3-319-46976-8_19

7. Everingham, M., et al.: The pascal visual object classes challenge: a retrospective. Int. J. Comput. Vis. **111**(1), 98–136 (2015). https://doi.org/10.1007/S11263-014-0733-5. ISSN: 15731405

8. Isensee, F., et al.: nnU-Net: a self-configuring method for deep learning-based biomedical image segmentation. Nat. Methods **18**(2), 203–211 (2020). https://doi.org/10.1038/s41592-020-01008-z. ISSN: 1548-7105

9. Larrazabal A.J., et al.: Gender imbalance in medical imaging datasets produces biased classifiers for computer-aided diagnosis. In: Proceedings of the National Academy of Sciences of the United States of America, vol. 117, no. 23, pp. 12592–12594 (2020). https://doi.org/10.1073/pnas.1919012117. ISSN: 10916490

10. Lee T., et al.: A systematic study of race and sex bias in CNN-based cardiac MR segmentation. In: Lecture Notes in Computer Science (including subseries Lecture Notes in Artificial Intelligence and Lecture Notes in Bioinformatics), LNCS, vol. 13593, pp. 233–244. Springer Science and Business Media Deutschland GmbH (2022). https://doi.org/10.1007/978-3-031-23443-9_22. ISBN: 9783031234422

11. Mehrabi N., et al.: A survey on bias and fairness in machine learning. In: ACM Computing Surveys (2019). https://doi.org/10.1145/3457607. ISSN: 15577341

12. Petersen E., et al.: Feature robustness and sex differences in medical imaging: a case study in MRI-based Alzheimer's disease detection. In: Lecture Notes in Computer Science (including subseries Lecture Notes in Artificial Intelligence and Lecture

Notes in Bioinformatics), LNCS, vol. 13431, pp. 88–98 (2022). https://doi.org/10.1007/978-3-031-16431-6_9. ISSN: 16113349

13. Petersen, S.E., et al.: UK biobank's cardiovascular magnetic resonance protocol. J. Cardiovasc. Magn. Reson. **18**(1), 1–7 (2016). https://doi.org/10.1186/s12968-016-0227-4. ISSN: 1532429X

14. Puyol-Antón, E., et al.: Fairness in cardiac magnetic resonance imaging: assessing sex and racial bias in deep learning-based segmentation. Front. Cardiovasc. Med. **9**, 859310 (2022). https://doi.org/10.3389/FCVM.2022.859310. ISSN: 2297–055X

15. Puyol-Antón E., et al.: Fairness in cardiac MR image analysis: an investigation of bias due to data imbalance in deep learning based segmentation. In: Medical Image Computing and Computer Assisted Intervention – MICCAI 2021, LNCS, vol. 12903, pp. 413–423. Springer International Publishing (2021). https://doi.org/10.1007/978-3-030-87199-4_39. ISBN: 9783030871987

16. Ronneberger, O., Fischer, P., Brox, T.: U-net: convolutional networks for biomedical image segmentation. In: Lecture Notes in Computer Science (including subseries Lecture Notes in Artificial Intelligence and Lecture Notes in Bioinformatics), vol. 9351, pp. 234–241 (2015). https://doi.org/10.1007/978-3-319-24574-4_28/COVER. ISSN: 16113349

17. Seyyed-Kalantari, L., et al.: Underdiagnosis bias of artificial intelligence algorithms applied to chest radiographs in under-served patient populations. Nat. Medicine **27**(12), 2176–2182 (2021). https://doi.org/10.1038/s41591-021-01595-0. ISSN: 1078–8956

An Investigation into Race Bias in Random Forest Models Based on Breast DCE-MRI Derived Radiomics Features

Mohamed Huti[1]([✉]), Tiarna Lee[1], Elinor Sawyer[2], and Andrew P. King[1]

[1] School of Biomedical Engineering and Imaging Sciences, King's College London, London, UK
mohamed.huti@kcl.ac.uk
[2] School of Cancer and Pharmaceutical Sciences, King's College London, London, UK

Abstract. Recent research has shown that artificial intelligence (AI) models can exhibit bias in performance when trained using data that are imbalanced by protected attribute(s). Most work to date has focused on deep learning models, but classical AI techniques that make use of hand-crafted features may also be susceptible to such bias. In this paper we investigate the potential for race bias in random forest (RF) models trained using radiomics features. Our application is prediction of tumour molecular subtype from dynamic contrast enhanced magnetic resonance imaging (DCE-MRI) of breast cancer patients. Our results show that radiomics features derived from DCE-MRI data do contain race-identifiable information, and that RF models can be trained to predict White and Black race from these data with 60–70% accuracy, depending on the subset of features used. Furthermore, RF models trained to predict tumour molecular subtype using race-imbalanced data seem to produce biased behaviour, exhibiting better performance on test data from the race on which they were trained.

Keywords: Bias · AI · Fairness · Radiomics · Breast · DCE-MRI

1 Introduction

The potential for artificial intelligence (AI) models to exhibit bias, or disparate performance for different protected groups, has been demonstrated in a range of computer vision and more recently medical imaging applications. For example, biased performance has been reported in AI models for diagnostic tasks from chest X-rays [9,21], cardiac magnetic resonance (MR) image segmentation [10, 18,19], brain MR image analysis [7,16,22,24] and dermatology image analysis [1,6]. In response, the field of *Fair AI* has emerged to address the challenge of making AI more trustworthy and equitable in its performance for protected groups [14].

S. Wesarg et al. (Eds.): CLIP/FAIMI/EPIMI 2023, LNCS 14242, pp. 225–234, 2023.
https://doi.org/10.1007/978-3-031-45249-9_22

A common cause of bias in AI model performance is the combination of a distributional shift between the data of different protected groups and demographic imbalance in the training set. For example, in chest X-rays there is a distributional shift between sexes due to the presence of breast tissue lowering the signal-to-noise ratio of images acquired from female subjects [9]. However, more subtle distributional shifts can also exist which cannot be perceived by human experts, and recent work has shown that race-based distributional shifts are present in a range of medical imaging modalities, including breast mammography [5]. This raises the possibility of race bias in AI models trained using imbalanced data from these modalities.

Most work on AI bias to date has focused on deep learning techniques, in which the features used for the target task are optimised as part of the training process. In the presence of distributional shift and training set imbalance this learning process can lead to bias in the features and potentially in model performance. Classical AI approaches are trained using fixed hand-crafted features such as radiomics, and so might be considered to be less susceptible to bias. However, despite these approaches still being widely applied, little experimental work has been performed to assess the potential for, and presence of, bias in these features and the resulting models.

In this paper, we investigate the potential for bias in a classical AI model (Random Forest) based on radiomics features. Our chosen application is potential race bias in Random Forest models trained using radiomics features derived from dynamic contrast enhanced magnetic resonance imaging (DCE-MRI) of breast cancer patients. This application is of interest because there have been reported differences in breast density and composition between races [12,15], as well as tumour biology [11], indicating a possible distributional shift in (imaging) data acquired from different races, and hence the possibility of bias in AI models trained using these data. Our target task is the prediction of tumour molecular subtype from the radiomics features. This is a clinically useful task because different types of tumour are commonly treated in different ways (e.g. surgery, chemotherapy), and tumour molecular subtype is normally determined through an invasive biopsy. Therefore, development and validation of an AI model to perform this task from imaging data would obviate the need for such biopsies.

This paper makes two key contributions to the field of Fair AI. First, we present the first thorough investigation into possible bias in AI models based on radiomics features. Second, we perform the first investigation of bias in AI models based on features derived from breast DCE-MRI imaging.

2   Materials

In our experiments we employ the dataset described in [20][1], which features pre-operative DCE-MRI images acquired from 922 female patients with invasive breast cancer at Duke Hospital, USA, together with demographic, clinical, pathology, genomic, treatment, outcome and other data. From the DCE-MRI

---

[1] The dataset is publicly available at: https://doi.org/10.7937/TCIA.e3sv-re93.

images, 529 radiomics features have been derived which are split into three (partially overlapping) categories: whole breast, fibroglandular tissue (FGT) only and tumour only. The full dataset consists of approximately 70% White subjects, 22% Black subjects and 8% other races. We refer the reader to [20] for a full summary of patient characteristics and the data provided.

3   Methods

For all experiments we employed a Random Forest (RF) classifier as our AI model, similar to the work described in [20]. For each model, we performed a grid search hyperparameter optimisation using a 5-fold cross validation on the training set. Following this, the final model was trained with the selected hyperparameter values using all training data and applied to the test set. The hyperparameters optimised were the number of trees (50, 100, 200, 250), the maximum depth of the trees (10, 15, 30, 45) and the splitting criterion (entropy, Gini). Our model training differed from that described in [20] in three important ways:

1. We used only Black and White subjects to enable us to analyse bias in a controlled environment. Data from all other races were excluded from both the training and test sets. This meant that our dataset comprised 854 subjects (651 White and 203 Black).
2. To simplify our analysis, we focused on just one of the binary classification problems reported in [20]: prediction of *Luminal A* vs *non-Luminal A* tumour molecular subtype. Based on these labels, the numbers of positive (*Luminal A*) and negative (*non-Luminal A*) subjects for each race are summarised in Table 1. As can be seen, there is a higher proportion of *non-Luminal A* tumours in the Black patients, which is consistent with prior studies on relative incidence of tumour subtypes by race [2,8].
3. We did not perform feature selection prior to training and evaluating the RF classifiers. We chose to omit this step because one of our objectives was to analyse which specific radiomics features (if any) could lead to bias in the trained models, so we did not want to exclude any features prior to this analysis.

Table 1. Summary of positive (*Luminal A*) and negative (*non-Luminal A*) labels in the dataset overall and broken down by race.

Label	White	Black	All
Positive (*Luminal A*)	442	107	549
Negative (*non-Luminal A*)	209	96	305

4   Experiments and Results

4.1   Race Classification

In the first experiment, our aim was to determine if the radiomics features contain race-identifiable information. The presence of such information is a known potential cause of bias in trained models as it would be indicative of a distributional shift in the data between races, not just in the imaging data but in the derived (hand-crafted) radiomics features. To investigate this, we trained RF classifiers to predict race (White or Black) from the entire radiomics feature set, and also for the whole breast, FGT and tumour features individually. For these experiments, to eliminate the effect of class (i.e. race) imbalance, we randomly sampled from the dataset to create race-balanced training and test sets, each consisting of 100/100 White/Black subjects.

Results are reported as percentage classification accuracy in Table 2 for all subjects in the test set and also separately for each race. We can see that it is possible to predict race from radiomics features with around 60–70% accuracy. The results are similar for both White and Black subjects and do not differ significantly for the category of radiomics features used. It should be noted that the whole breast, FGT and tumour categories are partially overlapping, hence the similar performance for the different radiomics categories. Specifically, a set of features related to breast and FGT volume is included in both the whole breast and FGT categories, and another set related to FGT and tumour enhancement is present in both the FGT and tumour categories [20].

Table 2. Race classification accuracy from radiomics features derived from breast DCE-MRI. Results are presented as percentage classification accuracy and reported for whole test set as well as broken down by race. Classification was performed from all radiomics features as well as just those derived from the whole breast, fibroglandular tissue (FGT) and tumour only.

Radiomics features	Whole test set	White subjects only	Black subjects only
All	63%	64%	66%
Whole breast only	62%	70%	57%
FGT only	61%	65%	60%
Tumour only	62%	62%	66%

4.2   Bias Analysis

Having established one of the key conditions for the presence of bias in AI models, i.e. a distributional shift between the data of different protected groups, we next investigated whether training with highly imbalanced training sets can lead to bias in performance.

For these experiments we split the dataset into a training set of 426 subjects and a test set of 428 subjects. The split was random under the constraints that the White and Black subjects and the *Luminal A* and *non-Luminal A* subjects were evenly distributed between train and test sets. The training set consisted of 325/101 White/Black subjects and 274/152 *Luminal A/non-Luminal A* subjects, and the test set consisted of 326/102 White/Black subjects and 275/153 *Luminal A/non-Luminal A* subjects.

In addition, we curated two additional training sets consisting of only the White subjects and only the Black subjects from the combined training set described above. Due to the racial imbalance in the database, these training sets consisted of 325/101 subjects for White/Black subjects. Using all three training sets (i.e. all, White-only and Black-only), we trained RF models for the task of classifying *Luminal A* vs *non-Luminal A* tumour molecular subtype and evaluated their performance for the entire test set as well as for the White subjects and the Black subjects in the test set individually. Class (i.e. molecular subtype) imbalance was addressed by applying a weighting to training samples that was inversely proportional to the class frequency.

Results are presented in Table 3, in which performance is quantified using the percentage classification accuracy. We performed this experiment using all radiomics features, just the whole breast features, just the FGT features and just the tumour features. We can see that in terms of overall performance, the models trained using all data and the White-only data had higher accuracy than the models trained using Black-only data, reflecting the impact of different training set sizes. Regarding race-specific performance, the models trained using all training data (i.e. 325/101 White/Black subjects) performed slightly better on White subjects, likely reflecting the effect of training set imbalance. The difference in performance in favour of White subjects varied from 3–11% (mean 6.25%), depending on the subset of features used. The models trained using White-only data had a larger performance disparity in favour of White subjects, varying between 6–11% (mean 9%). The models trained using Black-only data showed generally better performance on Black subjects (mean 3.5% difference), although the model trained using all radiomics features was 1% better for White subjects. In contrast, the model trained using whole breast radiomics features performed 10% better for Black subjects. With the exception of this last result, in general there was not a noticeable difference in bias between the models trained using all radiomics features, just whole breast features, just FGT features and just tumour features, which is consistent with the similar race classification results reported in Sect. 4.1.

4.3 Covariate Analysis

Next we investigated a range of covariates to test for the presence of confounding variable(s) that could be leading to the observed bias. From the full set of patient data available within the dataset we selected those variables that could most plausibly have associations with both race and model performance. These variables are summarised in Table 4. For the continuous variable (i.e. age), the

Table 3. Tumour molecular subtype classification accuracy for *Luminal A* vs. *non-Luminal A* task. Results presented as percentage accuracy and reported for training/testing using all subjects, White subjects only and Black subjects only. From top-to-bottom: results computed using all radiomics features, just whole breast features, just fibroglandular tissue (FGT) features and just tumour features.

ALL FEATURES	Train		
Test	All	White	Black
All	65%	65%	60%
White	68%	67%	60%
Black	57%	58%	59%
WHOLE BREAST	Train		
Test	All	White	Black
All	61%	62%	53%
White	62%	63%	51%
Black	57%	57%	61%
FGT	Train		
Test	All	White	Black
All	67%	64%	61%
White	68%	67%	60%
Black	62%	56%	62%
TUMOUR	Train		
Test	All	White	Black
All	67%	65%	59%
White	68%	67%	58%
Black	65%	57%	61%

table shows the median and lower/upper quartiles for White and Black patients separately. For categorical variables (i.e. all other variables), counts and percentages are provided. The p-values were computed using a Mann-Whitney U test for age and Chi-square tests for independence for all other variables. We can see that three of the covariates showed significant differences (at 0.05 significance) in their distributions between White and Black subjects: age, estrogen receptor status and neoadjuvant chemotherapy.

As stated earlier, non-luminal breast cancer, which is generally estrogen receptor negative, is more commonly seen in Black subjects than White subjects [2,8]. In addition, this cancer is more commonly treated with neoadjuvant chemotherapy, whereas luminal breast cancer is treated with surgery, followed by endocrine therapy and chemotherapy [23] [4]. This may contribute to the statistically significant differences seen in the covariates.

Table 4. Distributions of covariates in the dataset by race (White and Black subjects only). Continuous variables are reported as median (M), lower (L) and upper (U) quartiles. Categorical variables are reported as count (N) and percentage (%). p-values calculated using Mann Whitney U tests for continuous variables and Chi Square tests for independence for categorical variables.

Covariate		White	Black	p-value
Age at diagnosis (years, M(L,U))		53.3(45.9, 61.8)	50.5(44.0, 58.5)	0.012
Scanner (N/%):	GE	451/69.3%	134/ 66.0%	0.430
	Siemens	200/30.7%	69/34.0%	
Field strength (N/%):	1.5T	315/48.4%	111/54.7%	0.258
	2.89T	1/0.1%	0/0.0%	
	3T	335/ 51.5%	92/45.3%	
Menopause at diagnosis	Pre	276/42.4%	94/46.3%	0.574
(N/%):	Post	364/55.9%	105/51.7%	
	N/A	11/1.7%	4/2.0%	
Estrogen receptor status	Positive	510/ 78.3%	123/60.6%	7.430e−07
(N/%):	Negative	141/21.7%	80/39.4%	
Human epidermal growth	Positive	111/17.1%	36/17.7%	0.906
factor 2 receptor status (N/%):	Negative	540/82.9%	167/82.3%	
Adjuvant radiation	Yes	434/67.7%	144/71.0%	0.341
therapy (N/%):	No	210/32.3%	58/29.0%	
Neoadjuvant radiation	Yes	13/2.0%	7/ 3.4%	0.358
therapy (N/%):	No	632/98.0%	7/96.6%	
Adjuvant chemotherapy	Yes	391/ 63.1%	108/57.1%	0.167
(N/%):	No	229/36.9%	81/42.9%	
Neoadjuvant chemotherapy	Yes	178/ 28.1%	91/46.9%	1.593e−06
(N/%):	No	455 /71.9 %	103/53.1%	

5 Discussion and Conclusions

The main contribution of this paper has been to present the first investigation focused on potential bias in AI models trained using radiomics features. The work described in [20] also reported performance of their AI models based on radiomics features broken down by race. However, in our work we have performed a more controlled analysis to investigate the potential for bias and its possible causes. As a second key contribution, our paper represents the first investigation into bias in AI models based on breast DCE-MRI imaging.

Our key findings are that: (i) radiomics features derived from breast DCE-MRI data contain race-identifiable information, leading to the potential for bias in AI models trained using such data, and (ii) RF models trained to predict tumour molecular subtype seem to exhibit biased behaviour when trained using race-imbalanced training data.

These findings show that the process of producing hand-crafted features such as radiomics features does not remove the potential for bias from the imaging data, and so further investigation of the performances of other similar models is warranted. However, an unanswered question is whether the production of hand-crafted features *reduces* the potential for bias. To investigate this, in future work we will compare bias in radiomics-based AI models to similar image-based AI models.

Our analysis of covariates did highlight several possible confounders, so we emphasise that the cause of the bias we have observed remains to be established. In future work we will perform further analysis of these potential confounders, including of interactions between multiple variables, to help determine this cause.

Interestingly, the work described in [20], which included the same *Luminal A* vs. *non-Luminal A* classification task using the same dataset did not find a statistically significant difference in performance between races. However, there are a number of differences between our work and [20]. First, [20] used all training data (half of the full dataset) when training their RF models, i.e. they did not create deliberately imbalanced training sets as we did. Therefore, their race distribution was presumably similar to that of the full dataset (i.e. 70% White, 22% Black, 8% other races). It may be that this was not a sufficient level of imbalance to result in biased performances, and/or that the presence of other races (apart from White and Black) in the training and test sets reduced the bias effect. Second, we also note that the comparison performed in [20] was between White and other races, whereas we compared White and Black races. Third, in [20] a feature selection step was employed to optimise performance of their models. It is possible that this reduced the potential for bias by removing features that contained race-specific information, although our race classification results (see Sect. 4.1) suggest that this information is present across all categories of feature.

In this work we have focused on distributional shift in imaging data (and derived features) as a cause of bias, but bias can also arise from other sources, such as bias in data acquisition, annotations, and use of the models after deployment [3,13]. We emphasise that by focusing on this specific cause of bias we do not believe that others should be neglected, and we argue for the importance of considering possible bias in all parts of the healthcare AI pipeline.

Finally, this paper has focused on highlighting the *presence* of bias, and we have not addressed the important issue of what should be *done* about it. Bias mitigation techniques have been proposed and investigated in a range of medical imaging problems [19,25,26], and approaches such as these may have a role to play in addressing the bias we have uncovered in this work. However, when attempting to mitigate bias one should bear in mind that the classification tasks of different protected groups may have different levels of difficulty, making it challenging to eliminate bias completely. Furthermore, one should take care to ensure that the performances of the protected groups are 'levelled up' rather than 'levelled down' [17] to avoid causing harm to some protected groups.

Acknowledgements. This work was supported by the National Institute for Health Research (NIHR) Biomedical Research Centre at Guy's and St Thomas' NHS Foundation Trust and King's College London, United Kingdom. Additionally this research was funded in whole, or in part, by the Wellcome Trust, United Kingdom

WT203148/Z/16/Z. The views expressed in this paper are those of the authors and not necessarily those of the NHS, the NIHR or the Department of Health and Social Care.

References

1. Abbasi-Sureshjani, S., Raumanns, R., Michels, B.E.J., Schouten, G., Cheplygina, V.: Risk of training diagnostic algorithms on data with demographic bias. In: Cardoso, J., et al. (eds.) IMIMIC/MIL3ID/LABELS -2020. LNCS, vol. 12446, pp. 183–192. Springer, Cham (2020). https://doi.org/10.1007/978-3-030-61166-8_20
2. Abd El-Rehim, D.M., et al.: Expression of luminal and basal Cytokeratins in human breast carcinoma. J. Pathol. **203**(2), 661–671 (2004)
3. Chen, I.Y., Pierson, E., Rose, S., Joshi, S., Ferryman, K., Ghassemi, M.: Ethical machine learning in healthcare. Annu. Rev. Biomed. Data Sci. **4**(1), 123–144 (2021)
4. Domergue, C., et al.: Impact of her2 status on pathological response after neoadjuvant chemotherapy in early triple-negative breast cancer. Cancers **14**(10), 2509 (2022)
5. Gichoya, J.W., et al.: AI recognition of patient race in medical imaging: a modelling study. Lancet Digit. Health **7500**(22) (2022)
6. Guo, L.N., Lee, M.S., Kassamali, B., Mita, C., Nambudiri, V.E.: Bias in, bias out: underreporting and underrepresentation of diverse skin types in machine learning research for skin cancer detection - a scoping review. J. Am. Acad. Dermatol. **87**(1), 157–159 (2021)
7. Ioannou, S., Chockler, H., Hammers, A., King, A.P.: A study of demographic bias in CNN-based brain MR segmentation. In: Abdulkadir, A., et al. (eds.) Machine Learning in Clinical Neuroimaging. Lecture Notes in Computer Science, vol. 13596. Springer, Cham (2022). https://doi.org/10.1007/978-3-031-17899-3_2
8. Jones, V.C., Kruper, L., Mortimer, J., Ashing, K.T., Seewaldt, V.L.: Understanding drivers of the black: White breast cancer mortality gap: A call for more robust definitions. Cancer **128**(14), 2695–2697 (2022)
9. Larrazabal, A.J., Nieto, N., Peterson, V., Milone, D., Ferrante, E.: Gender imbalance in medical imaging datasets produces biased classifiers for computer-aided diagnosis. Proc. Natl. Acad. Sci. U.S.A. **117**(23), 12592–12594 (2020)
10. Lee, T., Puyol-Antón, E., Ruijsink, B., Shi, M., King, A.P.: A systematic study of race and sex bias in CNN-based cardiac MR segmentation. In: Camara, O., et al. (eds.) Statistical Atlases and Computational Models of the Heart. Regular and CMRxMotion Challenge Paper. Lecture Notes in Computer Science, vol. 13593. Springer, Cham (2022)
11. Martini, R., et al.: African ancestry-associated gene expression profiles in triple-negative breast cancer underlie altered tumor biology and clinical outcome in women of African descent. Cancer Discov. **12**(11), 2530–2551 (2022)
12. McCarthy, A.M., et al.: Racial differences in quantitative measures of area and volumetric breast density. J. Natl. Cancer Inst. **108**(10) (2016)
13. McCradden, M.D., Joshi, S., Mazwi, M., Anderson, J.A.: Ethical limitations of algorithmic fairness solutions in health care machine learning. Lancet Digit. Health **2**(5), e221–e223 (2020)
14. Mehrabi, N., Morstatter, F., Saxena, N., Lerman, K., Galstyan, A.: A survey on bias and fairness in machine learning. ACM Comput. Surv. **54**(6), 1–35 (2021)
15. Moore, J.X , Han, Y., Appleton, C., Colditz, G., Toriola, A.T.: Determinants of mammographic breast density by race among a large screening population. JNCI Cancer Spectr. **26**(4) (2020)

16. Petersen, E., et al.: Feature robustness and sex differences in medical imaging: a case study in MRI-based Alzheimer's disease detection. In: Wang, L., Dou, Q., Fletcher, P.T., Speidel, S., Li, S., et al. (eds.) Medical Image Computing and Computer Assisted Intervention - MICCAI 2022. Lecture Notes in Computer Science, Springer, Cham (2022). https://doi.org/10.1007/978-3-031-16431-6_9

17. Petersen, E., Holm, S., Ganz, M., Feragen, A.: The path toward equal performance in medical machine learning. Patterns **4**(7), 100790 (2023)

18. Puyol-Antón, E., et al.: Fairness in cardiac magnetic resonance imaging: assessing sex and racial bias in deep learning-based segmentation. Front. Cardiovasc. Med. **9**, 859310 (2022)

19. Puyol-Antón, E., et al.: Fairness in cardiac MR image analysis: an investigation of bias due to data imbalance in deep learning based segmentation. In: de Bruijne, M., et al. (eds.) MICCAI 2021. LNCS, vol. 12903, pp. 413–423. Springer, Cham (2021). https://doi.org/10.1007/978-3-030-87199-4_39

20. Saha, A., et al.: A machine learning approach to radiogenomics of breast cancer: a study of 922 subjects and 529 DCE-MRI features. Br. J. Cancer **119**, 508–516 (2018)

21. Seyyed-Kalantari, L., Zhang, H., McDermott, M.B.A., Chen, I.Y., Ghassemi, M.: Underdiagnosis bias of artificial intelligence algorithms applied to chest radiographs in under-served patient populations. Nat. Med. **27**(12), 2176–2182 (2021)

22. Stanley, E.A.M., Wilms, M., Mouches, P., Forkert, N.D.: Fairness-related performance and explainability effects in deep learning models for brain image analysis. J. Med. Imaging **9**(6), 061102 (2022)

23. Uchida, N., Suda, T., Ishiguro, K.: Effect of chemotherapy for luminal a breast cancer. Yonago Acta Med. **56**(2), 51–56 (2013)

24. Wang, R., Chaudhari, P., Davatzikos, C.: Bias in machine learning models can be significantly mitigated by careful training: evidence from neuroimaging studies. Proc. Natl. Acad. Sci. U.S.A. **120**(6), e2211613120 (2023)

25. Zhang, H., Dullerud, N., Roth, K., Oakden-Rayner, L., Pfohl, S., Ghassemi, M.: Improving the fairness of chest X-ray classifiers. In: Proceedings of Conference on Health, Inference, and Learning, pp. 204–233 (2022)

26. Zong, Y., Yang, Y., Hospedales, T.: MEDFAIR: benchmarking fairness for medical imaging. In: Proceedings of International Conference on Learning Representations (ICLR) (2023)

How You Split Matters: Data Leakage and Subject Characteristics Studies in Longitudinal Brain MRI Analysis

Dewinda J. Rumala(✉) ®

Institut Teknologi Sepuluh Nopember, Surabaya, Indonesia
dewinda.207022@mhs.its.ac.id

Abstract. Deep learning models have revolutionized the field of medical image analysis, offering significant promise for improved diagnostics and patient care. However, their performance can be misleadingly optimistic due to a hidden pitfall called 'data leakage'. In this study, we investigate data leakage in 3D medical imaging, specifically using 3D Convolutional Neural Networks (CNNs) for brain MRI analysis. While 3D CNNs appear less prone to leakage than 2D counterparts, improper data splitting during cross-validation (CV) can still pose issues, especially with longitudinal imaging data containing repeated scans from the same subject. We explore the impact of different data splitting strategies on model performance for longitudinal brain MRI analysis and identify potential data leakage concerns. GradCAM visualization helps reveal shortcuts in CNN models caused by identity confounding, where the model learns to identify subjects along with diagnostic features. Our findings, consistent with prior research, underscore the importance of subject-wise splitting and evaluating our model further on hold-out data from different subjects to ensure the integrity and reliability of deep learning models in medical image analysis.

Keywords: Data Leakage · Deep Learning · MRI · Alzheimer's Disease

1 Introduction

Medical image analysis has witnessed remarkable advancements with the integration of deep learning models, particularly Convolutional Neural Networks (CNNs), which have demonstrated great potential in enhancing diagnostics and patient care. Deep learning models have shown remarkable accuracy in classifying medical images, including brain MRI scans, and have become invaluable tools for medical professionals. However, the reliability of these models can be

Supplementary Information The online version contains supplementary material available at https://doi.org/10.1007/978-3-031-45249-9_23.

jeopardized by different biases [5], especially a hidden challenge known as 'data leakage'.

Data leakage occurs when information from the test set unintentionally leaks into the training process [12], leading to over-optimistic performance during model evaluation [3]. This misleading optimism can result in a false perception of the model's efficacy, potentially compromising clinical decision-making and patient outcomes. Data leakage can arise from various sources, including improper data splitting strategies during cross-validation (CV) procedures [4, 22].

In this study, we focus on investigating data leakage in the context of 3D medical imaging, specifically using 3D CNNs for longitudinal brain MRI analysis. Longitudinal brain MRI data is particularly challenging due to repeated scans from the same subjects over time, introducing complex temporal dependencies that can exacerbate data leakage. While 3D CNNs have been considered less prone to data leakage than their 2D counterparts [22], the impact of different data splitting strategies on model performance in this domain remains relatively unexplored.

Our primary objective is to examine the influence of data splitting strategies, specifically subject-wise, record-wise, and late-wise, on the performance of 3D CNN models in longitudinal brain MRI when employed for Alzheimer's Disaese (AD) classification. We aim to uncover potential issues related to data leakage and identity confounding as argued by [4], where the model learns to identify individual subjects rather than focusing solely on the diagnostic features of interest. To achieve this, we employ GradCAM visualization to gain insights into the attention patterns of the CNN models during classification.

This study contributes valuable insights into the impact of data leakage on deep learning models in the domain of medical image analysis, with a specific focus on longitudinal brain MRI data. Our findings are expected to shed light on the importance of proper data splitting strategies and the significance of subject-wise splitting in mitigating data leakage concerns. Ultimately, our research aims to enhance the integrity and reliability of deep learning models, promoting their seamless integration into clinical practice for longitudinal brain MRI classification.

Related Work. When employing deep learning models for Alzheimer's disease (AD) diagnosis using MR images, many studies have traditionally relied on two-dimensional (2D)-based analysis using 2D CNNs. However, the slice-level analysis is prone to bias, including data leakage, as highlighted by Yagis et al. [22]. In response to these limitations, recent works have shifted towards using 3D CNNs for volume-based analysis, which has shown to be more superior compared to 2D-based analysis [15]. Several studies have demonstrated the advantages of 3D CNNs in capturing spatial information and improving the accuracy of medical image analysis [8, 9, 13, 15, 23].

Despite significant advancements and high accuracies achieved by deep learning models, many studies overlook the importance of proper data splitting procedures to avoid data leakage. Specifically, in longitudinal data, the occurrence of identity confounding should be avoided by using subject-wise split, as advocated by Neto et al. [4] and Saeb et al. [19]. Additionally, the commonly overlooked practice of late-wise split (where data splitting is performed after data transformation) has been identified as a potential source of data leakage, as suggested by Yagis et al. [22]. Thus, it should be avoided to ensure a reliable evaluation of deep learning models.

2 Methods

2.1 Data Collection and Processing

We obtained and used the preprocessed MRI data from the Alzheimer's Disease Neuroimaging Initiative (ADNI) [10] (Full details about the preprocessing step can be found at the ADNI website[1]), where we specifically selected T1-weighted and T2-weighted images acquired using a 3T scanner from the same demographic. To ensure data consistency, we applied strict inclusion and exclusion criteria and included only subjects with complete 3T MRI scans for both modalities. The collected dataset consists of longitudinal data from multiple visits for each subject, comprises 25 patients of AD, 41 patients of healthy controls (CN), and 45 patients with mild cognitive impairment (MCI). For each class, there are 150 scans in total, except for AD, where we collected only 50 scans.

The data was then further analyzed using the Computational Anatomy Toolbox 12 (CAT12) [6] implemented in the Statistical Parametric Mapping 12 (SPM12) software. In CAT12, MRI images were processed using the usual VBM pipeline. As a result, all images have been rescaled to $113 \times 137 \times 113$ and the intensity values are in range between 0 and 1. And to increase the number of data, we performed 3D augmentation techniques using the libraries provided by [20]. The augmentation techniques performed include flipping and $5°C$ rotation as previously practiced by [9]. Thus, the final size of our data collection is 300 volume image scans for every class. Further detail about data statistics can be found in Table S1.

2.2 Training Setup

We employed the widely recognized CNN architecture, DenseNet121, which was originally designed for 2D image analysis. To adapt it for 3D medical image analysis, we utilized the 3D architecture version of DenseNet121 as proposed by Soleovyel et al. [20]. Based on their extensive investigation, 3DDenseNet121 demonstrated superior performance compared to other architectures when applied for 3D medical image analaysis task. In addition, we made a few modifications to the architecture to better suit our needs.

Following the last 3D convolutional layer in the architecture, we incorporated a 3D global average pooling layer to reduce the spatial dimensions and obtain a fixed-length feature vector. This feature vector is then connected to

---

[1] http://adni.loni.usc.edu/methods/mri-analysis.

fully connected layers with 2 units {128, 3}, enabling us to perform three-way classification for the target classes of CN, MCI, and AD.

Between the fully connected layers, we incorporated rectified linear unit (ReLU) activation layers to introduce non-linearity and enhance the network's expressiveness. The final layer is a softmax classification layer, which provides probability scores for each class. We utilized the categorical cross-entropy loss function to train the model for the multi-class classification task. More information about the training setup is summarized in the supplemental Table S2.

2.3   Evaluation Scheme

Our main goal is to assess the impact of data leakage during CV by considering three different splitting strategies, each adjusted to our longitudinal data conditions, as detailed in Sect. 2.1. To better illustrate the implementation, we provide a toy example of the three splitting strategies in Fig. S1. In this study, the dataset is then divided into train/validation/test sets with a rough ratio of 70/10/20%, respectively. We conduct CV across 5-folds to thoroughly evaluate the performance and generalizability of our deep learning models under various splitting strategies.

To ensure an unbiased evaluation of our models and maintain a balanced representation of each class in each fold, we employ a stratified splitting strategy. This involves partitioning the images per class into k-folds, ensuring an equal distribution of images for each class across the folds. By doing so, we prevent any class-specific biases during training and evaluation, thus providing a fair assessment of our deep learning model's performance.

These three evaluation schemes allow us to comprehensively investigate and understand the potential impact of data leakage on the model's performance under different data splitting conditions. This provides a holistic assessment of the deep learning model's reliability and generalizability.

Subject-Wise Split. In this scheme, all image scans of each subject are assigned as a group in a fold, regardless of their visit times before any data transformation is applied (early splitting). This approach ensures that the scans from the same subject are not spread across different folds but rather kept together, allowing the model to be evaluated on unseen subjects during each fold. This helps to avoid any potential data leakage and identity confounding between subjects as suggested by Neto, et al. [16].

Record-Wise Split. In this scheme, image scans from all subjects are grouped together into a fold based on the records or visits time in an early split manner. With this strategy, data from different visits of the same subject may appear in different folds, allowing the model to be trained and evaluated on the same subjects during each fold. While some studies [14] believe that this strategy might be conditionally appropriate and can lead to better results compared to subject-wise fashion data split, it may not be the best approach and can lead to overconfident performance due to data leakage, specifically caused by the occurrence of identity confounding as argued by Neto et al. [16].

Late Split. Data splitting is performed after the augmentation process in this scheme, by using a sequential numbering approach (more information described in Fig. S1). This split allows the augmented image scans of the same subjects from a certain visit to appear in different folds. According to previous studies by Varoquaux et al. [21] and Yagis et al. [22], this splitting strategy has been shown to cause data leakage and should be avoided.

3 Result

Evaluation Results on 5-Fold Data. We first investigated the impact of different data splitting strategies towards model performance on the 5-fold data. Table 1 reports the results for the experiments described in Sect. 2.3 on different MRI sequences of T1-weighted and T2-weighted MRI. As seen in the table, the performance of the models varied across the different splitting strategies.

Upon comparing the performance across various data splitting strategies, a significant difference was observed (P=0.0389), with the record-wise strategy achieving the highest mean accuracy across all MRI sequences. Conversely, the subject-wise strategy obtained the lowest mean accuracy when evaluated on both sequences.

Furthermore, our investigation revealed no statistically significant difference between the T1-weighted and T2-weighted MRI sequences in terms of model performance (P=0.7921). This finding suggests that both sequences are equally suitable for the three-way classification task of CN, MCI, and AD. Consequently, the choice of MRI sequence does not significantly influence the overall classification performance.

Table 1. Training and testing on longitudinal T1-weigted and T2-weighted MRI. Values are presented as mean and standard deviation across folds, expressed in percentages.

MRI Sequence	Scheme	Acc	Prec	Rec	F1-score
T1-weighted	Subject-wise	67.11 ± 6.11	69.38 ± 6.02	67.11 ± 6.12	68.28 ± 5.63
	Record-wise	97.33 ± 1.86	97.54 ± 1.66	97.33 ± 1.86	97.34 ± 1.85
	Late split	81.33 ± 12.37	89.45 ± 8.31	79.31 ± 13.29	89.44 ± 77.64
T2-weighted	Subject-wise	61.56 ± 5.23	63.81 ± 9.56	62.56 ± 5.59	61.55 ± 5.77
	Record-wise	93.33 ± 1.55	$95.53 \pm 1,45$	95.33 ± 1.55	95.42 ± 1.77
	Late split	88.89 ± 6.30	89.32 ± 6.34	88.89 ± 6.30	88.75 ± 6.37

Further Evaluation on More Subjects. We evaluated the robustness of 3D CNN models trained with different splitting strategies using a separate hold-out dataset of longitudinal brain MRI. The hold-out data consists of image scans from subjects with different visits, including 8 CN, 11 MCI, and 7 AD subjects, totaling 30 image scans per class (see Table S1 for details). Due to data availability constraints, the evaluation focused on T1-weighted MRI images, but future research could benefit from including T2-weighted MRI images for a more comprehensive assessment.

From the previous training and evaluation of the model on 5-fold data, we obtained five different models for each data splitting strategy that were directly evaluated on the hold-out data. The results of this robustness evaluation are presented in Table 2. From this investigation, we discovered notable discrepancies between the 5-fold and hold-out data evaluation for all splitting strategies.

For instance, the record-wise split, which demonstrated an impressive mean accuracy of 97.33% during CV, experienced a substantial drop to 38.71% when tested on the hold-out data. Similarly, the late split, initially yielding a mean accuracy of 81.33% during CV, exhibited a significant decline to 40.43%. Conversely, subject-wise split displayed a least drastic drop in accuracy compared to the other splitting strategies, from 67.12% to 42.15%.

We observed no statistically significant difference between data split strategies on hold-out data ($P = 0.8235$). However, it is worth noting that subject-wise splitting strategy achieved the highest mean accuracy in this evaluation, while record-wise split obtained the lowest mean accuracy. These results contrast with the model's performance on 5-fold data.

Table 2. Robustness evaluation of trained models on hold-out data of T1-weighted MRI. Values are presented as mean and standard deviation obtained from models trained previously on different folds, expressed in percentages.

Splitting Scheme	Acc	Prec	Rec	F1-score
Subject-wise	42.15 ± 5.45	38.71 ± 7.54	42.12 ± 5.50	38.57 ± 4.99
Record-wise	38.71 ± 7.75	37.48 ± 9.20	38.63 ± 7.72	35.68 ± 7.37
Late-wise	40.43 ± 8.95	$37,62 \pm 13.31$	40.43 ± 8.95	39.92 ± 4.80

Grad-CAM Visualization. We present in Fig. 1 examples of the gradient class activation maps (Grad-CAM) for the CN and AD classes extracted and fused from all layers of the 3D CNN models. GradCAM visualization was performed on the hold-out data to gain insights into the attention patterns during classification for the 3D CNN models trained using different splitting strategies.

The first row in Fig. 1 displays GradCAM examples of correctly classified CN and AD images from different splitting strategies. We observe that the patterns for achieving correct predictions on CN are relatively similar across all

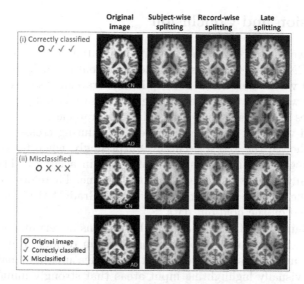

Fig. 1. GradCAM visualization examples for Normal Control (CN) and Alzheimer's Disease (AD) classes under different data split schemes on axial slices of T1-weighted MRI. Each pair of rows depicts (i) correctly classified images, and (ii) misclassified images across all splitting strategies

splitting strategies, where some activations are shown around middle area of the brain. However, there are notable differences between the attention patterns for correctly classified AD images under each splitting strategy.

The Grad-CAM results from the subject-wise split show activations around the peripheral and middle regions of the brain, contributing to the correct identification of AD. However, when employing the record-wise split, the activations become more scattered compared to using subject-wise split. While some activated areas remain shown in the left hemisphere, there are more irrelevant activated regions outside of the brain (in the background).

The same pattern is observed for the model trained using the late split, with even worse results. In this case, the activated areas are heavily concentrated in the corner of the right brain. The results from the record-wise and late splitting strategies may indicate that the models have learned hidden patterns that might not be easily understood by radiologists, rather than focusing on the expected brain patterns for detecting AD.

Moving to the second row, which showcases misclassified images, we can observe similarities in the attention patterns across the different data splitting strategies. Notably, the heatmaps tend to drift off to the background and do not heavily highlight regions in the brain, which suggests challenges in accurately classifying these images.

4   Discussion and Conclusion

In this study, we conducted a comprehensive evaluation of the effect of data leakage on a 3D CNN model for classifying longitudinal brain MRI data into CN, MCI, and AD when using three data splitting strategies: subject-wise, record-wise, and late-wise. Our investigation began with 5-fold data, revealing significant differences in accuracy across the splitting strategies.

The record-wise split performed impressively during cross-validation, while subject-wise showed the lowest accuracy. Surprisingly, when tested on hold-out data, all strategies experienced significant drops in accuracy. The record-wise strategy, initially the best, suffered the worst decline. Therefore, to gain further insights into the model's behavior, we performed GradCAM visualization of the model evaluated on the hold-out data.

Acknowledging GradCAM's potential limitations in certain contexts [1], our study does not rely on precise localization of brain regions. Rather, we utilize GradCAM to understand the model's attended features and patterns during classification, visually highlighting input areas that strongly influence decision-making. This approach offers qualitative insights into potential areas of interest.

The activation heatmaps for correctly classified images showed that there might be shortcut learning in record-wise and late split to correctly identify AD. This means that the models learned unintended patterns during CV, which do not generalize well to new, unseen data [7,11]. This might explain why there is significant drop in record-wise and late-wise strategies. As argued by Neto, et al. [4], such record-wise strategy is prone for identity confounding, where the model try to learn identifying the subject rather than target class. Meanwhile, late split was also found to cause data leakage [22] and might also experience identity confounding.

Meanwhile, in terms of subject-wise split, the model exhibits more interpretable heatmaps, indicating its focus on more relevant regions for AD diagnosis and a reduction in reliance on identity shortcut. However, there is a decrease in performance when tested on hold-out data compared to CV data, suggesting that the model might still be facing some challenges, which according to Little, et al. [14], subject-wise split can lead to model under-fitting and larger classification errors. The occurrence of under-fitting for this strategy is sensible, given that the model displayed unnecessary activations in irrelevant areas of misclassified images. This outcome suggests that the model may not be learning all the essential features required for accurate diagnosis.

However, the less significant drop in performance compared to record-wise and late splitting strategies indicates that subject-wise split might be more robust and less prone to data leakage. As suggested by Neto, et al. [4], overcoming model under-fitting when applying subject-wise split can be addressed by having more subjects in the dataset. With a larger and more diverse dataset, the model can learn from a broader range of samples, enabling it to capture more complex patterns and improve its performance and generalization on both CV and hold-out data.

We acknowledge certain limitations in this study, particularly the issue of under-fitting due to the limited number of subjects in the subject-wise approach. Additionally, we recognize the impact of sensitive attributes, such as age and gender imbalance, wherein certain classes in the dataset include a higher number of female participants compared to male participants, and vice versa. These sensitive attributes could also be the cause for the model's limited generalization on the hold-out data in different splits, which exhibit significant variations.

To enhance the impartiality of our analysis, future research should strive for a more comprehensive and balanced representation of participants, encompassing variables such as gender, age, and other demographic factors [2,17,18]. Additionally, exploring the effects of different splitting strategies on various demographic groups remains crucial to ensure higher level of fairness.

In summary, this study highlights the significant impact of data leakage on model performance, leading to over-optimistic results. Our findings align with previous research advocating for the use of subject-wise splitting approach [4,16,19] and early-split [22] to avoid data leakage and identity confounding. Moreover, it is advisable to incorporate hold-out data from different subjects whenever possible, in accordance with the recommendations of Varoquaux et al. [21]. Implementing these suggestions can improve deep learning model robustness and reliability in medical image analysis, particularly for longitudinal brain MRI analysis.

Acknowledgement. This research was funded by the Ministry of Education and Research Technology, Indonesia through the PMDSU scholarship. Special thanks to Prof. I Ketut Eddy Purnama, the author's PhD supervisor, for securing the research funding and for his valuable ideas and insights, and Prof. Tae-Seong Kim, whose insightful perspectives inspired the development of this paper. Additionally, the author would like to express gratitude to the Bio Imaging Laboratory at Kyung Hee University, South Korea, where the data collection for this study was conducted.

References

1. Arun, N., et al.: Assessing the trustworthiness of saliency maps for localizing abnormalities in medical imaging. radiology. Artif. Intell. **3**(6), e200267 (2021). https://doi.org/10.1148/ryai.2021200267
2. Brown, A., Tomasev, N., Freyberg, J., Liu, Y., Karthikesalingam, A., Schrouff, J.: Detecting shortcut learning for fair medical AI using shortcut testing. Nat. Commun. **14**(1), 4314 (2023). https://doi.org/10.1038/s41467-023-39902-7
3. Bussola, N., Marcolini, A., Maggio, V., Jurman, G., Furlanello, C.: AI slipping on tiles: data leakage in digital pathology. In: Del Bimbo, A., et al. (eds.) ICPR 2021. LNCS, vol. 12661, pp. 167–182. Springer, Cham (2021). https://doi.org/10.1007/978-3-030-68763-2_13
4. Chaibub Neto, E., et al.: Detecting the impact of subject characteristics on machine learning-based diagnostic applications. npj Digit. Med. **2**(1), 99 (2019). https://doi.org/10.1038/s41746-019-0178-x
5. Drukker, K., et al.: Toward fairness in artificial intelligence for medical image analysis: identification and mitigation of potential biases in the roadmap from

data collection to model deployment. J. Med. Imaging **10**(06) (2023). https://doi.org/10.1117/1.JMI.10.6.061104

6. Gaser, C., Dahnke, R., Thompson, P.M., Kurth, F., Luders, E.: Alzheimer's disease neuroimaging initiative: CAT – a computational anatomy toolbox for the analysis of structural MRI data. Neuroscience (2022). https://doi.org/10.1101/2022.06.11.495736

7. Geirhos, R., et al.: Shortcut learning in deep neural networks. Nat. Mach. Intell. **2**(11), 665–673 (2020). https://doi.org/10.1038/s42256-020-00257-z

8. Ghazal, M.: Alzheimer RSQUO s disease diagnostics by a 3D deeply supervised adaptable convolutional network. Front. Biosci. **23**(2), 584–596 (2018). https://doi.org/10.2741/4606

9. Goenka, N., Tiwari, S.: AlzVNet: a volumetric convolutional neural network for multiclass classification of alzheimer's disease through multiple neuroimaging computational approaches. Biomed. Sig. Process. Control **74**, 103500 (2022). https://doi.org/10.1016/j.bspc.2022.103500

10. Jack, C.R., et al.: ADNI study: the Alzheimer's disease neuroimaging initiative (ADNI): MRI methods. J. Magn. Reson. Imaging **27**(4), 685–691 (2008). https://doi.org/10.1002/jmri.21049

11. Jiménez-Sánchez, A., Juodelyte, D., Chamberlain, B., Cheplygina, V.: Detecting shortcuts in medical images - a case study in chest x-rays (2022)

12. Kaufman, S., Rosset, S., Perlich, C., Stitelman, O.: Leakage in data mining: formulation, detection, and avoidance. ACM Trans. Knowl. Discov. Data **6**(4) (2012). https://doi.org/10.1145/2382577.2382579

13. Korolev, S., Safiullin, A., Belyaev, M., Dodonova, Y.: Residual and plain convolutional neural networks for 3D brain MRI classification. In: 2017 IEEE 14th International Symposium on Biomedical Imaging (ISBI 2017), pp. 835–838. IEEE, Melbourne, Australia, April 2017. https://doi.org/10.1109/ISBI.2017.7950647

14. Little, M.A., et al.: Using and understanding cross-validation strategies. Perspectives on Saeb et al. GigaScience **6**(5) (2017). https://doi.org/10.1093/gigascience/gix020

15. Narazani, M., Sarasua, I., Pölsterl, S., Lizarraga, A., Yakushev, I., Wachinger, C.: Is a PET all you need? a multi-modal study for alzheimer's disease using 3D CNNs. In: Wang, L., Dou, Q., Fletcher, P.T., Speidel, S., Li, S. (eds.) Medical Image Computing and Computer Assisted Intervention – MICCAI 2022, pp. 66–76. Springer Nature Switzerland, Cham (2022). https://doi.org/10.1007/978-3-031-16431-6_7

16. Neto, E.C., Perumal, T.M., Pratap, A., Bot, B.M., Mangravite, L., Omberg, L.: On the analysis of personalized medication response and classification of case vs control patients in mobile health studies: the mpower case study (2017)

17. Petersen, E., Feragen, A., da Costa Zemsch, M.L., Henriksen, A., Wiese Christensen, O.E., Ganz, M.: Alzheimer's disease neuroimaging initiative: feature robustness and sex differences in medical imaging: a case study in MRI-based Alzheimer's disease detection. In: Wang, L., Dou, Q., Fletcher, P.T., Speidel, S., Li, S. (eds.) Medical Image Computing and Computer Assisted Intervention – MICCAI 2022, vol. 13431, pp. 88–98. Springer Nature Switzerland, Cham (2022). https://doi.org/10.1007/978-3-031-16431-6_9

18. Ricci Lara, M.A., Echeveste, R., Ferrante, E.: Addressing fairness in artificial intelligence for medical imaging. Nat. Commun. **13**(1), 4581 (2022). https://doi.org/10.1038/s41467-022-32186-3

19. Saeb, S., Lonini, L., Jayaraman, A., Mohr, D.C., Kording, K.P.: The need to approximate the use-case in clinical machine learning. GigaScience **6**(5) (2017). https://doi.org/10.1093/gigascience/gix019

20. Solovyev, R., Kalinin, A.A., Gabruseva, T.: 3D convolutional neural networks for stalled brain capillary detection. Comput. Biol. Med. **141**, 105089 (2022). https://doi.org/10.1016/j.compbiomed.2021.105089

21. Varoquaux, G., Cheplygina, V.: Machine learning for medical imaging: methodological failures and recommendations for the future. npj Digit. Med. **5**(1), 48 (2022). https://doi.org/10.1038/s41746-022-00592-y

22. Yagis, E., et al.: Effect of data leakage in brain MRI classification using 2D convolutional neural networks. Sci. Rep. **11**(1), 22544 (2021). https://doi.org/10.1038/s41598-021-01681-w

23. Zhang, J., Zheng, B., Gao, A., Feng, X., Liang, D., Long, X.: A 3D densely connected convolution neural network with connection-wise attention mechanism for Alzheimer's disease classification. Magn. Reson. Imaging **78**, 119–126 (2021). https://doi.org/10.1016/j.mri.2021.02.001

Revisiting Skin Tone Fairness
in Dermatological Lesion Classification

Thorsten Kalb[1], Kaisar Kushibar[1], Celia Cintas[2], Karim Lekadir[1],
Oliver Diaz[1], and Richard Osuala[1]

[1] Departament de Matemàtiques i Informàtica, Universitat de Barcelona, Barcelona,
Spain
richard.osuala@ub.edu
[2] IBM Research Africa, Nairobi, Kenya

Abstract. Addressing fairness in lesion classification from dermatological images is crucial due to variations in how skin diseases manifest across skin tones. However, the absence of skin tone labels in public datasets hinders building a fair classifier. To date, such skin tone labels have been estimated prior to fairness analysis in independent studies using the Individual Typology Angle (ITA). Briefly, ITA calculates an angle based on pixels extracted from skin images taking into account the lightness and yellow-blue tints. These angles are then categorised into skin tones that are subsequently used to analyse fairness in skin cancer classification. In this work, we review and compare four ITA-based approaches of skin tone classification on the ISIC18 dataset, a common benchmark for assessing skin cancer classification fairness in the literature. Our analyses reveal a high disagreement among previously published studies demonstrating the risks of ITA-based skin tone estimation methods. Moreover, we investigate the causes of such large discrepancy among these approaches and find that the lack of diversity in the ISIC18 dataset limits its use as a testbed for fairness analysis. Finally, we recommend further research on robust ITA estimation and diverse dataset acquisition with skin tone annotation to facilitate conclusive fairness assessments of artificial intelligence tools in dermatology. Our code is available at https://github.com/tkalbl/RevisitingSkinToneFairness.

Keywords: Dermatology · Fairness · Deep Learning · Skin Cancer

1   Introduction

Skin cancer is one of the most prevalent cancer types worldwide [23]. According to [2], early detection increases the survival rate to 99% compared to 32% in late stage detection. The 5-year survival rates after surgical removal of melanoma have been shown to be lower for black patients (73%) compared to white (88%), although melanoma is 23 times more prevalent in white patients [8]. Dick et al. [10] found that black patients were significantly more likely to present with advanced-stage disease, even after adjusting for tumour characteristics and

S. Wesarg et al. (Eds.): CLIP/FAIMI/EPIMI 2023, LNCS 14242, pp. 246–255, 2023.
https://doi.org/10.1007/978-3-031-45249-9_24

demographic factors. Deep learning models have demonstrated a remarkable performance in skin lesion analysis and classification [11]. Therefore, they are promising tools to detect skin cancer earlier and, thus, in theory, bear the potential to reduce the aforementioned disparities. In practice, however, deep learning models have been shown to be prone to and exacerbate existing societal biases [6,16,24]. Although bias and fairness assessment in skin lesion classification has been an active research area [4,17,24], there is a substantial limitation on developing an unbiased classifier. That is, many publicly available datasets lack information about ethnicity or skin types [1]. Hence, such labels are usually obtained using different automated methods. One of the common approaches is based on Individual Topology Angle (ITA) (see Sect. 2.3) that have been used in several studies [5,9,14,15,17,18]. For instance, Kinyanjui et al. [15] estimated skin tones to assess their effect on lesion classification performance, while Bevan et al. [5] labelled skin tones for model debiasing. These previous works [5,15,17,18] form the basis of our analysis and are further described in detail in Sect. 2.3.

These existing studies utilise ITA-based estimation of skin tones as a proxy to ground truth. However, in this work, we show that there is a large disagreement in the assigned skin tones on the same dataset, which suggests that the reported results may be inconclusive. Therefore, we investigate the causes and extent of the discrepancies between the estimated skin tones in previous studies. We uncover the complexities, pitfalls, and differences across ITA-based skin tone estimation techniques that question the conclusions derived in previous studies. In summary, our contributions are as follows:

- We compare different ITA-based automatic skin tone estimation algorithms and highlight common pitfalls.
- We demonstrate the impact of different skin tone estimations on the outcome of fairness analyses of the same model.
- We show the limitations of the commonly used ISIC18 dataset [22] to assess fairness of skin lesion classifiers.

2   Methods and Materials

2.1   Dataset

The ISIC18 dataset [22] includes 10015 dermatoscopic images from Austria and Australia. The present work focuses on classification fairness of seven skin lesion types distributed as: 1113 melanoma (MEL), 6705 melanocytic nevus (NV), 514 basal cell carcinoma (BCC), 327 actinic keratosis (AKIEC), 1099 benign keratosis (BKL), 115 dermatofibroma (DF), and 142 vascular lesions (VASC).

2.2   Evaluation of Skin Lesion Classification

To build a skin lesion classification model, we used a MobileNetV2 [21] initialised with ImageNet weights [12]. Only the last layer was replaced to correspond with the seven skin lesion classes. The choice of this model was driven by its universal applicability [16] due to its lightweight architecture that facilitates deployment across a wide range of devices including portable devices in low-resource settings. A grid search was performed for a combination of (a) a constant learning rate ranging from 1e-3 to 1e-6 in steps of 10 on a logarithmic scale and (b) the batch size, where 12, 24, 32, 48, and 64 were tested. Based on validation loss, the grid search yielded a learning rate of 1e-5 and a batch size of 16. The network was optimised using Adam to minimise the sparse categorical cross entropy loss for a maximum of 60 epochs with an early stopping policy to terminate training if the validation loss did not improve for 30 epochs. All experiments were performed on an NVIDIA GeForce RTX 2080 SUPER with 8GiB memory. All evaluations were conducted three times with different seeds for splitting into training (57%), validation (14%) and test (29%) sets, stratified by lesion, with a slightly different split for the RP data shift experiment (see Sect. 3.3).

2.3   Skin Tone Estimation

Individual Topology Angle (ITA). In the absence of dermatologist-confirmed skin type labels, researchers proposed automatic skin tone and skin type estimations [5,9,14,15,17,18], which are commonly based on the ITA [7]. The ITA is defined within the L*-b*-plane of the CIELab colour space, according to Eq. 1, and illustrated in Fig. 1a.

$$\text{ITA} = \arctan\left(\frac{L-50}{b}\right) \cdot \frac{180°}{\pi} \tag{1}$$

ITA values from different studies only become comparable when the lighting conditions and measurement devices are known and corrected for [19]. Categorical skin types are obtained by binning ITA values [14,15,19] as shown in Fig. 1b. We note that there is no consensus for ITA binning thresholds and high uncertainty for any ITA (colour) to Fitzpatrick [13] (sun reactivity) mapping. Given our analysis of four existing ITA-based automatic skin tone estimation methods [5,15,17,18], we note the following key issues that such methods need to address.

I.1 *Lighting conditions*: The ITA is sensitive to illumination, especially to brightness or lack of yellow chroma.
I.2 *Non-skin imaging artefacts*: The ITA is defined for any colour, but only meaningful for skin. Estimating ITA, in part, for hair, artefacts and dark borders can create misleading results.
I.3 *Lesion to skin contrast*: The pigmented lesion is not representative for the skin colour of the patient and needs to be excluded.
I.4 *From pixel to image-level*: The ITA is defined for each pixel. As one representative ITA value needs to be assigned to an image, there is no consensus how to address variance and outliers of the ITA distribution.

In the following, we describe the aforementioned four existing skin tone estimation methods [5,15,17,18], which are empirically analysed in Sect. 3.

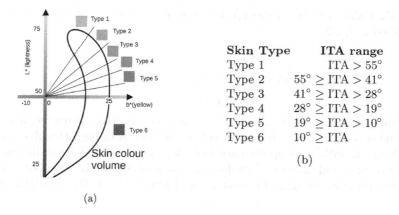

Fig. 1. (a) Skin colour volume in the $L*$ - $b*$ plane of CIELab colour space with ITA thresholds [5]. (b) Skin types from ITA thresholds, as in [5].

Method 1: Deep Learning-Based Skin Segmentation. We adopt a skin segmentation model kindly provided by the authors of [15], which is a Mask R-CNN trained with manually segmented ISIC18 images to address I.2 and I.3. The segmented pixels of healthy skin are converted to CIELab colour space, for which the median within one standard deviation of the mean of ITA's $L*$ and $b*$ components is calculated. Based on these $L*$ and $b*$ median values, an image's ITA is calculated according to Eq. 1. Selecting $L*$ and $b*$ median values separately can help to avoid outliers, but may also lead to less precise ITAs, as $L*$ and $b*$ medians likely do not correspond to the same pixels.

Method 2: Colour-Based Skin Segmentation. This method follows the skin segmentation algorithm proposed in [18]. Input images are converted to grayscale before applying Otsu binarisation and thresholding [20] to detect pixels that are non-lesion. For original values of healthy skin pixels, different thresholds in HSV and YCrCb colour spaces are applied to define potential skin colours. For the pixels within these thresholds, the mean values for red, green and blue are computed and define a representative skin colour. This skin colour is then converted into CIELab space before applying Eq. 1 to calculate the ITA.

Method 3: Random Patch Algorithms. Next method is based on semi-random patches proposed in [5]. It is assumed that a lesion is in the centre of the image and healthy skin is presumably found in at least one of eight patches of 20×20 pixels in the periphery. Before patch extraction, input images are centre-cropped, resized and small dark artefacts such as hair are removed via black-head morphology to address I.2. For each patch, the mean ITA is calculated. The ITA of the patch is selected that corresponds to the brightest skin type, which arguably has the effect of having excluded the pigmented lesion. In contrast to previous methods, this method used *arctan* in Eq. 1 for the ITA instead of

arctan2. Unlike *arctan*, *arctan2* takes into account the signs of the catheti as
described in Eq. 2.

$$\text{arctan2}(x, y) = \begin{cases} \text{arctan}(y/x), & \text{if } x > 0 \\ \text{arctan}(y/x) - \text{sgn}(y) * \pi, & \text{if } x < 0 \\ \text{sgn}(y) * \pi & \text{if } x = 0 \end{cases} \quad (2)$$

Using *arctan* in Eq. 1 assumes $b*$ cannot be negative. However, a negative
$b*$ value encodes blue or absence of yellow chroma. Although the blue colour in
skin is not intuitive, its appearance may depend on variations in illumination,
dermatoscope, and camera. Therefore, in our analysis, we include both versions
of ITA estimation using *arctan* and *arctan2* referred to as RP and RP2, respectively.

**Method 4: Histogram Thresholding with Grey-World White Balanc-
ing.** The next method is adopted from [17]. After centre-cropping and resizing,
a grey-world white balance algorithm is applied to correct for light influence
addressing I.1. Next, non-lesion skin is segmented using the Generalized Histogram Threshold [3] algorithm. The segmented area is transformed from RGB
to CIELab-space to calculate the ITA. However, in [17], reproducibility was
limited regarding how one representative ITA is obtained from the multiple segmented pixels. A further limitation is the assumption that skin images are grey
on average.

3   Experiments and Results

3.1   Comparison of ITA Estimation Methods

A comparison of the estimated skin tone distributions is shown in Fig. 2.
Although the same thresholds (see Fig. 1b) are applied to all four skin tone estimation methods for ITA binning, their estimates on the ISIC18 dataset clearly
differ. A detailed comparison of the agreement among the ITA estimation methods is shown in Fig. 3, where a diagonal matrix would represent perfect agree-
ment.There is little agreement between RP and DLHSS, especially for dark skin.
Most type 6 images in RP appear as type 2 in DLHSS and most type 2 images
in DLHSS are classified as type 1 in RP. All type 5 and 6 images in DLHSS are
labelled as type 1 in RP. Comparing the RP to RP2, the entries of the matrix
lay on the diagonal and in the first row. This implies that in most cases, RP and
RP2 agree on the same skin type. However, approximately 21.6% of the samples
are classified as type 1 by RP2 and darker by RP. Thus, using the *arctan* can
account for the over-estimation of dark skin images and affect all except for type
1. Comparing DLHSS and RP2, both agree that approximately 99.0% of the
samples show skin at least as light as type 3 (ITA > 28°), while only eight out
of 10015 samples are classified at least as dark as type 4 (ITA ≤ 28°).

These eight samples are shown in row 1 of Fig. 4 and indicate that images
labelled as dark skin are actually dark images of skin.

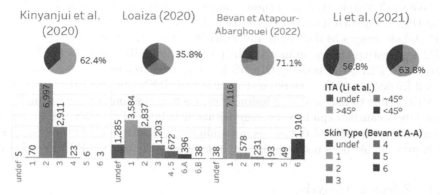

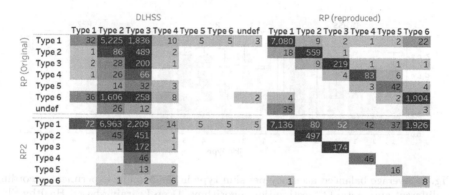

Fig. 2. Comparison of the four different ITA-based skin type estimation methods based on values reported in the literature [5,15,17,18]. Note that Li et al. [17] only reported darker (ITA < 45°, 43.2%) and lighter (ITA > 45°, 63.8%) skin and that their sum equals 107%; it is assumed that one of these values is correct and both possible distributions are shown.

Fig. 3. Comparison of ITA estimates of Deep Learning-based Healthy Skin Segmentation (DLHSS) [15], Random Patch [5] (RP) and Random Patch with *arctan2* (RP2).

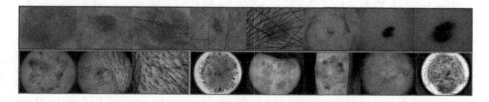

Fig. 4. Blue and red: All samples where both DLHSS and RP estimate dark skin tones (types 4–6, ITA ≤ 28°). Blue: Dark skin according to DLHSS and RP2. Green: Samples with darkest skin according to DLHSS (type 6, ITA ≤ 10°). (Color figure online)

We qualitatively evaluated these images with a dermatologist who confirmed that they likely correspond to Fitzpatrick (FST) [13] skin type III or lower. When RP2 labels images as dark skin and DLHSS does not, this can be explained by RP2 segmenting the lesion in all eight patches. Since RP2 reports the brightest patch and DLHSS a median, DLHSS is likely more reliable in these cases. On visual inspection of the five darkest images (DLHSS), this algorithm appears to be susceptible to hair and rare lesion sites, such as the tongue or ears; and dark skin labels can be explained by segmentation failures of DLHSS. These results suggest that the ISIC18 dataset is not sufficiently diverse for a fairness analysis, as it presumably does not contain any images of dark skin (i.e. FST IV-VI [13]).

3.2   Fairness Analysis

The question arises as to whether and how the different ITA estimation methods impact the result of a downstream fairness analysis. To this end, we analyse the balanced accuracy per skin type for the baseline experiment without data shift for DLHSS, RP, and RP2. As shown in Fig. 5, we obtain different fairness results per method despite using the exact same classification model and test set.

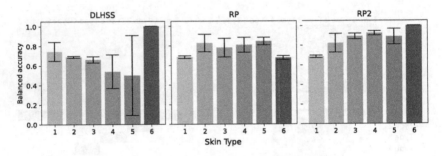

Fig. 5. Average balanced accuracy per skin type in the baseline experiment according to different automatic ITA estimation algorithms: Deep Learning-based Healthy Skin Segmentation (DLHSS), Random Patch algorithm with *arctan* (RP) and with *arctan2* (RP2). All bar charts stem from the same predictions.

For DLHSS ITA values, we note a decline in average balanced accuracy for types 1–4 (according to Fig. 1b). In contrast, the balanced accuracy of very light skin (type 1) and very dark skin (type 6) according to RP is lower than for intermediate skin tones (types 2–5). According to RP2, the light skin samples (type 1) show the worst performance, and the balanced accuracy appears to increase overall for darker skin types. Thus, analysing the fairness of the same lesion predictions, the outcome changes depending on the applied ITA estimation method.

3.3   Simulated Data Shifts

To simulate data shift, light skin images are used during training (ITA > 41°) while dark skin images are used in testing (ITA ≤ 41°). To assess whether the ITA estimation method impacts classification fairness in the presence of data shift, reproduced DLHSS and original RP ITA values are used. For comparison, the size of the baseline test set was defined equal to the DLHSS test set size and the train-validation ratio was 80:20 for all experiments. The results of the data shift experiments are visualised in Fig. 6. Despite a higher baseline accuracy, the weighted precision, recall, and f1-score, remain similar between the baseline and the data shift experiment with DLHSS labels. The balanced accuracy is the *unweighted* macro-averaged recall, hence the difference in balanced accuracy and *weighted* (macro-averaged) recall suggests that the classification performance differs per lesion type. This difference appears to depend on the train-test split induced by the ITA estimation method, yielding different lesion distributions. Thus, not only the choice of evaluation metric, and the lesion type distribution, but also the ITA estimations can alter the conclusions of the fairness analysis.

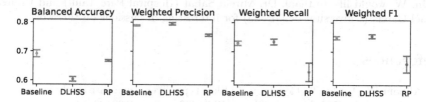

Fig. 6. Metrics for baseline and data shift experiments with the test set of the latter containing exclusively skin types 3–6 (ITA ≤ 41 according to Fig. 1b). ITAs are obtained from DLHSS and RP estimation methods. Error bars show the standard deviation between three experiment repetitions with different random train-validation (DLHSS, RP) or train-validation-test (Baseline) splits.

4   Conclusions

We compared ITA-based skin tone estimation methods that revealed common pitfalls. Namely, susceptibility to lighting conditions, colour space calibration, presence of hair or dark edges. Moreover, we showed the effects of the differences in extracting the healthy skin, and differences in mapping ITA values from pixel to image-level. Further, we observed disagreements between ITA estimation methods, which did not necessarily refer to the same images as dark, and overestimation of dark samples in the ISIC18 dataset. A qualitative analysis with a dermatologist revealed that the images, where different ITA estimation methods agree on their dark skin tones (ITA ≤ 28°), do not represent moderate brown to black skin types (FST IV-VI). Furthermore, our skin lesion classification experiments demonstrated that the choice of ITA estimation method substantially impacts the results of classification fairness analyses. Our data shift experiments further confirmed that ITA estimation altered conclusions drawn from the fairness analysis. We illustrated the need for more diverse dermatological datasets with diligently annotated skin tones, lighting conditions, dermatoscope and camera information. Apart from improving ITA estimation, further avenues of research include the measurement of the differences in model calibration and in epistemic and aleatoric uncertainty per skin tone type. In the presence of limited datasets, synthetic samples with controllable skin tones and lesion types can be explored to quantify their effect on fairness. Overall, our work shows that current skin tone fairness assessments are inconclusive and further research is needed into unified and robust algorithms for automatic skin tone estimation to avoid carrying unwanted biases into dermatology practice when deploying deep learning models.

Acknowledgements. This study has received funding from the European Union's Horizon 2020 research and innovation programme under grant agreement No 952103. It was further partially supported by the project FUTURE-ES (PID2021-126724OB-I00) and by grant FJC2021-047659-I from the Ministry of Science and Innovation of Spain. We would like to thank Dr. Mireia Sábat (Hospital Parc Taulí) and Professor Rafael Garcia (Universitat de Girona) for interesting discussions on this work.

References

1. Alipour, N., Burke, T., Courtney, J.: Skin type diversity: a case study in skin lesion datasets, July 2023. https://doi.org/10.21203/rs.3.rs-3160120/v1
2. American cancer society: cancer facts & figures 2023 (2023). https://www.cancer.org/content/dam/cancer-org/research/cancer-facts-and-statistics/annual-cancer-facts-and-figures/2023/2023-cancer-facts-and-figures.pdf
3. Barron, J.T.: A generalization of Otsu's method and minimum error thresholding. In: Vedaldi, A., Bischof, H., Brox, T., Frahm, J.-M. (eds.) ECCV 2020. LNCS, vol. 12350, pp. 455–470. Springer, Cham (2020). https://doi.org/10.1007/978-3-030-58558-7_27
4. Bevan, P.J., Atapour-Abarghouei, A.: Skin deep unlearning: artefact and instrument debiasing in the context of melanoma classification (2021). https://doi.org/10.48550/ARXIV.2109.09818
5. Bevan, P.J., Atapour-Abarghouei, A.: Detecting melanoma fairly: skin tone detection and debiasing for skin lesion classification. In: Kamnitsas, K., et al. (eds.) Domain Adaptation and Representation Transfer, pp. 1–11. Springer Nature Switzerland, Cham (2022). https://doi.org/10.1007/978-3-031-16852-9_1
6. Birhane, A., Prabhu, V., Han, S., Boddeti, V.N.: On hate scaling laws for data-swamps. arXiv preprint arXiv:2306.13141 (2023)
7. Chardon, A., Cretois, I., Horseau, C.: Skin colour typology and suntanning pathways. Int. J. Cosmet. Sci. **13**(4), 191–208 (1991). https://doi.org/10.1111/j.1467-2494.1991.tb00561.x
8. Collins, K.K., Fields, R.C., Baptiste, D., Liu, Y., Moley, J., Jeffe, D.B.: Racial differences in survival after surgical treatment for melanoma. Ann. Surg. Oncol. **18**(10), 2925–2936 (2011). https://doi.org/10.1245/s10434-011-1706-3

9. Corbin, A., Marques, O.: Exploring strategies to generate fitzpatrick skin type metadata for dermoscopic images using individual typology angle techniques. Multimedia Tools Appl. **82**, 23771–23795 (2022). https://doi.org/10.1007/s11042-022-14211-1

10. Dick, M., Aurit, S., Silberstein, P.: The odds of stage iv melanoma diagnoses based on socioeconomic factors. J. Cutaneous Med. Surg. **23**(4), 421–427 (2019)

11. Esteva, A., et al.: Dermatologist-level classification of skin cancer with deep neural networks. Nature **542**(7639), 115–118 (2017)

12. Fei-Fei, L., Deng, J., Li, K.: ImageNet: Constructing a large-scale image database. J. Vis. **9**(8), 1037–1037 (2010). https://doi.org/10.1167/9.8.1037

13. Fitzpatrick, T.B.: The validity and practicality of sun-reactive skin types I through VI. Arch. Dermatol. **124**(6), 869–871 (1988). https://doi.org/10.1001/archderm.1988.01670060015008

14. Groh, M., et al.: Evaluating deep neural networks trained on clinical images in dermatology with the Fitzpatrick 17k dataset (2021)

15. Kinyanjui, N.M., et al.: Fairness of classifiers across skin tones in dermatology. In: Martel, A.L., et al. (eds.) MICCAI 2020. LNCS, vol. 12266, pp. 320–329. Springer, Cham (2020). https://doi.org/10.1007/978-3-030-59725-2_31

16. Lekadir, K., et al.: FUTURE-AI: guiding principles and consensus recommendations for trustworthy artificial intelligence in medical imaging. arXiv preprint arXiv:2109.09658 (2021). https://doi.org/10.48550/arXiv.2109.09658

17. Li, X., Cui, Z., Wu, Y., Gu, L., Harada, T.: Estimating and improving fairness with adversarial learning. arXiv (2021). https://doi.org/10.48550/arXiv.2103.04243

18. Loaiza, K.: The skin tone problem in artificial intelligence. In: 1st Congress of Women in Bioinformatics and Data Science Latin America, September 2020. https://doi.org/10.13140/RG.2.2.20564.63361/1

19. Ly, B.C.K., Dyer, E.B., Feig, J.L., Chien, A.L., Bino, S.D.: Research techniques made simple: cutaneous colorimetry: a reliable technique for objective skin color measurement. J. Invest. Dermatol. **140**(1), 3–12.e1 (2020). https://doi.org/10.1016/j.jid.2019.11.003

20. Otsu, N.: A threshold selection method from gray-level histograms. IEEE Trans. Syst. Man Cybern. **9**(1), 62–66 (1979)

21. Sandler, M., Howard, A., Zhu, M., Zhmoginov, A., Chen, L.C.: MobileNetV2: Inverted residuals and linear bottlenecks. In: 2018 IEEE/CVF Conference on Computer Vision and Pattern Recognition, IEEE, June 2018. https://doi.org/10.1109/cvpr.2018.00474

22. Tschandl, P., Rosendahl, C., Kittler, H.: The HAM10000 dataset, a large collection of multi-source dermatoscopic images of common pigmented skin lesions. Sci. Data **5**(1) (2018). https://doi.org/10.1038/sdata.2018.161

23. World Health Organization: Skin cancer - IARC (2023). https://www.iarc.who.int/cancer-type/skin-cancer/. Accessed 28 July 2023

24. Wu, Y., Zeng, D., Xu, X., Shi, Y., Hu, J.: FairPrune: achieving fairness through pruning for dermatological disease diagnosis. In: Wang, L., Dou, Q., Fletcher, P.T., Speidel, S., Li, S. (eds.) Medical Image Computing and Computer Assisted Intervention - MICCAI 2022, pp. 743–753. Springer, Springer Nature Switzerland, Cham (2022). https://doi.org/10.1007/978-3-031-16431-6_70

A Study of Age and Sex Bias in Multiple Instance Learning Based Classification of Acute Myeloid Leukemia Subtypes

Ario Sadafi[1,2], Matthias Hehr[1,3], Nassir Navab[2,4], and Carsten Marr[1(✉)]

[1] Institute of AI for Health, Helmholtz Zentrum München - German Research Center for Environmental Health, Neuherberg, Germany
carsten.marr@helmholtz-munich.de
[2] Computer Aided Medical Procedures (CAMP), Technical University of Munich, Munich, Germany
[3] Laboratory of Leukemia Diagnostics, Department of Medicine III, University Hospital, Ludwig-Maximilian University Munich, Munich, Germany
[4] Computer Aided Medical Procedures, Johns Hopkins University, Baltimore, USA

Abstract. Accurate classification of Acute Myeloid Leukemia (AML) subtypes is crucial for clinical decision-making and patient care. In this study, we investigate the potential presence of age and sex bias in AML subtype classification using Multiple Instance Learning (MIL) architectures. To that end, we train multiple MIL models using different levels of sex imbalance in the training set and excluding certain age groups. To assess the sex bias, we evaluate the performance of the models on male and female test sets. For age bias, models are tested against underrepresented age groups in the training data. We find a significant effect of sex and age bias on the performance of the model for AML subtype classification. Specifically, we observe that females are more likely to be affected by sex imbalance dataset and certain age groups, such as patients with 72 to 86 years of age with the RUNX1::RUNX1T1 genetic subtype, are significantly affected by an age bias present in the training data. Ensuring inclusivity in the training data is thus essential for generating reliable and equitable outcomes in AML genetic subtype classification, ultimately benefiting diverse patient populations.

Keywords: Acute Myeloid Leukemia · Multiple Instance Learning · Sex Bias · Age Bias · Fairness

1 Introduction

Long before the advent of modern deep learning algorithms, fairness had already been a concern within the medical community. For instance, disparities in symptoms between men and women experiencing heart attacks led to differing treatment approaches [1]. The machine learning solutions that the community is developing for countless medical applications is not aware of bias-related challenges that persist within medicine. Healthcare practitioners adhere to an oath

S. Wesarg et al. (Eds.): CLIP/FAIMI/EPIMI 2023, LNCS 14242, pp. 256–265, 2023.
https://doi.org/10.1007/978-3-031-45249-9_25

that mandates unbiased treatment regardless of age, disease or disability, creed, ethnic origin, or sex [15]. Fairness remains an ever-present concern in healthcare systems as computer-aided diagnosis systems play an increasingly prominent role in medical decision-making, leading to a heightened awareness of potential biases in training data. Biased algorithms exhibit varying performance when assessed within sub-groups defined by attributes such as sex, gender, age, ethnicity, socioeconomic status, and other related factors [11]. Fairness evaluation has become essential to establish the credibility of the developed systems and the availability of data and models provides an avenue for identifying and rectifying these biases.

Acute Myeloid Leukemia (AML) is a critical hematopoietic malignancy that demands accurate subtype classification for effective clinical decision-making and optimized patient care. The ability to precisely identify different AML subtypes can significantly impact treatment strategies and prognosis. For instance, acute promyelocytic leukemia, a genetic subtype of AML with a PML::RARA gene fusion is considered an oncological emergency, where rapid and appropriate therapy is crucial and curative [13].

Several automated methods are developed for single cell analysis in peripheral blood [7,19,20,23]. Earlier works for the AML subtype classification [14,22] mostly focus on single white blood cell classification. Sidhom et al. [24] suggest a method of combining single cell level and patient level information to provide a rapid, accurate physician-aid for diagnosing AML with PML::RARA fusion using peripheral smear images. In contrast, more recent methods use multiple instance learning (MIL) that enables training a model with patient-level diagnosis rather than single-cell labels. Attention-based MIL [8] combines MIL with a trainable attention module, allowing the model to focus on specific instances within a patient's set of single-cell images for diagnosis [21]. Hehr et al. [5] suggest an explainable MIL model to accurately classify AML genetic subtypes from blood smears with a publicly available dataset [6].

Recently, various publications demonstrated the bias in medical image computing. For example Lee et al. [12] studied the impact of race and sex bias in segmentation of cardiac magnetic resonance (MR) images. Ioanno et al. [9] demonstrate the demographic bias in segmentation of brain MR images. Puyol-Anton et al. [16] investigated bias in AI-based cine cardiac MR segmentation, finding statistically significant differences in segmentation performance between racial groups.

However, to the best of our knowledge, fairness studies on hematology datasets or multiple instance learning architectures are scarce. In this paper, we are investigating possible sex and age bias in the crucial task of AML genetic subtype classification. We use the dataset and methodology of Hehr et al. [5] to conduct experiments with sub-groups varying in terms of sex and age representation during the training process. We examine their impact on both male and female test sets in a sex bias study, as well as the underrepresented age groups in our age bias study.

2   Materials and Methods

2.1   Data

We conducted our investigation on AML using the dataset published by Hehr et al. [5]. The cohort selection process involved meticulous curation of 242 samples from the Munich Leukemia Laboratory blood smear archives, with a primary focus on four distinct AML genetic subtypes and healthy controls. The blood smear images were scanned and cell detection was performed with the aid of the Metasystems Metafer software. Subsequently, a deep neural network (DNN) was employed to assign quality levels to each gallery image. The dataset consisted of a total of 101,947 single-cell images, with each patient contributing 99 to 500 images, ensuring a comprehensive representation of the disease's characteristics.

To ensure data quality, a comprehensive data cleaning process was implemented. This involved excluding blurry images using Canny edge detection and filtering out AML samples with a combined blast percentage of myeloblasts, promyelocytes, and myelocytes less than 20%. Furthermore, an expert hematologist assessed sub-samples of 96 cells per patient to eliminate data artifacts. The final, filtered dataset comprised 81,214 single-cell images from 189 individuals, with each AML subtype represented as follows: APL with PML::RARA (n = 24), AML with NPM1 mutation (n = 36), AML with CBFB::MYH11 fusion (n = 37), and AML with RUNX1::RUNX1T1 fusion (n = 32), and a healthy control group of 60 stem cell donors. Every patient sample is a single training data point and consists of 500 single cell images. Age and sex information of the patients in the dataset is also reported. Figure 1 provides an overview of the dataset details.

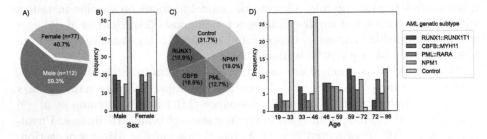

Fig. 1. Dataset composition. A) Distribution of sex in the dataset. B) AML genetic subtype distribution among male and female individuals. C) Overall distribution of AML genetic subtypes among the individuals in the dataset. D) Age distribution across five age groups, alongside the genetic subtype distribution within each age group.

2.2   Multiple Instance Learning

The proposed approach consists of two steps: (i) A feature extractor previously trained on a similar task is used to extract features from every single cell image. (ii) Using the extracted features, an attention based MIL model is used for AML subtype classification. We briefly describe the method in the following:

Feature Extraction. With a dataset of over 300,000 single-cell images from 2205 blood smears, a ResNet34 [4] model is trained to extract single-cell features. The dataset was annotated into 23 single-cell classes by experienced cytologists, excluding two classes due to their small size. Data augmentation techniques such as flipping, rotation, and random erasing were applied to the single-cell images, along with probabilistic oversampling to address class imbalance. The model was initialized with weights trained on ImageNet [17] and optimized using categorical cross-entropy. With the trained model a feature vector of 12,800 features is extracted from the 34th layer of the model before final pooling and fully connected layers. This feature vector represented the characteristics of the input image. Training continued until no significant improvement in validation loss was observed for 10 consecutive epochs. The best-performing model on the test set from the cross-validation folds was chosen for feature extraction on a separately scanned multiple instance learning dataset.

Multiple Instance Learning. A permutation invariant method is used to analyze a set of single-cell images from a patient and predict the associated AML subtype [5]. The method returns attention scores, indicating the importance of each cell in the bag for classifying different AML subtypes. The attention based MIL model works by computing embedded feature vectors from the initial feature vectors of the single-cell images. These vectors are then used to predict the patient's AML subtype based on the calculated attention scores. The class-wise attention matrix is designed to ensure that each attention value directly influences the prediction of its corresponding class, eliminating interclass competition of attention values. This way, the model can make more accurate predictions for each AML subtype.

The training procedure was identical to that of [5], resulting in similar metrics as reported. Training employed a learning rate of $5 \times 10 - 5$, and stopped if no improvement in validation loss was observed for 20 epochs, with a maximum of 150 epochs.

3 Experiments

In this section we describe the experiments we performed to investigate possible age and sex bias of the MIL model for AML subtype classification.

3.1 Sex Bias

From the total of 189 individuals, 77 are females and 112 are males. We designed 5 experiments to train a MIL model with different distributions of sex in the dataset and evaluate all of them on two hold out test sets of only male and only female individuals ensuring a stratified split to have all of the genetic subtypes present. Table 1 shows the sex proportions in different experiments both in training and test splits. A 10% validation set was sampled later from the training split. Holdout test sets are 20% of all of the data.

Table 1. Training and test sets for sex bias experiments. The table presents the number and percentages of male and female individuals in the entire training set. The mean and standard deviation of individuals' age is reported for each experiment. The test sets are comprised of 20% of the total respective female and male populations.

Section	Female (%)	Male (%)	Average age	Description	n
Training set	0	89 (100%)	46.8 ± 17.5	0% female	89
	30 (25.2%)	89 (74.8%)	49.1 ± 18.2	25% female	119
	61 (50%)	61 (50%)	52.3 ± 18.1	50% female	122
	61 (75.3%)	20 (24.6%)	54.9 ± 17.6	75% female	81
	61 (100%)	0	56.7 ± 17.4	100% female	61
Test set (Male)	–	23 (20%)	46.5 ± 20.6	All male testset	
Test set (Female)	16 (20%)	–	52.7 ± 17.5	All female testset	

3.2 Age Bias

To assess age bias, patients were divided into 5 distinct age groups (see Fig. 1c). Each experiment involved holding out one of the age groups as the test set while training the model on all other age groups. Table 2 displays the age groups along with the sex distribution within each group.

Table 2. Different age groups used in age bias experiments. The dataset is divided into 5 different age groups. Number of male and females in every group and percentage of males is reported for every age group.

Section	Age	Male - female (%)
1. Young adults	[19.8, 33.1]	28 - 11 (72%)
2. Middle adulthood	[33.1, 46.3]	37 - 10 (79%)
3. Midlife	[46.3, 59.5]	16 - 20 (44%)
4. Senior adults	[59.5, 72.8]	16 - 22 (42%)
5. Elderly	[72.8, 86.1]	15 - 14 (51%)

4 Results

We use the area under the precision-recall curve to evaluate the performance of the models for every class. Every experiment is repeated 5 times with different seeds and the standard deviation is reported. Statistical testing was performed using Kruskal-Wallis test [10] due to its suitability for comparing independent groups with non-normally distributed data. Subsequently, we performed post hoc Dunn test [3] with Bonferroni correction [2] to identify significantly diverse groups.

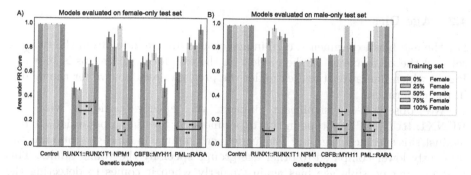

Fig. 2. Area under precision-recall (PR) curve for each class, representing models trained with varying proportions of females versus males in the training set. The models were tested on two separate test sets: A) the female-only test set, and B) the male-only test set. Statistical significance was determined using The statistical analysis involved the Kruskal-Wallis test followed by post hoc pairwise Dunn test with Bonferroni correction, indicated by asterisks: * $(0.1 \geq p > 0.05)$, ** $(0.05 \geq p > 0.01)$, *** $(0.01 \geq p > 0.001)$

4.1   Sex Bias

Figure 2A illustrates how all models perform on the holdout female test set under various training set conditions. Interestingly, we observed that a larger representation of females in the training data positively influences the classification of the PML::RARA genetic subtype. In Fig. 2B, we examined the performance of the same models on the male test set, and once again, we noticed that a higher proportion of females in the training set contributes to improved classification of PML::RARA samples. This is evident as the area under the precision-recall curve for both genetic subtypes is significantly higher when female patients are present. Statistical tests also support these observations.

For the RUNX1::RUNX1T1 genetic subtype, the best results are obtained when a balanced proportion of males and females are present in the training set. Conversely, when only male patients are present in the training set, a significantly lower performance is observed in the evaluation on both test sets.

In the female test set, NPM1 and CBFB::MYH11 subtypes show significantly better performance with a balanced training set, whereas in the male test set, NPM1 does not seem to depend much on the patients' sex in the training set. On the other hand, CBFB::MYH11 shows significantly better performance with more female representation in the training set.

Across both the male and female test sets, the models consistently achieve accurate classifications for control samples from healthy stem cell donors.

The presence of female patients in the training set has a more significant impact on the models' performance on the female test set compared to the male test set. While the male test set is generally affected to a lower degree, in most cases, a higher ratio of female to male presence in the training set positively affects the performance.

4.2   Age Bias

For the age bias experiments, certain age groups are excluded from the training set and the performance of the models on the excluded age groups is shown in Fig. 3, which illustrates the area under the precision-recall curve for the models in the experiment.

We observed a significant impact on elderly patients with the RUNX1::RUNX1T1 subtype in this experiment, suggesting a potential age bias against this specific age group. The model performance in this age group was significantly lower compared to other age groups specially the young adults. This is indicating possible age bias against elderly when it comes to detecting the RUNX1::RUNX1T1 subtype of AML.

Additionally, the absence of young adults with NPM1 and CBFB::MYH11 subtypes in the training set had a significant effect on the models' performance. In both cases, the classifiers' ability to accurately classify NPM1 and CBFB::MYH11 subtypes was compromised.

Since no elderly individuals were represented in the control samples of the dataset (Fig. 1c), their performance is not reported in Fig. 3.

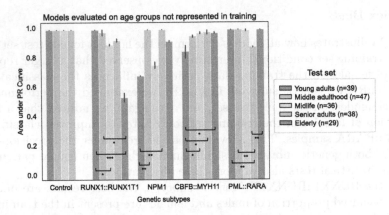

Fig. 3. Area under precision-recall (PR) curve for every class is reported for models trained with different age groups represented in the training set. The test set age group is always held out. Statistical significance was determined using Kruskal-Wallis test followed by post hoc pairwise Dunn test with Bonferroni correction, indicated by asterisks: * $(0.1 \geq p > 0.05)$, ** $(0.05 \geq p > 0.01)$, *** $(0.01 \geq p > 0.001)$

5   Discussion

We have demonstrated the possibility of age and sex bias in AML subtype classification with MIL architectures. Our results suggest that the absence or presence of certain age groups and sexes in the training data can significantly impact the classification performance of specific AML subtypes. In elderly, diagnosis of RUNX1::RUNX1T1 is more challenging if it is underrepresented in training. Sample quality is also worse in elderly. Interestingly, RUNX1::RUNX1T1 genetic subtype is also less common among patients (see Fig. 1).

Moreover, we have pointed out the possible influence of sex bias in the classification process. The variation in performance between male and female test sets indicates that certain AML subtypes may be classified more accurately when trained on balanced dataset representing both sexes. Specially, females are seen to be affected by a sex imbalanced training set more than males. This raises concerns about potential disparities in diagnosis and treatment recommendations based on sex.

Although MIL methods are providing explainability [5,18] for better clinical applicability, addressing and mitigating biases in AML subtype classification models is important for both the machine learning community and medical researchers. There is a pressing need to collect more diverse and well-balanced multicentric datasets, encompassing various age groups, sex, and ethnicities, in order to enhance the models' generalizability and fairness. Additionally, incorporating fairness-aware learning techniques and conducting comprehensive bias analysis during model development can play a significant role in reducing biases.

Naturally, the data distribution is different between our experiments as a result of sampling and this needs to be considered for the interpretation of our results. It also raises questions about fair data acquisition: Data bias does not start upon image acquisition, but needs to be considered during patient selection and study design. In the medical domain, almost every disease is distributed unevenly alongside various dependent variables such as age, sex, ethnicity or social status. Thus, methodological solutions must become part of every machine learning scientists' repertoire to cater the oath of physicians in pursuit of fair and just medical care.

Acknowledgments. C.M. has received funding from the European Research Council (ERC) under the European Union's Horizon 2020 research and innovation programme (Grant agreement No. 866411).

References

1. Ayanian, J.Z., Epstein, A.M.: Differences in the use of procedures between women and men hospitalized for coronary heart disease. N. Engl. J. Med. **325**(4), 221–225 (1991)

2. Bonferroni, C.: Teoria statistica delle classi e calcolo delle probabilita. Pubblicazioni del R Istituto Superiore di Scienze Economiche e Commericiali di Firenze **8**, 3–62 (1936)

3. Dunn, O.J.: Multiple comparisons using rank sums. Technometrics **6**(3), 241–252 (1964)

4. He, K., Zhang, X., Ren, S., Sun, J.: Deep residual learning for image recognition. In: Proceedings of the IEEE Conference on Computer Vision and Pattern Recognition, pp. 770–778 (2016)

5. Hehr, M., et al.: Explainable AI identifies diagnostic cells of genetic AML subtypes. PLOS Digit. Health **2**(3), e0000187 (2023)
6. Hehr, M., et al.: A morphological dataset of white blood cells from patients with four different genetic AML entities and non-malignant controls (AML-Cytomorphology_MLL_Helmholtz) (version 1) [data set]. Cancer Imaging Arch. (2023). https://doi.org/10.7937/6PPE-4020
7. Hiremath, P., Bannigidad, P., Geeta, S.: Automated identification and classification of white blood cells (leukocytes) in digital microscopic images. IJCA special issue on "recent trends in image processing and pattern recognition" RTIPPR, pp. 59–63 (2010)
8. Ilse, M., Tomczak, J., Welling, M.: Attention-based deep multiple instance learning. In: International Conference on Machine Learning, pp. 2127–2136. PMLR (2018)
9. Ioannou, S., Chockler, H., Hammers, A., King, A.P., Initiative, A.D.N.: A study of demographic bias in CNN-based brain MR segmentation. In: Abdulkadir, A., et al. (eds.) MLCN 2022. LNCS, vol. 13596, pp. 13–22. Springer, Cham (2022). https://doi.org/10.1007/978-3-031-17899-3_2
10. Kruskal, W.H., Wallis, W.A.: Use of ranks in one-criterion variance analysis. J. Am. Stat. Assoc. **47**(260), 583–621 (1952)
11. Lara, M.A.R., Mosquera, C., Ferrante, E., Echeveste, R.: Towards unraveling calibration biases in medical image analysis. arXiv preprint arXiv:2305.05101 (2023)
12. Lee, T., Puyol-Antón, E., Ruijsink, B., Shi, M., King, A.P.: A systematic study of race and sex bias in CNN-based cardiac MR segmentation. In: Camara, O., et al. (eds.) STACOM 2022. LNCS, vol. 13593, pp. 233–244. Springer, Cham (2022). https://doi.org/10.1007/978-3-031-23443-9_22
13. Lo-Coco, F., et al.: Front-line treatment of acute promyelocytic leukemia with AIDA induction followed by risk-adapted consolidation for adults younger than 61 years: results of the AIDA-2000 trial of the gimema group. Blood J. Am. Soc. Hematol. **116**(17), 3171–3179 (2010)
14. Matek, C., Schwarz, S., Spiekermann, K., Marr, C.: Human-level recognition of blast cells in acute myeloid leukaemia with convolutional neural networks. Nat. Mach. Intell. **1**(11), 538–544 (2019)
15. Parsa-Parsi, R.W.: The revised declaration of Geneva: a modern-day physician's pledge. JAMA **318**(20), 1971–1972 (2017)
16. Puyol-Antón, E., et al.: Fairness in cardiac magnetic resonance imaging: assessing sex and racial bias in deep learning-based segmentation. Front. Cardiovasc. Med. **9**, 859310 (2022)
17. Russakovsky, O., et al.: ImageNet large scale visual recognition challenge. Int. J. Comput. Vision **115**, 211–252 (2015)
18. Sadafi, A., et al.: Pixel-level explanation of multiple instance learning models in biomedical single cell images. In: Frangi, A., de Bruijne, M., Wassermann, D., Navab, N. (eds.) IPMI 2023. LNCS, vol. 13939, pp. 170–182. Springer, Cham (2023). https://doi.org/10.1007/978-3-031-34048-2_14
19. Sadafi, A., Bordukova, M., Makhro, A., Navab, N., Bogdanova, A., Marr, C.: RedTell: an AI tool for interpretable analysis of red blood cell morphology. Front. Physiol. **14**, 1058720 (2023)
20. Sadafi, A., et al.: Multiclass deep active learning for detecting red blood cell subtypes in brightfield microscopy. In: Shen, D., et al. (eds.) MICCAI 2019, Part I. LNCS, vol. 11764, pp. 685–693. Springer, Cham (2019). https://doi.org/10.1007/978-3-030-32239-7_76

21. Sadafi, A., et al.: Attention based multiple instance learning for classification of blood cell disorders. In: Martel, A.L., et al. (eds.) MICCAI 2020. LNCS, vol. 12265, pp. 246–256. Springer, Cham (2020). https://doi.org/10.1007/978-3-030-59722-1_24

22. Salehi, R., et al.: Unsupervised cross-domain feature extraction for single blood cell image classification. In: Wang, L., Dou, Q., Fletcher, P.T., Speidel, S., Li, S. (eds.) MICCAI 2022. LNCS, vol. 13433, pp. 739–748. Springer, Cham (2022). https://doi.org/10.1007/978-3-031-16437-8_71

23. Sharma, S., et al.: Deep learning model for the automatic classification of white blood cells. Comput. Intell. Neurosci. **2022** (2022)

24. Sidhom, J.W., et al.: Deep learning for diagnosis of acute promyelocytic leukemia via recognition of genomically imprinted morphologic features. NPJ Precis. Oncol. **5**(1), 38 (2021)

Unsupervised Bias Discovery in Medical Image Segmentation

Nicolás Gaggion[(⊠)], Rodrigo Echeveste, Lucas Mansilla, Diego H. Milone, and Enzo Ferrante

Research Institute for Signals, Systems and Computational Intelligence, sinc(i) (CONICET, Universidad Nacional del Litoral), Santa Fe, Argentina
ngaggion@sinc.unl.edu.ar

Abstract. It has recently been shown that deep learning models for anatomical segmentation in medical images can exhibit biases against certain sub-populations defined in terms of protected attributes like sex or ethnicity. In this context, auditing fairness of deep segmentation models becomes crucial. However, such audit process generally requires access to ground-truth segmentation masks for the target population, which may not always be available, especially when going from development to deployment. Here we propose a new method to anticipate model biases in biomedical image segmentation in the absence of ground-truth annotations. Our unsupervised bias discovery method leverages the reverse classification accuracy framework to estimate segmentation quality. Through numerical experiments in synthetic and realistic scenarios we show how our method is able to successfully anticipate fairness issues in the absence of ground-truth labels, constituting a novel and valuable tool in this field.

Keywords: unsupervised bias discovery · fairness · medical image segmentation · reverse classification accuracy

1 Introduction

An ever growing body of work has shown that machine learning systems can be systematically biased against certain sub-populations based on attributes like race or gender in a variety of settings [17]. When it comes to machine learning systems analyzing health data, this topic becomes extremely relevant, particularly in medical image computing (MIC) tasks like computed assisted diagnosis [12] or anatomical segmentation [16]. It is hence vital to audit the models considering fairness metrics to assess potential disparate performance of various types among subgroups [13]. A usual way to detect such fairness issues consists in employing the standard *group fairness* framework to audit bias in machine learning models. This framework usually requires access to a dataset of images with two important pieces of information: the *demographic attributes* used to define the groups of analysis, and the *ground truth labels* for the task of interest. However, in many situations, we may not have access to them. Medical images, for example, may lack the necessary metadata due to privacy concerns. Moreover, obtaining expert label annotations for extensive image databases can be a time-consuming and costly endeavor, significantly limiting the availability of these annotations. More importantly, in real life scenarios, we may be interested in auditing existing deployed systems in a new target population for which we do not have ground-truth labels to ensure these models are still fair. To tackle these issues, here we propose an unsupervised bias discovery (UBD) method, which to our knowledge is the first one specifically designed for biomedical image segmentation.

The term UBD refers to the process of identifying and uncovering biases in a machine learning system without using ground-truth labels or demographic information. Two primary UBD scenarios can be considered. The first one, when we have access to ground-truth (such as classification labels or anatomical masks for image segmentation) but lack knowledge about the specific sub-population metadata for sub-group analysis. Methods addressing this issue have been recently proposed in fairness literature for machine learning [10,11]. The second scenario, of main interest to us, appears when we possess demographic attributes at an individual level to construct pre-defined analysis groups (e.g. gender, sex, age, ethnicity) but lack ground-truth annotations for the target population. Here we focus in this last case, i.e. performing UBD in the absence of ground-truth to anticipate fairness issues in unseen subjects, particularly in the context of medical image segmentation. Given recent research indicating how fairness properties may not transfer under distribution shifts in healthcare [19], understanding how to anticipate fairness issues in new populations with unforeseen distribution shifts, especially when ground-truth labels are unavailable, becomes of paramount importance.

2 Related Work

Anatomical segmentation is a fundamental task in medical image analysis, with applications ranging from computational anatomy studies and patient follow-up, to radiotherapy planning [3,6]. Several methods have been proposed in the literature to estimate segmentation performance in the absence of ground-truth annotations, usually in the context of automatic quality control pipelines. Some of these methods leverage the concept of predictive uncertainty [3,4] under the hypothesis that highly uncertain predictions will correlate with erroneous pixels. However, these methods heavily rely on the quality of the uncertainty estimates provided by the segmentation model. In [5], the authors proposed an alternative learning-based approach where a convolutional neural network (CNN) is trained to predict the Dice-Sørensen coefficient (or DSC, a commonly used metric for segmentation quality) from pairs of images with the corresponding predicted segmentation. This approach is agnostic to the segmentation model, but it does require to train an extra CNN to predict the DSC score. Here we build on top of a different approach introduced by Valindria et al. [21], entitled reverse classification accuracy (RCA). This method proposes to construct a *reverse classifier*, using only one test image and its predicted segmentation as pseudo ground truth. Subsequently, this classifier is assessed on a reference dataset containing available segmentations, and its performance is used as a proxy of the quality of the predicted segmentation. Different variants of RCA were originally proposed, including atlas-forest, CNN-based RCA and atlas-based label propagation RCA via classic image registration. Here we implement our own deep learning variant of atlas-based label propagation RCA, which has shown to be fast and reliable in previous work [14,15] (see details in Sect. 3), and use it to perform UBD.

Contributions: To the best of our knowledge, this is the first study to explore unsupervised bias discovery in biomedical image segmentation. To this end, we propose a deep learning variant of atlas-based label propagation RCA and demonstrate its capacity to reveal hidden biases in segmentation models that exhibit disparate performance across specific sub-populations. The proposed method is applied to the task of chest X-ray anatomical segmentation, revealing sex biases in situations where ground-truth masks are unavailable. This is fundamentally important when implementing anatomical segmentation systems in real and dynamic clinical scenarios, as it may help to monitor that fairness is maintained when models are deployed into production.

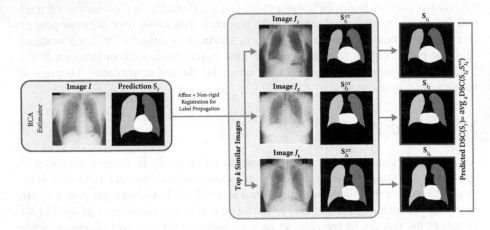

Fig. 1. Outline of the RCA framework (based on [21]). The DSC-score for the predicted segmentation S_I (for which we do not have ground-truth) is estimated by registering image I to the top-k similar images from the reference database, and computing the DSC-score of the propagated labels with respect to the corresponding segmentation masks. The mean DSC computed for the k reference masks is then used as a DSC estimate of the predicted segmentation S_I.

3   Unsupervised Bias Discovery via Reverse Classification Accuracy

Let us briefly start by describing the atlas based label-propagation RCA framework [21] adopted in our work to estimate individual DSC coefficients per image. The process starts with an image I and its corresponding predicted segmentation S_I for which quality needs to be assessed (see Fig. 1). We will refer to this image as the *atlas*. Additionally, there is a reference database containing images J_i with known ground-truth segmentation masks $S_{J_i}^{GT}$.

Non-rigid registration is then applied to align the atlas with the reference images in the database, deforming it via a deformation field D_i to match their anatomy. Through label propagation, the predicted segmentation mask under evaluation (S_I) is transferred to the corresponding target images by warping it with the resulting deformation fields, providing a candidate segmentation mask $\hat{S}_{J_i} = S_I \circ D_i$ per reference image J_i. Subsequently, we evaluate the quality of the propagated labels $\hat{S}_{J_i}$ by comparing them with the corresponding ground-truth segmentation masks $S_{J_i}^{GT}$ using DSC (or any other metric of interest). Thus, given an image I, the prediction S_I and the reference database $\{J, S_J^{GT}\}_k$, we estimate the quality of S_I as the average:

$$DSC^{RCA}(I, S_i, \{J, S_J^{GT}\}_k) = \frac{\sum_k DSC(\hat{S}_{J_k}, S_{J_k}^{GT})}{k} \qquad (1)$$

The underlying assumption is that if the segmentation quality of the new image is high, the RCA estimator will have a high performance on the reference images; conversely, poor segmentation quality will result in lower performance on the reference images. By aggregating the predictions of the RCA estimator on those images, one can assess the quality of the segmentation of a new image without requiring ground truth data. Note that even though RCA stands for reverse *classification* accuracy, here there is no actual classifier, as the process is fully based on image registration. However, we keep the name following the original publication [21].

Deep Registration Networks for Atlas-Based RCA. We propose a variant of the original atlas-based label propagation RCA framework which employs deep registration networks and top-k image selection to reduce computational time. Our method has four main stages:

1. First, the top-k images, which are most similar to the atlas, are identified within our reference database, to avoid registering the atlas with all the references.
2. Subsequently, a deep CNN is employed to learn the affine registration parameters for every reference image. This network is implemented following a Siamese architecture, with two CNN encoders whose output are then processed by 2 fully connected layers to produce the affine parameters of a 2D transformation.

3. Then, a dense deformation field is estimated by an anatomically constrained deformable registration network (ACNN-RegNet) for every reference image which is finally used to warp the atlas mask (following Ref. [15]).[1]
4. Finally, the estimated DSC^{RCA} is computed following Eq. 1, i.e. by averaging over all estimates. Note that in the original work [21] the maximum is computed instead of the average of the DSC scores for the reference images. In our experiments, we found that taking the average was more robust than the maximum. We believe this difference may be due to the fact that we perform a top-k selection where only the top-k similar images are chosen, reducing the chance of having really different images in our reference set.

Unsupervised Bias Discovery. Let us say we have a trained model $\mathbf{M}$ which, given an image I, produces a segmentation mask $S_I = \mathbf{M}(I)$. The idea is to audit $\mathbf{M}$ for fairness with respect to a demographic attribute $\mathbf{A}$. Following previous work [16], we will study fairness in terms of the DSC coefficient. In particular, we consider the difference in terms of DSC between demographic groups as a measure of bias (the closer to zero, the fairer the model). However, since we do not have access to segmentation masks for the target population, we will use the difference in terms of DSC^{RCA} as a proxy for this performance gap. For example, let us assume A refers to the biological sex. We then divide our target population according to whether they are male or female, and analyze the gap between the corresponding $DSC^{RCA}_{A=M}$ and $DSC^{RCA}_{A=F}$. Note that throughout the experimental section we will use the signed difference $\Delta DSC^{RCA} = DSC^{RCA}_{A=M} - DSC^{RCA}_{A=F}$ to be sure that our estimator is capturing the biases in the right direction. In what follows, we present synthetic and real experiments to validate the proposed UBD framework.

4 Experiments and Discussion

We study the task of lung and heart segmentation in x-ray images, and measure fairness taking sex as the protected attribute. We employed four different x-ray datasets (comprising a total of 911 images) including JSRT [20], Montgomery [2], Shenzhen [9], and a minor subset of the Padchest dataset [1] for which we had access to ground-truth annotations[2].

All experiments were conducted using a UNet model [18] trained via a compound soft Dice and cross-entropy loss. In an effort to combine heterogeneous annotations –given that some images contain only lung annotations while others contain both lung and heart annotations, we adopted the approach outlined in [8]. Here, each organ is treated as an independent output channel using binary versions of the loss functions, and gradients are back-propagated on a specific channel only when annotations for that channel are available.

---

[1] Code for the full RCA pipeline based on deep registration networks is publicly available at https://github.com/ngaggion/UBD_SourceCode.

[2] For JSRT, Montgomery and Shenzhen we used the original annotations. For PadChest, we used the annotations released in the Chest X-ray Landmark Database [7] publicly available at https://github.com/ngaggion/Chest-xray-landmark-dataset.

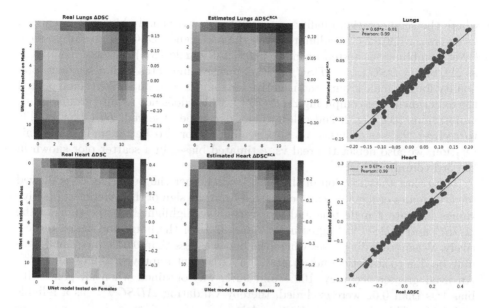

Fig. 2. Synthetic experiment. Heatmaps illustrating the distribution of real bias as measured by ΔDSC (1st column) and the estimated ΔDSC^{RCA} (2nd column) across all combinations of models for male and female subjects. The third column shows scatter plots depicting the correlation between the real ΔDSC and the estimated ΔDSC^{RCA} bias. Each point represents a combination of models for male and female subjects. The positive correlation indicates that ΔDSC^{RCA} serves as a valid estimator of biases in the absence of ground-truth. Top/bottom row shows results for lung/heart segmentation.

4.1 Synthetic Experiment: Validating RCA for UBD

We start with a synthetic experiment to validate the effectiveness of RCA as a bias detector. To obtain networks of varying performance levels, we trained UNet models until convergence, and used 12 different versions from intermediate saving checkpoints $M_{1 \leq i \leq 12}$, one after each epoch from the 1st to the 11th and a final checkpoint at convergence. Note that since performance improves over training, in general the performance of model M_i will tend to be higher than that of M_j if $i > j$. We used balanced training and test sets in terms of sex (50% male and 50% female) (80–20% split). To asses biases the test subset was analyzed in terms of the two sex-specific subgroups.

We simulated a scenario where the segmentation quality varies based on sex, with either male or female patients exhibiting superior performance. This was achieved by selecting pairs of UNet models $(M_i, M_j) \forall i, j : 1 \leq i, j \leq 12$, and segmenting the male test set with M_i and the female set with M_j. Note that even though masks are coming from different models, in this synthetic experiment we consider them to be generated by a single fictitious model whose fairness we would aim to audit. This led to 144 possible combinations (fictitious models), where the segmentation quality in the male set was usually higher if $i > j$ and vice-versa. Figure 2 (1st column) shows the real signed DSC gap for all possible UNet combinations tested on male (M_i) and female (M_j) patients, computed as $\Delta DSC = DSC_{A=M} - DSC_{A=F}$ (i.e. positive indicates higher segmentation quality for the male group, and negative for the female group). We then repeated the experiment but computing ΔDSC^{RCA}, which does not require ground-truth annotations (see Fig. 2, 2nd column). As it can be observed, DSC^{RCA} is able to recover the same biases, up to a scaling factor. To better understand this relation, we also depict the real vs. estimated biases in a scatter plot shown in the 3rd column of Fig. 2.

Note that the direction of the bias, i.e., whether the performance is skewed towards males or females, can be inferred from the sign of the difference; positive values indicate a male bias and vice versa. We highlight that the signs of the ΔDSC and ΔDSC^{RCA} agree in 90% and 92% for the lungs and heart, respectively. When excluding cases where the bias was less than 0.01 (since small gaps may be due to statistical noise), the sign agreement improved to 96% and 92%. An even greater accuracy, 100% and 96%, was obtained when all pairs with bias less than 0.02 were excluded, thereby validating ΔDSC^{RCA} as a reliable estimator of bias in segmentation models.

4.2 Real Experiment: Auditing Chest X-Ray Segmentation Models for Sex Bias

We now proceed to validate ΔDSC^{RCA} in a more realistic scenario, auditing the bias of two concrete chest x-ray segmentation models. Data imbalance in terms of protected attributes in the training set is known to result in models which are biased in terms of those attributes [12]. We exploit this fact to produce biased models by training two sets of 5-fold cross validation UNet models: one using only the male subset of the training set, and the other on the female subset. These models were then evaluated separately on the Male and Female partitions of the test set. Here again we employed an 80–20% split for the complete train and test sets respectively. Intriguingly, when measuring biases using the available ground-truth annotations, we discovered that independently of the training set, models tend to perform better on female patients, with a slightly more pronounced gap when the model is trained on female patients (Fig. 3, left and center column boxplots).

Crucially, estimated performance gaps via RCA strongly correlate with true gaps both for lung and heart segmentation also in this scenario (Fig. 3, right column top and bottom). However, it should be noted that the slope of the correlation is not exactly one and varies between tasks (cf. Fig. 3 top vs bottom row). This suggests that while ΔDSC^{RCA} provides reliable estimates of bias up to a scaling factor, as in the synthetic experiment, calibration may be necessary to accurately estimate the magnitude of bias when transitioning from development to deployment scenarios.

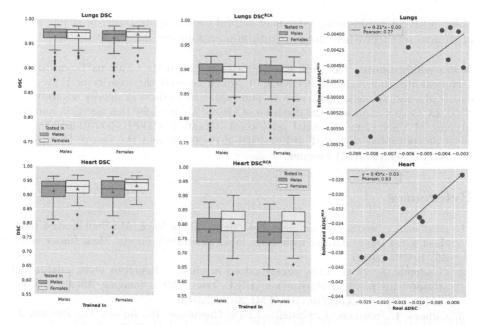

Fig. 3. Real experiment Boxplots depict the distribution of real bias as measured by Real DSC (1st column) and the estimated DSC^{RCA} (2nd column) across all combinations of models trained and tested on male and female subjects separately. The third column shows scatter plots depicting the correlation between the real ΔDSC and the estimated ΔDSC^{RCA} bias. Each point represents a model. The positive correlation indicates that ΔDSC^{RCA} serves as a valid estimator of biases in the absence of ground-truth. Top/bottom row shows results for lungs/heart segmentation.

5 Conclusion

The increasing adoption of automated methods in the context of MIC brings with it a responsibility to produce models which do not unfairly discriminate in terms of protected attributes. After development of a model, the reduced number of expert label annotations in the real-world dataset where the model is to be deployed may significantly limit the possibility to anticipate fairness issues. In order to estimate possible biases, here we propose a novel unsupervised bias discovery for biomedical image segmentation based on the RCA framework.

To evaluate our framework we employed x-ray datasets where ground truth annotations were actually available and true biases could hence be obtained. A first synthetic experiment, where the performance of a model could be easily manipulated, allowed us to validate our estimated bias score in a controlled scenario, showing the viability of our method. A second experiment performed in a more realistic scenario, where models were trained for high performance but in a data-imbalance setting resulting in biases, confirmed UBD was able to correctly capture these biases. Overall, these results show that UBD methods based on RCA could prove extremely helpful to anticipate biases at deployment in clinical settings where ground truth annotations are not yet available.

Acknowledgments. This work was supported by Argentina's National Scientific and Technical Research Council (CONICET), which covered the salaries of E.F., R.E. and D.M., and the fellowships of N.G. and L.M. The authors gratefully acknowledge NVIDIA Corporation with the donation of the GPUs used for this research, and the support of Universidad Nacional del Litoral (Grants CAID-PIC-50220140100084LI, 50620190100145LI), ANPCyT (PICT-PRH-2019-00009) and the Google Award for Inclusion Research (AIR) Program.

References

1. Bustos, A., Pertusa, A., Salinas, J.M., de la Iglesia-Vayá, M.: PadChest: a large chest x-ray image dataset with multi-label annotated reports. Med. Image Anal. **66**, 101797 (2020)
2. Candemir, S., et al.: Lung segmentation in chest radiographs using anatomical atlases with nonrigid registration. IEEE Trans. Med. Imaging **33**(2), 577–590 (2014). https://doi.org/10.1109/TMI.2013.2290491
3. Cubero, L., Serrano, J., Castelli, J., De Crevoisier, R., Acosta, O., Pascau, J.: Exploring uncertainty for clinical acceptability in head and neck deep learning-based oar segmentation. In: IEEE ISBI 2023. IEEE (2023)
4. Czolbe, S., Arnavaz, K., Krause, O., Feragen, A.: Is segmentation uncertainty useful? In: Feragen, A., Sommer, S., Schnabel, J., Nielsen, M. (eds.) IPMI 2021. LNCS, vol. 12729, pp. 715–726. Springer, Cham (2021). https://doi.org/10.1007/978-3-030-78191-0_55
5. Fournel, J., et al.: Medical image segmentation automatic quality control: a multi-dimensional approach. Med. Image Anal. **74**, 102213 (2021)
6. Fu, Y., Lei, Y., Wang, T., Curran, W.J., Liu, T., Yang, X.: A review of deep learning based methods for medical image multi-organ segmentation. Physica Med. **85**, 107–122 (2021)
7. Gaggion, N., Mansilla, L., Mosquera, C., Milone, D.H., Ferrante, E.: Improving anatomical plausibility in medical image segmentation via hybrid graph neural networks: applications to chest x-ray analysis. IEEE Trans. Med. Imaging (2022). https://doi.org/10.1109/tmi.2022.3224660. https://doi.org/10.1109
8. Gaggion, N., Vakalopoulou, M., Milone, D.H., Ferrante, E.: Multi-center anatomical segmentation with heterogeneous labels via landmark-based models. In: 20th IEEE International Symposium on Biomedical Imaging (ISBI). IEEE (2023)
9. Jaeger, S., et al.: Automatic tuberculosis screening using chest radiographs. IEEE Trans. Med. Imaging **33**(2), 233–245 (2014). https://doi.org/10.1109/TMI.2013.2284099
10. Krishnakumar, A., Prabhu, V., Sudhakar, S., Hoffman, J.: UDIS: unsupervised discovery of bias in deep visual recognition models. In: British Machine Vision Conference (BMVC), vol. 1, p. 3 (2021)
11. Lahoti, P., et al.: Fairness without demographics through adversarially reweighted learning. Adv. Neural. Inf. Process. Syst. **33**, 728–740 (2020)
12. Larrazabal, A.J., Nieto, N., Peterson, V., Milone, D.H., Ferrante, E.: Gender imbalance in medical imaging datasets produces biased classifiers for computer-aided diagnosis. Proc. Natl. Acad. Sci. **117**(23), 12592–12594 (2020)

13. Liu, X., Glocker, B., McCradden, M.M., Ghassemi, M., Denniston, A.K., Oakden-Rayner, L.: The medical algorithmic audit. Lancet Digit. Health **4**(5), e384–e397 (2022)
14. Mansilla, L., Ferrante, E.: Segmentación multi-atlas de imágenes médicas con selección de atlas inteligente y control de calidad automático. In: XXIV Congreso Argentino de Ciencias de la Computación (La Plata, 2018) (2018)
15. Mansilla, L., Milone, D.H., Ferrante, E.: Learning deformable registration of medical images with anatomical constraints. Neural Netw. **124**, 269–279 (2020)
16. Puyol-Antón, E., et al.: Fairness in cardiac MR image analysis: an investigation of bias due to data imbalance in deep learning based segmentation. In: de Bruijne, M., et al. (eds.) MICCAI 2021, Part III. LNCS, vol. 12903, pp. 413–423. Springer, Cham (2021). https://doi.org/10.1007/978-3-030-87199-4_39
17. Ricci Lara, M.A., Echeveste, R., Ferrante, E.: Addressing fairness in artificial intelligence for medical imaging. Nat. Commun. **13**(1), 1–6 (2022)
18. Ronneberger, O., Fischer, P., Brox, T.: U-Net: convolutional networks for biomedical image segmentation. In: Navab, N., Hornegger, J., Wells, W.M., Frangi, A.F. (eds.) MICCAI 2015, Part III. LNCS, vol. 9351, pp. 234–241. Springer, Cham (2015). https://doi.org/10.1007/978-3-319-24574-4_28
19. Schrouff, J., Chen, C., et al.: Diagnosing failures of fairness transfer across distribution shift in real-world medical settings. Adv. Neural. Inf. Process. Syst. **35**, 19304–19318 (2022)
20. Shiraishi, J., et al.: Development of a digital image database for chest radiographs with and without a lung nodule: receiver operating characteristic analysis of radiologists' detection of pulmonary nodules. Am. J. Roentgenol. **174**(1), 71–74 (2000)
21. Valindria, V.V., et al.: Reverse classification accuracy: predicting segmentation performance in the absence of ground truth. IEEE Trans. Med. Imaging **36**(8), 1597–1606 (2017)

Debiasing Counterfactuals in the Presence of Spurious Correlations

Amar Kumar[1,2]([⊠]), Nima Fathi[1,2], Raghav Mehta[1,2], Brennan Nichyporuk[1,2], Jean-Pierre R. Falet[2,3], Sotirios Tsaftaris[4,5], and Tal Arbel[1,2]

[1] Center for Intelligent Machines, McGill University, Montreal, Canada
amarkr@cim.mcgill.ca
[2] MILA (Quebec AI institute), Montreal, Canada
[3] Montreal Neurological Institute, McGill University, Montreal, Canada
[4] Institute for Digital Communications, School of Engineering, University of Edinburgh, Edinburgh, UK
[5] The Alan Turing Institute, London, UK

Abstract. Deep learning models can perform well in complex medical imaging classification tasks, even when basing their conclusions on spurious correlations (i.e. confounders), should they be prevalent in the training dataset, rather than on the causal image markers of interest. This would thereby limit their ability to generalize across the population. Explainability based on counterfactual image generation can be used to expose the confounders but does not provide a strategy to mitigate the bias. In this work, we introduce the first end-to-end training framework that integrates both (i) popular debiasing classifiers (e.g. distributionally robust optimization (DRO)) to avoid latching onto the spurious correlations and (ii) counterfactual image generation to unveil generalizable imaging markers of relevance to the task. Additionally, we propose a novel metric, *Spurious Correlation Latching Score (SCLS)*, to quantify the extent of the classifier reliance on the spurious correlation as exposed by the counterfactual images. Through comprehensive experiments on two public datasets (with the simulated and real visual artifacts), we demonstrate that the debiasing method: (i) learns generalizable markers across the population, and (ii) successfully ignores spurious correlations and focuses on the underlying disease pathology.

Keywords: Biomarker · Counterfactuals · Debiasing · Explainablity

1 Introduction

Deep learning models have shown tremendous success in disease classification based on medical images, given their ability to learn complex imaging markers

Supplementary Information The online version contains supplementary material available at https://doi.org/10.1007/978-3-031-45249-9_27.

across a wide population of subjects. These models can show good performance and still be *biased* as they may focus on spurious correlations in the image that are not causally related to the disease but arise due to confounding factors - should they be common across the majority of samples in the training dataset. As a result, the confounding predictive image markers may not generalize across the population. For example, a deep learning model was able to accurately detect COVID-19 from chest radiographs, but rather than relying on pathological evidence, the model latched on to spurious correlations such as medical devices or lettering in the image [3]. As a result, these image markers did not generalize across the population.

In order to safely deploy black-box deep learning models in real clinical applications, explainability should be integrated into the framework so as to expose the spurious correlations on which the classifier based its conclusions. Popular post-hoc explainability strategies, such as Grad-CAM [6,16,19], SHAP [10], LIME [11] are not designed to expose the precise predictive markers driving a classifier. Models that integrate counterfactual image generation, along with black-box classifers [2,21,23], permit exposing the predictive markers used by the classifier. However, should these methods discover that the markers are indeed simply visual artifacts there are no strategies to mitigate the resulting biases. Furthermore, although several debiasing methods have been successfully implemented to account for generalizability [1,8,17,26,27], they do not integrate explainability into the framework in order to provide reasons for improved performance.

Therefore, the important question to be answered is - *Can a model be trained to disregard spurious correlations and identify generalizable predictive disease markers?*

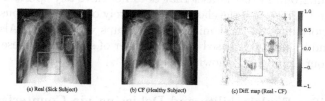

Fig. 1. Counterfactual (CF) image indicating that the classifier latched onto spurious correlations (medical devices) when correctly predicting that subject is sick (class: Pleural Effusion), due to their prevalence in the training dataset for this class. (a) Chest radiograph of a sick subject with several medical devices shown (cyan boxes), (b) Generated (CF) image, (c) Difference heat map shows maximum change around the medical devices, rather than indicating the correct markers for the disease.

In this paper, we propose the first end-to-end training framework for the explainability of classifier and debiasing via counterfactual image generation. We seek to discover imaging markers that reflect underlying disease pathology and that generalize across subgroups. Extensive experiments are performed on two different publicly available datasets - (i) *RSNA Pneumonia Detection Challenge* and (ii) *CheXpert* [5]. To illustrate the goal, Fig. 1 shows an example from the contrived CheXpert dataset, where most of the sick subjects have medical device(s) (e.g. a pacemaker) in their images while most of the healthy subjects do not. As such, there exists a spurious correlation between a confounding visual artifact (the medical devices) and the disease. A classifier based on a standard optimization technique, empirical risk minimization (ERM), incorrectly indicates the medical device as a disease marker, as depicted by the counterfactual (CF). In this work, we propose replacing ERM with a popular debiasing method, Group-DRO (distributional robust optimization). This permits the classifier to focus on the pathological image markers of the disease rather than on spurious correlation(s). Additionally, we show that Group-DRO ignores the visual artifact when making its decision, and generalizes across subgroups without the spurious correlation. Since standard metrics to evaluate counterfactuals do not indicate the region where the classifier focuses, we also propose a novel metric, the Spurious Correlation Latching Score (SCLS), to measure the degree to which the classifier latches onto spurious correlations. Our experiments indicate an improvement (in terms of differences in classifier outputs) of 0.68 and 0.54 in the SCLS using the Group-DRO classifier over the ERM for each of the two datasets.

2   Methodology

We propose an end-to-end training strategy to explain the output of a classifier. Here, we are considering a scenario where majority of the training data encompasses a spurious correlation with the target label. However, there is also a minority subgroup in the dataset that does not have any spurious correlation with the target label i.e., if the classifier was to rely onto the spurious correlation then the performance on these minority subgroups will be poor. Also, the term 'majority' and 'minority' is based on the number of samples in these groups. An overview of our approach is shown in Fig. 2.

2.1   Classifier Explainability and Debiasing via Counterfactual Image Generation

Disease Classification. Binary (e.g. "sick" or "healthy") classification of the images is performed using either a standard classifier (ERM [24]), or a classifier that mitigates biases across sub-groups (Group-DRO [18]). The ERM classifier (f_{ERM}) is expected to be affected by the spurious correlation present in the training dataset, as it minimizes the loss over the entire training dataset and latching onto spurious correlation is a shortcut to minimize the loss. Thus, it would not generalize across the minority subgroups of the dataset [12,20]. On the contrary, the DRO classifier (f_{DRO}) is not expected to learn the spurious correlation as it considers the majority and minority subgroups separately when optimizing the loss. Thus, it would generalize well across all subgroups.

Generative Model for Synthesizing Counterfactuals. We develop an explainability framework that integrates counterfactual image generation together with a classifier during training. We adapted Cycle-GAN [25] as the generative model for counterfactual image generation, chosen for its strong performance across a variety of domains [13,25]. A pre-trained, frozen binary classifier (f_{ERM} or f_{DRO}) provides supervision to the generator. The proposed architecture and optimization objectives (see Fig. 2) are designed to generate counterfactual images that adhere to the following common constraints [7,14,15]: (i) *Identity preservation*: The counterfactual images resemble the input images with minimal change; (ii) *Classifier consistency*: Counterfactual images belong to the target class; (iii) *Cycle consistency*: When counterfactual images are fed through the opposing generator, the output reverts to the original image (see Fig. 2).

During inference, based on the classifier's decision (i.e., f_{ERM} or f_{DRO}) for the input image, we generate counterfactual images and analyze the difference heatmap between the factual (input) and counterfactual (synthesized) images. This interpretable heat map indicates the image markers that contribute the most to changing the classifier's decision.

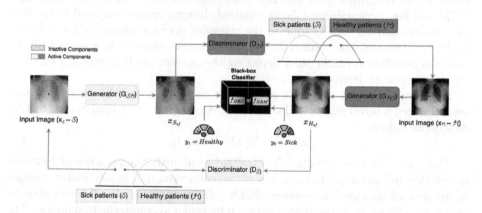

Fig. 2. Training procedure overview: The black-box classifier can be f_{ERM} or f_{DRO} and provides supervision to maintain the correct target class, y_t. Two U-Net generators, G_{SH} and G_{HS}, are employed to synthesize counterfactual images, namely $x_{S_{cf}}$ and $x_{H_{cf}}$. The discriminator D_H and D_S compares the counterfactual images with the domain of healthy H and sick S subjects respectively. Note, training a cycle-GAN requires simultaneous use of two input images from the two distributions.

2.2 Metrics for Evaluating Counterfactuals: Accounting for Spurious Correlations

Standard counterfactual evaluation metrics are structured so as to ensure that the generated images (a) preserve the subject identity and thus penalize generated counterfactual images that are significantly different from the factual (original) images and (b) result in a maximal change in the class label (e.g. from healthy to sick). Identity preservation is typically measured by *structural similarity index* (SSIM) [4] and *Actionability* [14,15], defined as $\mathbb{E}\left[\|x - x_{cf}\|_{L_1}\right]$ between factual (x) and counterfactual (x_{cf}) images. Here, a higher value for SSIM and a lower value for Actionability would indicate better counterfactuals. The *counterfactual prediction gain* (CPG) [15], defined as $|f(x) - f(x_{cf})|$, indicates the degree of change in the classifier's decision such that a higher value of CPG indicates better counterfactuals.

While such metrics are required to measure the validity of the generated counterfactuals, they do not assess whether the classifier latched onto spurious correlations. For example, consider an image of a sick subject in the presence of a spurious correlation. If the disease classifier, f_{ERM}, latched onto the spurious correlation when identifying the subject as sick, the corresponding counterfactual image (i.e., depicting a healthy subject) would show changes in the area of the spurious correlation. In this case, all three evaluation metrics mentioned above would determine that this is a valid counterfactual image, based on high SSIM and low Actionability (shows minimal changes made compared to the factual image) and high CPG (due to the classifier decision changing from sick to healthy). However, the counterfactual image shows changes in the area of the spurious correlation rather than depicting the correct predictive image markers for the disease as desired.

In order to indicate that the classifier is correct but for the wrong reasons, we introduce a novel metric called Spurious Correlation Latching Score (SCLS) defined as follows:

$$\text{SCLS} = |d(x) - d(x_{cf})|. \tag{1}$$

Here, $d(\cdot)$ is a separate classifier, trained to identify the presence of spurious correlation in the image. In cases where the counterfactual image makes changes in an area of spurious correlation, SCLS will be high, as the $d(\cdot)$ will show a maximum change in its prediction between factual and counterfactual images. On the other hand, if the counterfactual image does not make changes in the area of the spurious correlation then SCLS will have a low value. As such, this evaluation strategy will validate how well the counterfactuals can help to determine that the classifier latched onto spurious correlations.

3 Experiments and Results

3.1 Dataset and Implementation Details

We perform experiments on two publicly available datasets. The absence of ground truth makes the validation of counterfactual images particularly

challenging. Therefore, to directly evaluate the quality of the generated counterfactual images in the presence of spurious correlations, we modify a publicly available dataset (*RSNA Pneumonia Detection Challenge*) by adding a synthetic artifact to the majority of the sick images (90%). The majority of the sick
and few of the healthy subjects have an artifact in the image, whereas the majority of the healthy and a few sick subjects do not have this artifact. The spurious
correlation (artifact) is a black dot of radius 9 pixels at the center of the image.
Thus, there are a total of four subgroups ($majority_S$, $majority_H$, $minority_S$
and $minority_H$) in the dataset with varying number of images: $majority_S$ and
$majority_H$ are majority subgroups (sick with artifact and healthy without artifact), while $minority_S$ and $minority_H$ are minority subgroups (sick without
artifact and healthy with artifact). Henceforth, this dataset will be referred to
as Dataset 1.

Table 1. Implementation details for the two datasets

	Disease	Image size	Classifier	# samples [$majority_S$, $minority_S$, $minority_H$, $majority_H$]
Dataset 1	Pneumonia	512×512	AlexNet [13]	5413, 1526, 883, 7968
Dataset 2	Pleural Effusion	224×224	Resnet-50 [22] (pre-trained)	2600, 260, 350, 3456

We also show experiments on a subset of a publicly available dataset (*CheX-
pert* [5]) with medical devices (visual artifacts), spuriously correlated with the
disease. Specifically, we extract the subset of images that have labels "healthy"
or "pleural effusion" (subjects with the presence of other diseases are removed
from the dataset). This dataset will be referred to as Dataset 2. More details
about both datasets are provided in Table 1. Note that both the datasets are
divided into training/validation/testing with 70/10/20 random split. Example
images for both datasets and all four subgroups are shown in Fig. 3.

3.2   Results

Classifier Evaluation. For both datasets (Fig. 4), the DRO-based classifier
(f_{DRO}) performs better for the minority subgroups ($minority_S$ and $minority_H$);
indicating that it can better generalize to sub-populations that do not have the
same visual artifact as the majority subgroups. Both classifiers perform similarly
for the majority subgroups ($majority_S$ and $majority_H$).

Qualitative Counterfactual Evaluation. Pneumonia in chest radiograph
manifests as increased brightness in some regions of the lungs. In dataset 1,
when examining the majority subgroup of sick subjects, the ERM-based classifier latches onto the spurious correlation, as seen by the difference maps. On the

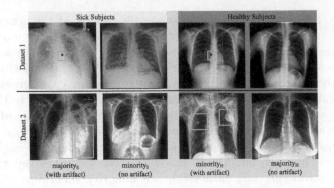

Fig. 3. Datasets 1 and 2 group division: The majority of the sick subjects [$majority_S$] and the minority of healthy subjects [$minority_H$] have visual artifacts (shown in cyan boxes). The majority of healthy subjects [$majority_H$] and the minority of sick subjects [$minority_S$] do not have visual artifacts. Top row: Simulated artifacts (black dots); Bottom row: Real artifacts (medical devices).

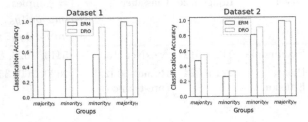

Fig. 4. Performance of ERM (f_{ERM}) and DRO (f_{DRO}) based classifier on a held out test set across all subgroups for both datasets. Notice that DRO has improved performance on minority subgroup [$minority_S$ and $minority_H$] showing improved generalizability across all subgroups.

other hand, a DRO-based classifier focuses on the pathology of the disease, indicated by darker intensity regions over the lungs, as shown in Fig. 5. The behavior of f_{ERM} is also evident in the minority subgroup, where the counterfactual for a healthy subject exhibits an enlarged artifact, wrongly suggesting that the visual artifact serves as a disease marker. Pleural effusion is characterized by the rounding of the costophrenic angle, augmented lung opacity, and reduced clarity of the diaphragm and lung fissures [9]. For the majority subgroup of sick subjects in Dataset 2, the counterfactual images based on ERM remove the medical device rather than focusing on the disease. In addition, for healthy subjects from the minority subgroup, maximum changes are observed around the medical device. On the other hand, for the majority subgroup, the DRO-based counterfactuals show changes around the expected areas while preserving the medical device.

Quantitative Counterfactual Evaluation. In Table 2, counterfactual images generated by ERM and DRO show similar scores according to standard metrics: SSIM, Actionability and CPG. As these metrics are not designed to quantify

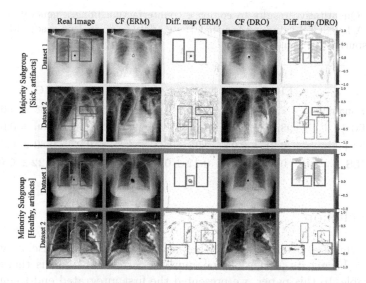

Fig. 5. Qualitative comparison of counterfactual (CF) images generated with ERM and DRO classifiers for both majority (top row) and minority (bottom row) subgroups. The ERM-based CFs show significant changes in the areas of spurious correlation (cyan boxes), whereas the DRO-based CFs show almost no changes in the same areas. In contrast, significant changes can be seen in the expected area of disease pathology (magenta boxes) in DRO-based CFs, while the ERM-based CFs show little to no changes in these areas. (Color figure online)

whether the generated counterfactuals are affected by spurious correlations (see Sect. 2.2), the quality of the counterfactuals is now examined based on the proposed SCLS metric. The AUC of the classifier, d, trained to detect the presence of artifacts is 1.0 and 0.82 for Dataset 1 and Dataset 2, respectively. As indicated by the last row of Table 2, the ERM-based classifier shows a high value (poor performance) for SCLS for both datasets. On the other hand, the DRO-based classifier has a low value (good performance) for SCLS for both datasets. These results corroborate the finding made by visual comparison of the counterfactual images generated by the ERM and DRO classifiers. Overall, both qualitative and quantitative evaluations indicate that an ERM-optimized classifier latches on to the spurious correlation prevalent in the dataset, while a DRO-optimized classifier can be trained to successfully ignore the spurious correlation.

Table 2. Quantitative results to compare counterfactual images generated for both datasets. A low SCLS value implies that the model (f_{DRO} in this case) did not latch onto the spurious correlation.

	Dataset 1		Dataset 2	
	ERM	DRO	ERM	DRO
Actionability ↓	7.68 ± 0.01	7.86 ± 0.01	4.93 ± 0.01	5.68 ± 0.04
SSIM ↑	98.03 ± 0.00	98.44 ± 0.01	98.21 ± 0.01	98.36 ± 0.01
CPG ↑	0.91 ± 0.04	0.96 ± 0.03	0.88 ± 0.07	0.89 ± 0.04
SCLS ↓	0.80 ± 0.08	$\mathbf{0.12 \pm 0.07}$	0.76 ± 0.09	$\mathbf{0.22 \pm 0.06}$

4   Conclusion

Safe deployment of black-box models requires explainability to disclose when the classifier is basing its predictions on spurious correlations and is therefore not generalizable. In this paper, we presented the first integrated end-to-end training strategy for generating unbiased counterfactual images, capitalizing on a DRO classifier to enhance generalization. Our experiments based on two datasets demonstrate that, unlike standard ERM classifiers which are susceptible to latching onto spurious correlations, the unbiased DRO classifier performs significantly better for minority subgroups in terms of- (a) the classifier performance and (b) the novel SCLS metric, which quantifies the degree to which the classifier latches on to the spurious correlation as depicted by the generated counterfactual images.

Current datasets typically do not provide the ground truth predictive markers of interest. Future work will require localizing the predictive markers (e.g. with bounding boxes) and determining the degree of overlap with the discovered markers. Further, we intend to explore the power of alternative debiasing techniques and their potential contribution to discovering generalizable image markers.

Acknowledgements. The authors are grateful for funding provided by the Natural Sciences and Engineering Research Council of Canada, the Canadian Institute for Advanced Research (CIFAR) Artificial Intelligence Chairs program, the Mila - Quebec AI Institute technology transfer program, Microsoft Research, Calcul Quebec, and the Digital Research Alliance of Canada. S.A. Tsaftaris acknowledges the support of Canon Medical and the Royal Academy of Engineering and the Research Chairs and Senior Research Fellowships scheme (grant RCSRF1819 / 8 / 25), and the UK's Engineering and Physical Sciences Research Council (EPSRC) support via grant EP/X017680/1.

References

1. Burlina, P., Joshi, N., Paul, W., Pacheco, K.D., Bressler, N.M.: Addressing artificial intelligence bias in retinal diagnostics. Transl. Vis. Sci. Technol. **10**(2), 13 (2021)
2. Cohen, J.P., et al.: Gifsplanation via latent shift: a simple autoencoder approach to counterfactual generation for chest X-rays. In: Medical Imaging with Deep Learning, pp. 74–104. PMLR (2021)
3. DeGrave, A.J., Janizek, J.D., Lee, S.I.: AI for radiographic COVID-19 detection selects shortcuts over signal. Nat. Mach. Intell. **3**(7), 610–619 (2021)
4. Hore, A., Ziou, D.: Image quality metrics: PSNR vs. SSIM. In: 2010 20th International Conference on Pattern Recognition, pp. 2366–2369. IEEE (2010)
5. Irvin, J., et al.: CheXpert: A large chest radiograph dataset with uncertainty labels and expert comparison. In: Proceedings of the AAAI Conference on Artificial Intelligence, vol. 33, pp. 590–597 (2019)
6. Jiang, H., et al.: A multi-label deep learning model with interpretable grad-cam for diabetic retinopathy classification. In: 2020 42nd Annual International Conference of the IEEE Engineering in Medicine & Biology Society (EMBC), pp. 1560–1563. IEEE (2020)
7. Kumar, A., et al.: Counterfactual image synthesis for discovery of personalized predictive image markers. In: Kakileti, S.T., et al. (eds.) MIABID AIIIMA 2022 2022. LNCS, vol. 13602, pp. 113–124. Springer, Cham (2022). https://doi.org/10.1007/978-3-031-19660-7_11
8. Larrazabal, A.J., Nieto, N., Peterson, V., Milone, D.H., Ferrante, E.: Gender imbalance in medical imaging datasets produces biased classifiers for computer-aided diagnosis. Proc. Natl. Acad. Sci. **117**(23), 12592–12594 (2020)
9. Light, R.W.: Pleural effusion. N. Engl. J. Med. **346**(25), 1971–1977 (2002)
10. Lundberg, S.M., Lee, S.I.: A unified approach to interpreting model predictions. In: Advances in Neural Information Processing Systems, vol. 30 (2017)
11. Magesh, P.R., Myloth, R.D., Tom, R.J.: An explainable machine learning model for early detection of Parkinson's disease using LIME on DaTSCAN imagery. Comput. Biol. Med. **126**, 104041 (2020)
12. Mehta, R., Shui, C., Arbel, T.: Evaluating the fairness of deep learning uncertainty estimates in medical image analysis. In: Medical Imaging with Deep Learning (2023)
13. Mertes, S., Huber, T., Weitz, K., Heimerl, A., André, E.: GANterfactual-counterfactual explanations for medical non-experts using generative adversarial learning. Front. Artif. Intell. **5**, 825565 (2022)
14. Mothilal, R.K., Sharma, A., Tan, C.: Explaining machine learning classifiers through diverse counterfactual explanations. In: Proceedings of the 2020 Conference on Fairness, Accountability, and Transparency, pp. 607–617 (2020)
15. Nemirovsky, D., Thiebaut, N., Xu, Y., Gupta, A.: CounteRGAN: Generating realistic counterfactuals with residual generative adversarial nets. arXiv preprint arXiv:2009.05199 (2020)
16. Panwar, H., Gupta, P., Siddiqui, M.K., Morales-Menendez, R., Bhardwaj, P., Singh, V.: A deep learning and grad-CAM based color visualization approach for fast detection of COVID-19 cases using chest X-ray and CT-scan images. Chaos, Solitons Fractals **140**, 110190 (2020)
17. Ricci Lara, M.A., Echeveste, R., Ferrante, E.: Addressing fairness in artificial intelligence for medical imaging. Nat. Commun. **13**(1), 4581 (2022)

18. Sagawa, S., Koh, P.W., Hashimoto, T.B., Liang, P.: Distributionally robust neural networks for group shifts: on the importance of regularization for worst-case generalization. In: International Conference on Learning Representations (2019)
19. Selvaraju, R.R., Cogswell, M., Das, A., Vedantam, R., Parikh, D., Batra, D.: Grad-CAM: Visual explanations from deep networks via gradient-based localization. In: Proceedings of the IEEE International Conference on Computer Vision, pp. 618–626 (2017)
20. Shui, C., Szeto, J., Mehta, R., Arnold, D., Arbel, T.: Mitigating calibration bias without fixed attribute grouping for improved fairness in medical imaging analysis. arXiv preprint arXiv:2307.01738 (2023)
21. Singla, S., Eslami, M., Pollack, B., Wallace, S., Batmanghelich, K.: Explaining the black-box smoothly-a counterfactual approach. Med. Image Anal. **84**, 102721 (2023)
22. Targ, S., Almeida, D., Lyman, K.: Resnet in resnet: Generalizing residual architectures. arXiv preprint arXiv:1603.08029 (2016)
23. Thiagarajan, J.J., Thopalli, K., Rajan, D., Turaga, P.: Training calibration-based counterfactual explainers for deep learning models in medical image analysis. Sci. Rep. **12**(1), 597 (2022)
24. Vapnik, V.: Principles of risk minimization for learning theory. In: Advances in Neural Information Processing Systems, vol. 4 (1991)
25. Zhu, J.Y., Park, T., Isola, P., Efros, A.A.: Unpaired image-to-image translation using cycle-consistent adversarial networks. In: Proceedings of the IEEE International Conference on Computer Vision, pp. 2223–2232 (2017)
26. Zong, Y., Yang, Y., Hospedales, T.: MEDFAIR: benchmarking fairness for medical imaging. In: International Conference on Learning Representations (2023)
27. Zou, J., Schiebinger, L.: AI can be sexist and racist-it's time to make it fair (2018)

EPIMI

On the Relationship Between Open Science in Artificial Intelligence for Medical Imaging and Global Health Equity

Raissa Souza[1,2,3,4], Emma A.M. Stanley[1,2,3,4(✉)], and Nils D. Forkert[1,2,3,4,5]

[1] Biomedical Engineering Graduate Program, University of Calgary, Calgary, Canada
{raissa.souzadeandrad,emma.stanley}@ucalgary.ca
[2] Department of Radiology, University of Calgary, Calgary, Canada
[3] Hotchkiss Brain Institute, University of Calgary, Calgary, Canada
[4] Alberta Children's Hospital Research Institute, University of Calgary, Calgary, Canada
[5] Department of Clinical Neurosciences, University of Calgary, Calgary, Canada

Abstract. Artificial intelligence (AI) holds tremendous promise for medical image analysis and computer-aided diagnosis, but various challenges must be addressed to enable its widespread adoption and impact in patient care. Open science, specifically through open-source code and public databases, brings multiple benefits to the progress of AI in medical imaging. It is expected to facilitate research output sharing, promote collaboration among researchers, improve the reproducibility of findings, and foster innovation. However, it is important to recognize that the current state of open-source research, particularly with respect to the large, public datasets commonly used in medical imaging AI, is inherently centered around high-income countries (HIC) and privileged populations. Low- and middle-income countries (LMIC) often face several limitations in contributing to and benefiting from open science research in this domain, such as inadequate digital infrastructure, limited funding for research and development, and a scarcity of healthcare and data science professionals. This may lead to further global disparities in health equity as AI-based clinical decision support systems continue to be implemented in practice. While transfer learning and distributed learning hold promise in addressing some challenges related to limited and non-public data in LMIC, practical obstacles arise when dealing with small, lower-quality datasets, resource constraints, and the need for tailored local implementation of these models. In this commentary, we explore the relationship between open-source models and public medical imaging data repositories in the context of transfer learning and distributed learning, specifically considering their implications for global health equity.

R. Souza and E.A.M. Stanley—Contributed equally.

1   Introduction

Artificial intelligence (AI) presents incredibly promising opportunities for medical image analysis and computer-aided diagnosis systems [18,19]. However, several challenges, including the complexity of data, data accessibility and management, patient privacy concerns, reproducibility and transferability of algorithms, and integration of these tools into the clinical workflow, must be effectively and carefully addressed to enable the widespread adoption of this technology and realize its potential to enhance patient care [30]. Many of these limitations motivated the open science movement in the field. In essence, open science involves making scientific research and data openly available to the public, allowing others to access, use, and build upon the knowledge generated [38]. Open science revolves around the principle of providing access to research findings, publications, and data for the scientific community and the general public. This inclusive approach has facilitated many successful collaborations, increased reproducibility, and advancement through collective efforts within the scientific community and funding organizations. In medical imaging specifically, significant efforts have been devoted to creating and curating large publicly (or academically) accessible databases such as UK Biobank [36], Alzheimer's Disease Neuroimaging Initiative (ADNI) [16], and the Multimodal Brain Tumor Segmentation (BraTS) dataset [22]. Furthermore, significant efforts have been made to foster advancement based on open science, as evidenced by the organization of workshops and competitions aimed at expediting technological and biomedical discoveries [30].

While open science has evolved, many major journals, conferences, and societies [5,9,28] have focused on defining guidelines to promote reproducibility and transferability of algorithms, for example through requiring a clear description of the mathematical setting, algorithm and/or model, study cohort, training code, and the details of train/validation/test splits, among others. Within the active research community studying the development and use of deep learning models for medical image analysis and computer-aided diagnosis, open-source models and the use of large public medical imaging datasets have become highly valued. These standards and new values are expected to encourage not only reproducibility and transferability but also transparency and international research collaboration.

In this commentary, we emphasize that while open-source research is highly valuable in many regards, it is intrinsically linked to high-income countries (HIC) and privileged populations, which can result in a globally imbalanced distribution of its positive impacts. We briefly discuss the relationship between open-source models and public medical imaging data repositories on global health equity, specifically in the context of two popular deep learning paradigms that seem useful in this context - transfer learning and distributed learning.

2   Open-Source Data in AI for Medical Imaging

The reproducibility crisis, or the ongoing phenomenon of scientific studies being difficult or impossible to replicate, has been well-documented in the domain of AI for healthcare [20]. The use of public datasets and open-source model code have been suggested as promising avenues for addressing this challenge [8]. Thus, recent years have seen a strong movement within the AI community toward valuing open-source data and models. Within the medical image computing and computer-assisted interventions (MICCAI) community, this open science model has become so important that reviewers for the annual MICCAI conference are now asked to consider whether the authors will release code and share links to datasets used as part of the acceptance criteria. An overarching goal of MICCAI research is eventual translation and application of AI-based medical image analysis in clinical practice. Open science values are assumed to play a vital role in achieving this objective by promoting transparency and reproducibility, which brings us closer to translating technical innovation and performance gains into tangible improvements in healthcare.

Some of the datasets commonly used in MICCAI research include UK Biobank, ADNI, and BraTS. Since the MICCAI community encompasses diverse research domains, additional datasets specific to certain areas like cardiac imaging, retinal imaging, and histopathology are also widely available and used. Such public datasets used in medical imaging AI papers and challenges typically become commonly used in the community, with new innovations and methods seeking to improve upon previously achieved performance [12]. The efforts of researchers and AI practitioners around the world using public medical imaging datasets in this way have led to more research and progress on radiological clinical decision support tasks. These models are often seen as a way to enable a higher level of global health equity, particularly in regions with a shortage of radiologists and other healthcare professionals [23]. However, the common characteristic of most data available in these public, open repositories is that they originate from HIC (see Fig. 1). This is due to the availability of advanced and accessible medical imaging technology, well-established digital infrastructure, investment into research funding, and higher prevalence of researchers and healthcare professionals. Despite LMIC producing accessible datasets such as Drishti-GS [33] and IDRID [29], both retinal fundus datasets gathered in India and comprised of 101 and 516 images, respectively, these datasets may be underutilized by researchers due to their limited scale. In contrast, the Kaggle EYE-PACS diabetic retinopathy retinal fundus dataset [3] from the United States, comprised of 88,702 images, is extensively used.

Evidently, the ability to open-source big data that is required for training deep learning models is a privilege of nations and institutions who have the resources to generate, collect, curate, annotate, and store those datasets in the first place. Many low- and middle-income countries (LMIC) lack the technological infrastructure and human resources required to generate large, high-quality biomedical datasets [4,23]. As a result, model research and development are disproportionately conducted using data from HIC, and are further biased toward those who voluntarily participate in research studies (which may be skewed on the basis of sociodemographic factors such as race, educational attainment, or socioeconomic status [7,13,14,41]). This realization leads to concerns related to another major challenge for AI research and development - domain shift and generalization. While we celebrate improvements over state-of-the-art performance in the medical image analysis domain, we may forget that those gains were realized on specific populations primarily within datasets from HIC and often fail to consider whether those models can generalize to reach even clinically acceptable performance on data from other populations and from LMIC [10]. Moreover, the ongoing research on developing and evaluating multimodal deep learning models exacerbates these challenges, as acquiring and curating multimodal data poses an additional significant barrier for LMIC.

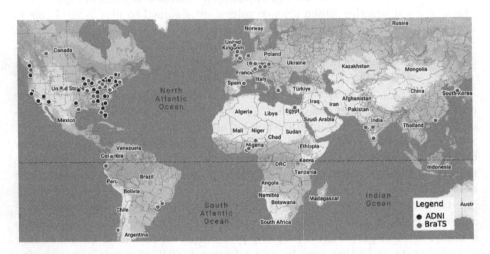

Fig. 1. Collaborating institutions globally are involved in contributing data to two of the most prevalent datasets used in medical imaging research: ADNI [1] (represented in blue) and BraTS [2] (represented in red). A visual representation of this map clearly indicates that a significant portion of the data originates from HIC. While ADNI solely had collaborators in North America, BraTS had 20% (16 out of 79) collaborators from middle-income countries. However, even with this inclusion, it is important to recognize that the representation of LMIC within the dataset may remain insufficient to capture the complete context. (Color figure online)

Furthermore, significant disparities in prevalence of some diseases exist between LMIC and HIC, particularly for communicable diseases such as tuberculosis, malaria, and HIV [11]. Consequently, conditions that are substantial concerns in LMIC but not in HIC have relatively fewer amounts of accessible, high-quality, public data available to study and to develop AI models for. Investment into research and development efforts may tend to be biased towards diseases and disorders that primarily impact HIC, as there is greater access to large, high-quality datasets from these regions [26].

The digital divide contributes to an imbalanced representation within open-source medical imaging data, not only between HIC and LMIC but also within middle income countries that have widespread digital infrastructure and within HIC. For example, Indigenous communities in Canada are less likely to gain access to appropriate health services and share their health data with the scientific community due to past and present systemic racism by the government and health institutions [15,25]. This group of people as well as others impacted by the digital and health access divides in affluent countries (*e.g.*, rural communities, undocumented people, aging population) may be less likely to be represented in public biomedical data repositories. Therefore, they may also fail to benefit from the focused development of medical imaging AI with open-source data. In this paper, we focus on the disparate impacts between HIC and LMIC, but it should be noted that most points made below can also be applied to marginalized and disadvantaged populations within HIC and in general.

Ultimately, the open science principles that we value so highly in our community may disproportionately benefit HIC and privileged populations. Deep learning applied to medical imaging is often seen as a way to improve global health, for example by making radiological screening more accessible [23]. However, such promises can only be achieved with the appropriate sociotechnical considerations and resources required to translate achievements made using data from countries like Canada, the U.S., and the U.K., to data from countries in South America, sub-Saharan Africa, and Southeast Asia, for example. Deep learning frameworks that could address some of these challenges related to domain shift and data availability in LMIC include transfer learning and distributed learning, which we describe and discuss in the following two sections.

3 Open Science and Transfer Learning

Transfer learning is a technique used in deep learning, which often utilizes a model pre-trained on a large public dataset, and then fine-tunes it on additional data for a new domain or for a task that shares similar features [39]. One such example would be a brain aging model trained using UK Biobank neuroimaging data that is then fine-tuned for dementia classification using neuroimaging data from a different cohort. In this way, it is expected that transfer learning can significantly reduce the required computational resources and data to achieve high performance in a model [40]. Transfer learning theoretically provides a solution to the problem that most medical imaging AI models have been trained on public data from HIC, because those models could simply be fine-tuned using data from LMIC to account for generalization issues and/or domain shift. However, it should be emphasized that we once again reach the problem of a lack of digital infrastructure and data collection resources in many LMIC - without high-quality, accessible data to fine-tune those models on, this solution is futile. Lower-quality medical imaging systems, scarcity of experts to properly label data, and presence of health co-morbidities in which prevalence differs between HIC and LMIC (*e.g.*, malnutrition) can further reduce the ability of transfer learning to perform well [40]. Furthermore, biases that have been learned by the original model may end up transferring to the new task [31]. These biases can emerge from intricate biological factors (*e.g.,* age, sex, pathological characteristics) and non-biological factors (*e.g.,* sample size, scanner type) that vary between cohorts [35]. In such situations, the model's performance may even deteriorate due to the extreme adjustments needed when the available data is limited. With all these considerations, it is unknown whether transfer learning is truly effective enough to allow models pre-trained on public data from HIC to perform with acceptable accuracy in the context of LMIC, which is challenging to investigate in detail due to a lack of available data.

It is also important to note that transfer learning merely accounts for shifts in the joint distribution of data and labels fed to a deep learning model, and it does not take into account how shifts in social context may alter the utilization and efficacy of the AI system [32, 40]. Considerations of social context need to be incorporated into the development of clinical AI systems early in the design process in order to ensure responsible and effective outcomes [4]. Even if the goals of transfer learning are successful and a model is indeed able to achieve state-of-the-art performance once fine-tuned on data from LMIC, this does not guarantee that it will lead to improved health outcomes. For instance, consider a deep learning-based radiological screening tool that is validated for implementation in LMIC. However, consider that it may only be possible to make this tool available in urban centers due to computational and connectivity requirements. In such urban centers, screening for a specific condition may already be widely available due to access to a radiologist who performs screening using traditional methods. On the contrary, it may be in rural areas where patient outcomes could be improved the most by applying this tool, but without the appropriate education, outreach, transportation services, or modification of the system to allow it to be used in remote areas, those who stand to benefit the most from this model would not necessarily be positively impacted by its implementation. Thus, even if transfer learning enables the creation of better fine-tuned models, it does not account for the problem that models developed on data from HIC are also primarily developed by and for people in HIC. Therefore, even though transfer learning may theoretically address issues related to the domain shift in applying models developed on large open data from HIC to data from LMIC, it cannot remove barriers of data availability, digital infrastructure, and local social context required for these models to be clinically effective and sustainable.

4   Open Science and Distributed Learning

Distributed learning is a rather new approach to training deep learning models that addresses data accessibility and privacy issues by avoiding sharing patient data outside of healthcare centers where the data was acquired. In this approach, instead of collecting data in a centralized database to train a model, each institution trains models locally, and a server is responsible for combining or orchestrating the locally trained models to create the global model, which is expected to leverage knowledge from data across all institutions. In theory, distributed learning approaches could overcome some barriers that LMIC face in the context of contributing to and benefitting from open science research in medical imaging AI. Instead of having to train large, computationally expensive models locally, they could contribute data to the distributed learning network and still benefit from the final trained model as it is more likely to be applicable to the local data. Additionally, LMIC with limited data and without the ability to generate large datasets could still contribute to and benefit from certain distributed learning frameworks designed for small samples sizes [34].

In the distributed learning scenario, an institution is any entity capable of collecting and storing data locally. Institutions are required to have technological resources available to train deep learning models locally and to ensure that confidential patient information does not leave the safety of the institution's health data management protocols, which should be aligned with the governmental protocols [37]. However, given this consideration, it is likely that many LMIC healthcare centers will not be able to contribute to a model trained in this decentralized approach. While HIC possess the capability to rapidly integrate AI into their healthcare systems, many healthcare centers in LMIC will encounter obstacles due to inadequate digital infrastructure within their institutions. For example, institutions in these countries often grapple with issues such as insufficient imaging equipment, limited internet connectivity, outdated computer systems, and a scarcity of skilled radiologists and data scientists, hindering their ability to effectively leverage and contribute to advancements in AI [23]. Furthermore, the implementation of distributed learning requires a solid foundation of data management considerations. These regulations encompass legal and ethical principles governing the appropriate data storage and access. Within this context, it is important to acknowledge that many LMIC still face challenges in establishing comprehensive and consistent regulatory systems for managing health data [6,23].

Federated learning (FL) has emerged as the leading method for implementing distributed deep learning across various domains, including healthcare [21]. In the FL setup, an initial global model is created at a central server, and each collaborating institution receives a replica of this model. The models undergo parallel training at each institution using locally available data, and their periodic aggregation takes place at the central server. This iterative process is typically repeated multiple times to enhance the performance of the global model. The effectiveness of the global model relies on the performance of the locally trained models, making it crucial for each institution to have access to diverse and substantial datasets to contribute valuable information to the global model [34]. However, the implementation of FL in LMIC poses several concerns. Capability to train models locally may be hindered by potentially limited access to powerful graphics processing units (GPUs) and available datasets [24]. Furthermore, if participating in a global distributed learning network, a relatively lower amount of data coming from LMIC may be overwhelmed by data from HIC. Thus, in the global model, features learned on data from LMIC may be overpowered by features learned on the majority data from HIC. Moreover, a consensual solution achieved by a global model may not always be desirable in all cases. Although the global model can improve performance and generalization, it is important to consider the local heterogeneity of datasets. Embracing this heterogeneity allows for the customization of the global solution to fit the specific context of LMIC, taking into account factors such as disease manifestation, cultural aspects, and patterns of medical practices. This consideration acknowledges the importance of tailoring models to address specific regional characteristics and needs, rather than solely relying on a uniform global approach [37].

The progress made in distributed learning in the medical domain has been heavily facilitated by the numerous public databases that have been established and curated as a result of the open science movement. Common databases used in this research, such as ADNI and BraTs, have played a crucial role in enabling technical exploration by offering data obtained from multiple healthcare institutions [37]. In particular, the MICCAI community has delved into distributed learning, specifically examining federated learning for brain tumor segmentation (FeTS) [27]. However, since most of this data originates from HIC (see Fig. 1), limited conclusions can be made regarding its potential for training deep learning models that are widely applicable.

While distributed learning has benefited greatly from open science, clinical translation of advancements made thus far will only be beneficial to countries that have the infrastructure and resources to participate in such networks. Furthermore, although frameworks such as FL could address barriers that prevent LMIC from participating in and benefiting from open science, the likely smaller and potentially lower-quality datasets from LMIC could lead to models that are unreliable and do not generalize well when considering the nature of parallel training and model aggregation. The resulting global models also fail to consider crucial differences in clinical experience, resource constraints, demographics, and public policies across different regions.

5   Conclusions

Open science – particularly open-source code and public databases – offers numerous advantages for advancing AI in medical imaging. It is expected to facilitate the sharing of research outputs, encourage collaboration among researchers, promote reproducibility of findings, and foster innovation. Open science is widely regarded as a transformative movement that seeks to make scientific research more inclusive, transparent, and ultimately beneficial for society [38]. However, considering the fundamental objective of AI for medical imaging, which is widespread clinical implementation to enhance health outcomes and equity, it is imperative to address the fact that the impressive and extensive technical advancements made thus far, largely facilitated by open science, tend to primarily benefit majority groups in HIC and privileged populations represented in public databases.

The well-known problems of domain shift and generalizability that affect many deep learning models developed by researchers render many state-of-the-art models being developed on UK Biobank, ADNI, and BraTS, to name a few, not directly applicable in healthcare centers in LMIC in which digital infrastructure, funding for research and development, and healthcare and data science professionals are often lacking. Transfer learning and distributed learning could theoretically address some of the concerns related to limited, non-public data in LMIC, but both frameworks come with practical problems when facing small amounts of low-quality data, insufficient resources for data governance and management, and local contextual implementation of these models. Additionally, neither of these frameworks seem to be viable solutions to the problem that data for diseases that are more prevalent in LMIC may not be available in large public repositories, and therefore overlooked by research in HIC.

An apparent solution would be an investment in data infrastructure, healthcare and data science professionals, and computational resources in LMIC so that these nations could generate their own open medical imaging databases and enable researchers globally to develop and validate models on this data. However, consideration needs to be made for data autonomy and sovereignty. Publicizing potentially sensitive biomedical data carries significant risks related to privacy, which must be addressed with appropriate data governance. Additionally, the mass generation and collection of personal data introduce concerns related to commercialization and surveillance capitalism [42]. Consequently, at a national, community, and individual level, autonomy should be granted with respect to the choice of whether to openly share data with the global research community. Should the choice to make medical imaging data publicly available be predetermined, perhaps as a stipulation of foreign investment in digital infrastructure, this would translate to further consequences of data colonialism. In other words, data may be compulsorily taken from LMIC with humanitarian intentions to enable researchers and developers in HIC to develop and validate clinical models. However, those models may then be sold at high costs back to LMIC, repeating patterns of colonialism and further widening the divide between those who collect and control data and those who provided it but have a lack of control over how it is used [4,17]. Thus, it is not only the digital infrastructure required to generate and manage data that should be prioritized in LMIC, but also the digital literacy and data governance that is necessary for empowering data sovereignty and individual data autonomy. Overall, investment into both the technological and human resources that enable end-to-end development of AI (from defining the clinical task, to data collection, to model training, to evaluation and implementation) in underrepresented nations and institutions can allow them not only to benefit from research efforts utilizing open-source models and public data, but also to develop systems that represent their own local populations.

References

1. ADNI — Acquisition Sites. https://adni.loni.usc.edu/about/centers-cores/study-sites/
2. The Federated Tumor Segmentation (FeTS) initiative. https://www.med.upenn.edu/cbica/fets/#FeTSCollaboratingSites6

3. Kaggle DR dataset (EyePACS). https://www.kaggle.com/datasets/mariaherrerot/eyepacspreprocess

4. Ethics and governance of artificial intelligence for health: WHO Guidance (2021). https://www.who.int/publications-detail-redirect/9789240029200

5. Reporting standards and availability of data, materials, code and protocols (2023). https://www.nature.com/nature-portfolio/editorial-policies/reporting-standards

6. Annas, G.J.: HIPAA regulations - a new era of medical-record privacy? N. Engl. J. Med. **348**(15), 1486–1490 (2003)

7. Ashford, M.T., et al.: Screening and enrollment of underrepresented ethnocultural and educational populations in the Alzheimer's Disease Neuroimaging Initiative (ADNI). Alzheimer's & Dementia **18**(12), 2603–2613 (2022)

8. Beam, A.L., Manrai, A.K., Ghassemi, M.: Challenges to the reproducibility of machine learning models in health care. JAMA **323**(4), 305–306 (2020)

9. Bosma, J., et al.: Reproducibility of Training Deep Learning Models for Medical Image Analysis. Medical Imaging with Deep Learning (2023). https://openreview.net/forum?id=MR01DcGST9

10. Ciecierski-Holmes, T., Singh, R., Axt, M., Brenner, S., Barteit, S.: Artificial intelligence for strengthening healthcare systems in low- and middle-income countries: a systematic scoping review. NPJ Digit. Med. **5**(1), 1–13 (2022)

11. Coates, M.M., et al.: Burden of disease among the world's poorest billion people: an expert-informed secondary analysis of global burden of disease estimates. PLoS ONE **16**(8), e0253073 (2021)

12. Diaz, O., et al.: Data preparation for artificial intelligence in medical imaging: a comprehensive guide to open-access platforms and tools. Physica Med. **83**, 25–37 (2021)

13. Fry, A., et al.: Comparison of sociodemographic and health-related characteristics of UK biobank participants with those of the general population. Am. J. Epidemiol. **186**(9), 1026–1034 (2017)

14. Ganguli, M., Lee, C.W., Hughes, T., Snitz, B.E., Jakubcak, J., Duara, R., Chang, C.C.H.: Who wants a free brain scan? Assessing and correcting for recruitment biases in a population-based sMRI pilot study. Brain Imaging Behav. **9**(2), 204–212 (2015)

15. Hyett, S., Marjerrison, S., Gabel, C.: Improving health research among Indigenous Peoples in Canada. CMAJ **190**(20), E616–E621 (2018)

16. Jack, C.R., et al.: The Alzheimer's disease neuroimaging initiative (ADNI): MRI methods. J Magn. Reson. Imaging **27**(4), 685–691 (2008)

17. Kwet, M.: Digital colonialism: US empire and the new imperialism in the Global South. Race Class **60**(4), 3–26 (2019)

18. Lo Vercio, L., et al.: Supervised machine learning tools: a tutorial for clinicians. J. Neural Eng. **17**(6), 062001 (2020)

19. MacEachern, S.J., Forkert, N.D.: Machine learning for precision medicine. Genome **64**(4), 416–425 (2021)

20. McDermott, M.B.A., Wang, S., Marinsek, N., Ranganath, R., Foschini, L., Ghassemi, M.: Reproducibility in machine learning for health research: still a ways to go. Sci. Transl. Med. **13**(586), eabb1655 (2021)

21. McMahan, B., Moore, E., Ramage, D., Hampson, S., Arcas, B.A.y.: Communication-efficient learning of deep networks from decentralized data. In: Proceedings of the 20th International Conference on Artificial Intelligence and Statistics, pp. 1273–1282. PMLR (2017)

22. Menze, B.H., et al.: The multimodal brain tumor image segmentation benchmark (BRATS). IEEE Trans. Med. Imaging **34**(10), 1993–2024 (2015)

23. Mollura, D.J., et al.: Artificial Intelligence in low- and middle-income countries: innovating global health radiology. Radiology **297**(3), 513–520 (2020)

24. Ng, D., Lan, X., Yao, M.M.S., Chan, W.P., Feng, M.: Federated learning: a collaborative effort to achieve better medical imaging models for individual sites that have small labelled datasets. Quant. Imaging Med. Surg. **11**(2), 852–857 (2021)

25. Nguyen, N.H., Subhan, F.B., Williams, K., Chan, C.B.: Barriers and mitigating strategies to healthcare access in indigenous communities of Canada: a narrative review. Healthcare **8**(2), 112 (2020)

26. Okolo, C.T.: Optimizing human-centered AI for healthcare in the Global South. Patterns **3**(2), 100421 (2022)

27. Pati, S., et al.: The federated tumor segmentation (FeTS) challenge, May 2021. arXiv:2105.05874 [cs, eess]

28. Pineau, J., et al.: Improving reproducibility in machine learning research (a report from the NeurIPS 2019 reproducibility program). J. Mach. Learn. Res. **22**(1), 164:7459–164:7478 (2021)

29. Porwal, P.: Indian Diabetic Retinopathy Image Dataset (IDRiD) (2018). https://ieee-dataport.org/open-access/indian-diabetic-retinopathy-image-dataset-idrid

30. Prevedello, M., et al.: Challenges related to artificial intelligence research in medical imaging and the importance of image analysis competitions. Radiol. Artif. Intell. **1**(1), e180031 (2019)

31. Salman, H., Jain, S., Ilyas, A., Engstrom, L., Wong, E., Madry, A.: When does bias transfer in transfer learning? (2022). arXiv:2207.02842 [cs]

32. Selbst, A.D., Boyd, D., Friedler, S.A., Venkatasubramanian, S., Vertesi, J.: Fairness and abstraction in sociotechnical systems. In: Proceedings of the Conference on Fairness, Accountability, and Transparency. pp. 59–68. FAT* 2019. Association for Computing Machinery, New York (2019)

33. Sivaswamy, J., Krishnadas, S.R., Datt Joshi, G., Jain, M., Syed Tabish, A.U.: Drishti-GS: retinal image dataset for optic nerve head(ONH) segmentation. In: 2014 IEEE 11th International Symposium on Biomedical Imaging (ISBI), pp. 53–56 (2014)

34. Souza, R., Mouches, P., Wilms, M., Tuladhar, A., Langner, S., Forkert, N.D.: An analysis of the effects of limited training data in distributed learning scenarios for brain age prediction. J. Am. Med. Inform. Assoc. **30**(1), 112–119 (2023)

35. Souza, R., et al.: Image-encoded biological and non-biological variables may be used as shortcuts in deep learning models trained on multi-site neuroimaging data. J. Am. Med. Inform. Assoc. (2023). https://doi.org/10.1093/jamia/ocad171

36. Sudlow, C., et al.: UK Biobank: an open access resource for identifying the causes of a wide range of complex diseases of middle and old age. PLoS Med. **12**(3), e1001779 (2015)

37. Tuladhar, A., Rajashekar, D., Forkert, N.D.: Distributed learning in healthcare. In: Sakly, H., Yeom, K., Halabi, S., Said, M., Seekins, J., Tagina, M. (eds.) Trends of Artificial Intelligence and Big Data for E-Health, pp. 183–212. Springer, Cham, Integrated Science (2022). https://doi.org/10.1007/978-3-031-11199-0_10

38. Wang, K.: Opportunities in open science with AI. Front. Big Data **2**, 26 (2019)

39. Weiss, K., Khoshgoftaar, T.M., Wang, D.: A survey of transfer learning. J. Big Data **3**(1), 9 (2016)

40. Williams, D., Hornung, H., Nadimpalli, A., Peery, A.: Deep learning and its application for healthcare delivery in low and middle income countries. Front. Artif. Intelli. **4**, 553987 (2021)

41. Zhou, Y., Elashoff, D., Kremen, S., Teng, E., Karlawish, J., Grill, J.D.: African Americans are less likely to enroll in preclinical Alzheimer's disease clinical trials. Alzheimer's Dementia Transl. Res. Clin. Intervent. **3**(1), 57–64 (2017)
42. Zuboff, S.: Big other: surveillance capitalism and the prospects of an information civilization. J. Inf. Technol. **30**(1), 75–89 (2015)

Gradient-Based Enhancement Attacks in Biomedical Machine Learning

Matthew Rosenblatt[1]([⊠]), Javid Dadashkarimi[2], and Dustin Scheinost[1,3]

[1] Department of Biomedical Engineering, Yale University, New Haven, USA
{matthew.rosenblatt,dustin.scheinost}@yale.edu
[2] Department of Computer Science, Yale University, New Haven, USA
javid.dadashkarimi@yale.edu
[3] Department of Radiology and Biomedical Imaging, Yale School of Medicine,
New Haven, USA

Abstract. The prevalence of machine learning in biomedical research is rapidly growing, yet the trustworthiness of such research is often overlooked. While some previous works have investigated the ability of adversarial attacks to degrade model performance in medical imaging, the ability to falsely improve performance via recently-developed "enhancement attacks" may be a greater threat to biomedical machine learning. In the spirit of developing attacks to better understand trustworthiness, we developed two techniques to drastically enhance prediction performance of classifiers with minimal changes to features: 1) general enhancement of prediction performance, and 2) enhancement of a particular method over another. Our enhancement framework falsely improved classifiers' accuracy from 50% to almost 100% while maintaining high feature similarities between original and enhanced data (Pearson's $r's > 0.99$). Similarly, the method-specific enhancement framework was effective in falsely improving the performance of one method over another. For example, a simple neural network outperformed logistic regression by 17% on our enhanced dataset, although no performance differences were present in the original dataset. Crucially, the original and enhanced data were still similar ($r = 0.99$). Our results demonstrate the feasibility of minor data manipulations to achieve any desired prediction performance, which presents an interesting ethical challenge for the future of biomedical machine learning. These findings emphasize the need for more robust data provenance tracking and other precautionary measures to ensure the integrity of biomedical machine learning research. The code is available at https://github.com/mattrosenblatt7/enhancement_EPIMI.

Keywords: machine learning · adversarial attacks · neuroimaging

1 Introduction

Machine learning has demonstrated great success across numerous fields. However, the success of these models is not immune to attacks. Adversarial

S. Wesarg et al. (Eds.): CLIP/FAIMI/EPIMI 2023, LNCS 14242, pp. 301–312, 2023.
https://doi.org/10.1007/978-3-031-45249-9_29

attacks, or data manipulations designed to alter the prediction [3], threaten real-world machine learning applications. Adversarial attacks include evasion attacks [2,4,12,30], where only test data are manipulated, or poisoning attacks [3,4,7,19], where the attacker may contribute manipulated test and/or training data. Understanding adversarial attacks and developing corresponding defenses is crucial to the integrity of machine learning applications.

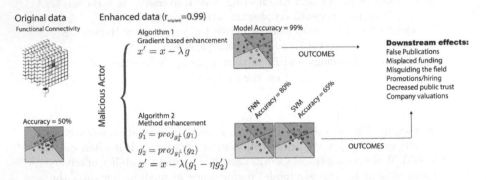

Fig. 1. Overview of enhancement methods used in this paper. Classification accuracy in the original dataset is 50%. After applying gradient-based enhancement attacks following Algorithm 1, classification accuracy in the enhanced dataset is 99%. Using method enhancement attacks (Algorithm 2), datasets are altered such that a specific method (e.g., feedforward neural network) outperforms another (e.g., support vector machine). In all cases, the changes between the original and enhanced datasets are minor. The "Downstream effects" box highlights possible implications of enhancement attacks.

Machine learning is becoming increasingly prevalent in biomedical research—including biomedical imaging. Previous studies of adversarial attacks in medical imaging have focused on clinical applications where a malicious party would be interested in altering the prediction outcomes for financial or other purposes. Most of these studies implemented evasion attacks [6,10,11,17,17,32], while a smaller subset used poisoning attacks [9,18,20,26].

An equally relevant yet understudied motivation in scientific machine learning is the feasibility of manipulating data to improve model performance falsely. In scientific research, data manipulations designed to make results seem more impressive are regarded as a major ethical issue. For example, a line of highly-cited Alzheimer's research was recently flagged for likely manipulation of biological images [22]. This paper is just one example of many where scientific data were manipulated in ways that harmed their respective fields [1,5]. The potential for data manipulation is not widely acknowledged in the scientific machine learning community. Traditional approaches to preventing and detecting academic fraud include data and code sharing. While data and code sharing are useful for improving the replicability, they do not necessarily ensure that the results can be trusted. Even if the data and code are shared, we will show in this work how scientific machine learning results can be modified through subtle and unnoticeable data manipulations, which could have major consequences.

For example, a malicious party might manipulate their data to improve model performance, falsely claim that they can classify a specific mental health condition, and make a paper more publishable or increase the valuation of a healthcare start-up (Fig. 1). If they manipulate the data in a subtle way, they could then publicly share this manipulated dataset, and other researchers would likely not notice that anything was wrong with the data. These data manipulations could waste grant money, misdirect future research directions, and potentially cause harmful public effects. One recent work showed that the performance of regression models using neuroimaging data could be falsely enhanced by injecting subtle associations into the data, labeled as "enhancement attacks" [25]. However, this enhancement framework relies on manually adding patterns to the data that are correlated with the prediction outcome of interest. Thus, the previous framework only works for regression problems and cannot generalize to other settings, such as classification. Given the prevalence of classification problems in biomedical machine learning, developing a general framework for enhancement is novel and important for understanding the feasibility of data manipulations in machine learning.

In this work, we first extend the enhancement attack to classification models with a gradient-based enhancement framework (GOAL #1). Then, we present an additional way in which data can be enhanced with only subtle manipulations: falsely demonstrating that a particular method (e.g., a type of machine learning model) outperforms another (GOAL #2). We found that both methods were successful in falsely improving classification accuracy with only minimal changes to the dataset. Finally, given the vulnerability of biomedical datasets to enhancement attacks, we discuss the implications and potential solutions.

2   Methods

In the following sections, we considered two separate attacker goals: falsely enhancing 1) overall classifier performance (GOAL #1), and 2) the performance of one method over another (GOAL #2) (Fig. 1). Whereas traditional adversarial attackers may have complete ("white-box") or limited ("black-box" or "gray-box") knowledge of the model and/or data [4], enhancement attackers have even greater knowledge than the "white-box" setting. Our methods can modify the entire dataset, which includes both training and test data. These enhancement attacker assumptions mimic a realistic setting, where a researcher could modify their data to falsely improve performance and then publicly release the data such that their results are computationally reproduced by others. We assess all methods with two main metrics: classification accuracy in the enhanced dataset and similarity between the original and enhanced datasets, as measured by Pearson's correlation. A highly effective enhancement attack is one that falsely increases the accuracy while the original and enhanced datasets remain highly similar.

Algorithm 1. Gradient-based enhancement attacks

$D \in \{X, y\}$: dataset
f: model
n_{folds}: folds for K-fold partitioning
λ: enhancement step size
for $k = 1 : n_{folds}$ **do**
    Establish $D_{tr}, D_{held-out}$
    Train f
    $g \leftarrow \nabla_x A$ where $A = L(f, x)$ or $DF(f, x)$
    $X_{held-out} \leftarrow X_{held-out} - \lambda g$
end for

GOAL #1: Gradient-Based Enhancement. The key idea behind gradient-based enhancement is to "push" the samples in the direction of a learned model to make the decision boundaries clearer and more consistent across all samples, thus improving performance. The method is similar to an adversarial evasion attack, except we are altering the entire dataset, not just the test data. For a single held-out point, one may optimally change the classification by perturbing the point in the direction of $g = \nabla_x A$, where A can be a decision function or loss function. For example, in the case of binary linear support vector machine (SVM) or logistic regression (LR):

$$X_{held-out,y=-1} \leftarrow X_{held-out,y=-1} - \epsilon * w \tag{1}$$

$$X_{held-out,y=1} \leftarrow X_{held-out,y=1} + \epsilon * w \tag{2}$$

where w is a vector of model coefficients and ϵ is a scaling factor. Equations 1 and 2 would move the corresponding held-out points toward the correct side of the decision boundary. As summarized in Algorithm 1, a model f is first trained by holding one or numerous points out with K-fold partitioning. Then, the held-out point(s) are updated with the attacker gradient $g = \nabla_x A$, such that the model will predict them correctly. This process repeats until all points in the dataset are held out. Empirically, we found that enhancement was most effective for complex models (i.e., neural networks) when updating the held-out data iteratively within the cross-validation loop, whereas updating the dataset once after the cross-validation loop was more effective for simpler models (i.e., support vector machine or logistic regression). Since learned model coefficients are similar when only holding out a small fraction of the points, this method pushes all points of a given class in a consistent direction. Eventually, when the enhanced dataset is released, an independent researcher would not notice any perturbations but would falsely find higher performance.

GOAL #2: Enhancement of a Particular Method. A roadblock to method-specific enhancement is that the gradients used in Eqs. 1 and 2 generally transfer well across model types [8], which would make this process ineffective in enhancing the performance of a specific method over another. Transferability

---

Algorithm 2. Method enhancement

$D \in \{X, y\}$: dataset
f_1: model to enhance
f_2: model to avoid enhancement
n_{folds}: number of folds for K-fold partitioning
λ: enhancement step size
η: enhancement suppression coefficient
for $k = 1 : n_{folds}$ **do**
     Establish $D_{train}, D_{held-out}$
     Train f_1, f_2
     $g_1 \leftarrow \nabla_x A(f_1, X_{held-out})$
     $g_2 \leftarrow \nabla_x A(f_2, X_{held-out})$
     $X_{held-out} \leftarrow X_{held-out} - \lambda(proj_{g_2^{\perp}}(g_1) - \eta \, proj_{g_1^{\perp}}(g_2))$
end for

---

of attacks from a base classifier f_1 to another classifier f_2 is defined by [8] as how well an attack designed for f_1 works on f_2. In this case, we do *not* wish for the attacks to transfer between models. We want to find a new direction g_1' that enhances the performance of f_1 but does not affect f_2. We achieve this by taking the component of g_1 that is orthogonal to g_2:

$$g_1' = proj_{g_2^{\perp}}(g_1) \tag{3}$$

Furthermore, Eq. 3 may not be sufficient to limit the performance of f_2, since f_2 can learn a new decision boundary after retraining. As such, we propose to include a term g_2' to suppress the performance of f_2:

$$g_2' = proj_{g_1^{\perp}}(g_2) \tag{4}$$

Then, for a held-out sample, we can update it as follows to attempt to improve the performance of f_1 but not f_2:

$$x' = x - \lambda(g_1' - \eta g_2') \tag{5}$$

where λ and the suppression coefficient η control the influence of g_1' and g_2'.

Similar to the model-based data enhancement, we split the data into k folds. For each partitioning, we train two models: 1) A model that we want to enhance (i.e., f_1), and 2) a second model that we do not want to enhance (i.e., f_2). Subsequently, Eqs. 3–5 are applied to update the held-out data, and the process is repeated until each sample is held out once.

3   Experiments

Datasets. Resting-state functional MRI (fMRI) data were obtained from the UCLA Consortium for Neuropsychiatric Phenomics (CNP) [23] dataset. For all data, we performed motion correction, registration to common space, regression of covariates of no interest, temporal smoothing, and gray matter masking. Participants were also excluded for lack of full-brain coverage. After these exclusion criteria, 245 participants remained. We parcellated the fMRI data for each participant into 268 nodes using the Shen atlas [29]. To form functional connectivity matrices, the time series data from each pair of nodes (regions of interest) was correlated using Pearson's correlation, and then Fisher's transform was applied. Before inputting the functional connectivity matrices into machine learning models, we first vectorized the upper triangle of each matrix to use as features. In all the following models, we classified participants in CNP based on their diagnoses, including no diagnosis (n = 117), schizophrenia (n = 46), bipolar disorder (n = 44), and ADHD (n = 38). In addition, all plots below were made with seaborn [14,31].

GOAL #1: Gradient-Based Enhancement. We enhanced the CNP dataset for a classification problem with the following four classes: participants with 1) no diagnosis, 2) bipolar disorder, 3) schizophrenia and 4) ADHD. We used linear support vector machine (SVM), logistic regression (LR) [21], and a feedforward neural network (FNN) as models. Our FNN consisted of three fully connected layers with the ReLU activation function. During the enhancement attack, all model hyperparameters were held constant. For SVM and LR, this included an L2 regularization parameter $C = 1$. For the FNN, we used cross entropy loss and the Adam [16] optimizer, with a learning rate of $\alpha = 0.001$ and batch size of 10 for 10 epochs.

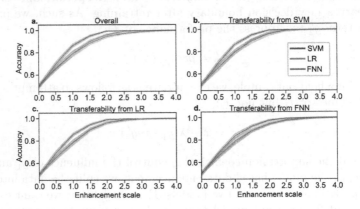

Fig. 2. a) Model-based enhancement of SVM, LR, and FNN models for various enhancement scales. The enhancement scale is multiplied by the unit norm direction of the perturbation for each sample, where an enhancement scale of 0 reflects the original dataset. b-d) Transferability of enhancement between the three models. All accuracies were evaluated with 5-fold cross-validation, with error bars showing standard deviation across 10 random seeds.

Gradients were computed as the model coefficients in SVM and LR (linear models), while Pytorch's autograd feature was used for the FNN. All gradients (*i.e.*, $\nabla_x A$ in Algorithm 1) were normalized to have a Frobenius norm of 1 and then multiplied by the corresponding enhancement scale in Fig. 2. After creating an "enhanced dataset," we evaluated enhanced performance with nested k-fold cross-validation, as one would do if receiving this dataset without any knowledge of the enhancement, with a grid search in SVM and LR for the L2 regularization parameter $C = \{1e\text{-}4, 1e\text{-}3, 1e\text{-}2, 1e\text{-}1, 1\}$ within each fold. Due to computational restraints, we did not perform a hyperparameter search for the FNN but used the same hyperparameters described above. Enhancement brought prediction performance from 50% to 99% in all three models (Fig. 2a), despite similar feature values between the original and enhanced datasets ($r = 0.99$).

Furthermore, we considered whether the enhancement attacks could transfer between models. After data were enhanced using Algorithm 1 for each of the three models (Fig. 2a), we re-trained the other two types of models to assess transferability. For example, in Fig. 2b, data are first enhanced using SVM and Algorithm 1; then, the performance of that enhanced dataset is evaluated using LR and FNN. Overall, we found that the enhancement attacks transferred between each of the three models (Fig. 2b-d).

GOAL #2: Enhancement of a Particular Method. Since the enhancement attacks above transferred between models, we investigated how enhancement may be targeted to a specific model. Although there are countless machine learning models, we selected three of the most common models for functional connectivity data to perform a case study: SVM, LR, and a FNN. We demonstrated the hypothetical scenario in which one may perturb a dataset such that a FNN outperforms simpler methods like SVM and LR.

CNP data were manipulated following Algorithm 2 to promote the performance of a particular method over another. Hyperparameters were held constant when enhancing the data with Algorithm 2 (as detailed in GOAL #1). After enhancing the data, we performed 10-fold cross-validation to evaluate accuracy with nested folds and a grid search for the L2 regularization parameter C in SVM and LR. We consider different enhancement scales λ for the classifier of interest ($f_1 = $ FNN) and different suppression values η for the classifier which we do not wish to perform well ($f_2 = $ SVM or LR). Despite no differences in the original dataset, FNN outperformed SVM and LR (Fig. 3a-b), while maintaining high feature similarities (Fig. 3c-d). The performance on f_2 generally increased, though less, as performance on f_1 increased, but increasing the suppression coefficient η limited performance improvements of f_2. Furthermore, attacks transferred between SVM and LR. For $f_2 = $ SVM, $\lambda = 0.4$, and $\eta = 0$, accuracies for SVM and LR were 76.8% and 77.8% vs. 90.6% for FNN. For $f_2 = $ LR, $\lambda = 0.4$, and $\eta = 0$, accuracies for SVM and LR were 61.8% and 62.3% vs. 77.0% for FNN. This method was less effective than that of Algorithm 1, which is likely because the enhancement is weakened by the suppression coefficient and the projection of the gradients.

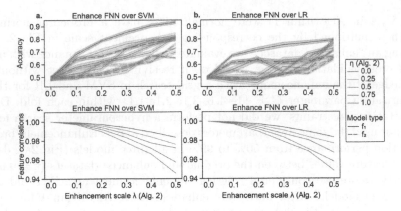

Fig. 3. Enhancement of a FNN over a,c) SVM and b,d) LR in CNP. In a,b), data are enhanced with increasing λ (see Algorithm 2). Solid lines represent the accuracy for f_1 (FNN), while dashed lines show the accuracy for f_2 (SVM in a and LR in b). Error bars reflect standard deviation across 10 random seeds of k-fold cross-validation and initialization seeds for FNN. Line color shows the suppression coefficient for f_2, η. In c,d), the correlation between original and enhanced features is shown with increasing λ. The original and enhanced features are highly correlated ($r's > 0.94$).

Proof-of-Concept Validation in Other Models and Datasets. While the focus of this work is enhancement of SVM, LR, and FNN models in the CNP dataset, we also wanted to demonstrate that other datasets and models were easily manipulated. To demonstrate generalizability to another dataset, we performed enhancement attacks with enhancement scale $\lambda = 2$ in the Philadelphia Neurodevelopmental Cohort [27,28] (N = 1126) dataset to predict self-reported sex using resting-state functional connectivity data. The accuracies were (baseline/after enhancement): SVM (79.40%/100%), logistic regression (79.13%/100%), and FNN (76.47%/99.91%).

Furthermore, we demonstrated our results in the CNP dataset with another model, BrainNetCNN [15], a deep learning model for brain connectivity data. BrainNetCNN consists of two edge-to-edge filters, one edge-to-node filter, one node-to-graph filter, and three dense layers. Further details of BrainNetCNN are in the original paper [15]. BrainNetCNN was trained with cross entropy loss and the Adam [16] optimizer, with a learning rate of $\alpha = 0.001$ and batch size of 10 for 20 epochs. For the gradient-based enhancement attack with enhancement scale $\lambda = 3$, the resulting accuracy was 79.31% (baseline: 41.10%). We enhanced BrainNetCNN over the FNN with $\lambda = 0.15$, $\eta = 0$; accuracy was 85.43% for BrainNetCNN and 52.33% for FNN. We enhanced the FNN over BrainNetCNN with $\lambda = 0.15$, $\eta = 0$; accuracy was 80.16% for BrainNetCNN and 99.27% for FNN. These results demonstrate that complex models are also susceptible to designed data manipulations.

4   Discussion

In this work, we developed a gradient-based framework for enhancement attacks and showed that a functional neuroimaging dataset could be modified to achieve essentially any desired prediction performance. The vulnerability of machine learning pipelines to possible fraud (enhancement) is integral to the trustworthiness of the field. Considering the prevalence of research fraud [1,5,22] and the increasing popularity of machine learning in biomedical research, we believe that enhancement could become a major issue, if it is not already.

For GOAL #1, we demonstrated that a four-way classification task went from near-chance performance, where chance is defined as the most frequently occurring group (47.76%), to over 99% accuracy while the original and enhanced data remained highly similar ($r > 0.99$). In a hypothetical scenario, a malicious researcher could collect data, enhance it to perform well in a specific classification or regression task, and release this data on a public repository. For example, if the enhanced CNP dataset in this paper was released, the scientific community would be excited and impressed by the near-perfect classification accuracy between participants with no diagnosis, schizophrenia, bipolar disorder, and ADHD. In the academic sector, these false results could lead to hiring and grant decisions or the distribution of future grant funding under false pretenses. In the industrial sector, these false results may cause increased investments in certain companies. Furthermore, enhancement attacks could have additional downstream effects, such as the undermining of public trust or the wasted time and resources of other researchers attempting to build upon the false results.

For GOAL #2, we found that in the best case, a FNN outperformed LR and SVM by 17% in the enhanced dataset, despite no differences in original performance and high similarity between original and enhanced data ($r = 0.99$). The feasibility of modifying a dataset such that a specific type of model outperforms another is both powerful and potentially dangerous. For instance, a start-up company may demonstrate that their new model outperforms the current industry standard to increase their valuation. Alternatively, researchers may perturb a dataset so their novel method performs the best, leading to a (falsely) more exciting paper.

There were several limitations to our study. First, we investigated enhancement only in functional neuroimaging. Future work should expand these concepts to other disciplines, which may have different sample sizes or dimensionality. Second, the method enhancement algorithm (Algorithm 2) only allows for the suppression of a specific model, and it remains to be seen how this can be extended to improve the performance of one model over many other models. However, we demonstrated that the method-specific enhancement transferred across SVM and LR, which is promising for the extension of it to multiple models. Third, we evaluated enhancement attacks in processed connectome data, and applying enhancement earlier in the processing pipeline (e.g., raw fMRI data) may reduce its effectiveness. Fourth, future work should evaluate gradient-based enhancement of additional model types. By design, the gradient-based enhancement attack is generalizable and should be able to work for any machine learning model, including both classification and regression models, for which a gradient of the loss with respect to the input can be computed.

In conclusion, although our analysis was restricted to functional neuroimaging, these problems extend to the greater biomedical machine learning communities, where many view data and code sharing as the panacea for trustworthiness. Whereas adversarial attackers only have limited access to the model and data, enhancement attackers have unrestricted access, which makes developing defenses more difficult. Still, future work should explore whether existing adversarial defenses, including augmentation and input transformations [24], can be adapted to defend against enhancement attacks, though we expect these defenses will be less effective given the much greater capabilities of enhancement relative to adversarial attacks. Another possible defense is data provenance tracking, such as DataLad [13]. However, one caveat is that data could be manipulated before provenance tracking begins. Thus, an alternative solution could be the replication of studies in an independent dataset. Overall, we hope that this work sparks additional discussion about possible defenses against enhancement attacks and data manipulations to secure the integrity of the field.

Data use declaration and acknowledgment. The UCLA Consortium for Neuropsychiatric Phenomics (download: https://openneuro.org/datasets/ds000030/versions/00016) and the Philadelphia Neurodevelopmental Cohort (dbGaP Study Accession: phs000607.v1.p1) are public datasets that obtained consent from participants and supervision from ethical review boards. We have local human research approval for using these datasets.

References

1. Al-Marzouki, S., Evans, S., Marshall, T., Roberts, I.: Are these data real? Statistical methods for the detection of data fabrication in clinical trials. BMJ **331**(7511), 267–270 (2005)
2. Biggio, B., et al.: Evasion attacks against machine learning at test time. In: Blockeel, H., Kersting, K., Nijssen, S., Železný, F. (eds.) ECML PKDD 2013. LNCS (LNAI), vol. 8190, pp. 387–402. Springer, Heidelberg (2013). https://doi.org/10.1007/978-3-642-40994-3_25
3. Biggio, B., Nelson, B., Laskov, P.: Poisoning attacks against support vector machines (Jun 2012)
4. Biggio, B., Roli, F.: Wild patterns: ten years after the rise of adversarial machine learning. Pattern Recognit. **84**, 317–331 (2018)
5. Bik, E.M., Casadevall, A., Fang, F.C.: The prevalence of inappropriate image duplication in biomedical research publications. MBio **7**(3), 10–1128 (2016)
6. Bortsova, G., et al.: Adversarial attack vulnerability of medical image analysis systems: unexplored factors. Med. Image Anal. **73**, 102141 (2021)
7. Cinà, A.E., et al.: Wild patterns reloaded: a survey of machine learning security against training data poisoning (May 2022)
8. Demontis, A., et al.: Why do adversarial attacks transfer? Explaining transferability of evasion and poisoning attacks. In: USENIX Security Symposium 2019, pp. 321–338 (2019)

9. Feng, Y., Ma, B., Zhang, J., Zhao, S., Xia, Y., Tao, D.: FIBA: frequency-Injection based backdoor attack in medical image analysis. arXiv preprint arXiv:2112.01148 (2021)

10. Finlayson, S.G., Bowers, J.D., Ito, J., Zittrain, J.L., Beam, A.L., Kohane, I.S.: Adversarial attacks on medical machine learning. Science **363**(6433), 1287–1289 (2019)

11. Finlayson, S.G., Chung, H.W., Kohane, I.S., Beam, A.L.: Adversarial attacks against medical deep learning systems. arXiv preprint arXiv:1804.05296 (2018)

12. Goodfellow, I.J., Shlens, J., Szegedy, C.: Explaining and harnessing adversarial examples (Dec 2014)

13. Halchenko, Y., et al.: DataLad: distributed system for joint management of code, data, and their relationship. J. Open Source Softw. **6**(63), 3262 (2021)

14. Hunter, J.D.: Matplotlib: a 2D graphics environment. Comput. Sci. Eng. **9**(03), 90–95 (2007)

15. Kawahara, J., et al.: BrainNetCNN: convolutional neural networks for brain networks; towards predicting neurodevelopment. Neuroimage **146**, 1038–1049 (2017)

16. Kingma, D.P., Ba, J.: Adam: a method for stochastic optimization. arXiv preprint arXiv:1412.6980 (2014)

17. Ma, X., et al.: Understanding adversarial attacks on deep learning based medical image analysis systems. Pattern Recognit. **110**, 107332 (2021)

18. Matsuo, Y., Takemoto, K.: Backdoor attacks to deep neural Network-Based system for COVID-19 detection from chest x-ray images. NATO Adv. Sci. Inst. Ser. E Appl. Sci. **11**(20), 9556 (2021)

19. Muñoz-González, et al.: Towards poisoning of deep learning algorithms with back-gradient optimization. In: Proceedings of the 10th ACM Workshop on Artificial Intelligence and Security, pp. 27–38 (Nov 2017)

20. Nwadike, M., Miyawaki, T., Sarkar, E., Maniatakos, M., Shamout, F.: Explainability matters: backdoor attacks on medical imaging. arXiv preprint arXiv:2101.00008 (2020)

21. Pedregosa, F., et al.: Scikit-learn: machine learning in python. J. Mach. Learn. Res. **12**, 2825–2830 (2011)

22. Piller, C.: Blots on a field? Science (New York, NY) **377**(6604), 358–363 (2022)

23. Poldrack, R.A., et al.: A phenome-wide examination of neural and cognitive function. Sci. Data **3**, 160110 (2016)

24. Ren, K., Zheng, T., Qin, Z., Liu, X.: Adversarial attacks and defenses in deep learning. Proc. Est. Acad. Sci. Eng. **6**(3), 346–360 (2020)

25. Rosenblatt, M., et al.: Connectome-based machine learning models are vulnerable to subtle data manipulations. Patterns **4**(7), 100756 (2023)

26. Rosenblatt, M., Scheinost, D.: Data poisoning attack and defenses in Connectome-Based predictive models. In: Workshop on the Ethical and Philosophical Issues in Medical Imaging,Multimodal Learning and Fusion Across Scales for Clinical Decision Support, and Topological Data Analysis for Biomedical Imaging. EPIMI ML-CDS TDA4BiomedicalImaging 2022 2022 2022. Lecture Notes in Computer Science, vol.13755, pp. 3–13. Springer Nature Switzerland (2022). https://doi.org/10.1007/978-3-031-23223-7_1

27. Satterthwaite, T.D., et al.: The Philadelphia neurodevelopmental cohort: a publicly available resource for the study of normal and abnormal brain development in youth. Neuroimage **124**(Pt B), 1115–1119 (2016)

28. Satterthwaite, T.D., et al.: Neuroimaging of the Philadelphia neurodevelopmental cohort. Neuroimage **86**, 544–553 (2014)

29. Shen, X., Tokoglu, F., Papademetris, X., Constable, R.T.: Groupwise whole-brain parcellation from resting-state fMRI data for network node identification. Neuroimage **82**, 403–415 (2013)
30. Szegedy, C., et al.: Intriguing properties of neural networks (Dec 2013)
31. Waskom, M.: seaborn: statistical data visualization. J. Open Source Softw. **6**(60), 3021 (2021)
32. Yao, Q., He, Z., Lin, Y., Ma, K., Zheng, Y., Zhou, S.K.: A hierarchical feature constraint to camouflage medical adversarial attacks. In: de Bruijne, M., et al. (eds.) MICCAI 2021. LNCS, vol. 12903, pp. 36–47. Springer, Cham (2021). https://doi.org/10.1007/978-3-030-87199-4_4

Correction to: De-identification and Obfuscation of Gender Attributes from Retinal Scans

Chenwei Wu[✉], Xiyu Yang, Emil Ghitman Gilkes, Hanwen Cui, Jiheon Choi, Na Sun, Ziqian Liao, Bo Fan, Mauricio Santillana, Leo Celi, Paolo Silva, and Luis Nakayama

Correction to:
Chapter 9 in: S. Wesarg et al. (Eds.): *Clinical Image-Based Procedures, Fairness of AI in Medical Imaging, and Ethical and Philosophical Issues in Medical Imaging*, **LNCS 14242, https://doi.org/10.1007/978-3-031-45249-9_9**

In the originally published version of chapter 9 it has not been made clear that the authors Chenwei Wu and Xiyu Yang contributed equally to the publication. This has been corrected and a footnote has been added.

The updated version of this chapter can be found at
https://doi.org/10.1007/978-3-031-45249-9_9

Author Index

© The Editor(s) (if applicable) and The Author(s), under exclusive license to Springer Nature Switzerland AG 2023
S. Wesarg et al. (Eds.): CLIP/FAIMI/EPIMI 2023, LNCS 14242, pp. 313–314, 2023.
https://doi.org/10.1007/978-3-031-45249-9

Printed in the United States
by Baker & Taylor Publisher Services